Chaos and Its Influence on Children's Development

Chaos and Its Influence on Children's Development
An Ecological Perspective

Edited by
Gary W. Evans and Theodore D. Wachs

American Psychological Association
Washington, DC

Copyright © 2010 by the American Psychological Association. All rights reserved. Except as permitted under the United States Copyright Act of 1976, no part of this publication may be reproduced or distributed in any form or by any means, including, but not limited to, the process of scanning and digitization, or stored in a database or retrieval system, without the prior written permission of the publisher.

Published by
American Psychological Association
750 First Street, NE
Washington, DC 20002
www.apa.org

To order
APA Order Department
P.O. Box 92984
Washington, DC 20090-2984
Tel: (800) 374-2721; Direct: (202) 336-5510
Fax: (202) 336-5502; TDD/TTY: (202) 336-6123
Online: www.apa.org/books/
E-mail: order@apa.org

In the U.K., Europe, Africa, and the Middle East, copies may be ordered from
American Psychological Association
3 Henrietta Street
Covent Garden, London
WC2E 8LU England

Typeset in New Century Schoolbook by Stephen McDougal, Mechanicsville, MD

Printer: United Book Press, Inc., Baltimore, MD
Cover Designer: Minker Design, Bethesda, MD
Technical/Production Editor: Devon Bourexis

The opinions and statements published are the responsibility of the authors, and such opinions and statements do not necessarily represent the policies of the American Psychological Association.

Library of Congress Cataloging-in-Publication Data

Chaos and its influence on children's development : an ecological perspective / edited by Gary W. Evans & Theodore D. Wachs.
  p. cm.
  Includes bibliographical references and index.
  ISBN-13: 978-1-4338-0565-3
  ISBN-10: 1-4338-0565-0
  1. Child development. 2. Noise. I. Evans, Gary W. II. Wachs, Theodore D., 1941–

  HQ772.C4123 2010
  305.231—dc22
                                                                            2009015694

British Library Cataloguing-in-Publication Data
A CIP record is available from the British Library.

Printed in the United States of America
First Edition

APA Science Volumes

Attribution and Social Interaction: The Legacy of Edward E. Jones

Best Methods for the Analysis of Change: Recent Advances, Unanswered Questions, Future Directions

Cardiovascular Reactivity to Psychological Stress and Disease

The Challenge in Mathematics and Science Education: Psychology's Response

Changing Employment Relations: Behavioral and Social Perspectives

Children Exposed to Marital Violence: Theory, Research, and Applied Issues

Cognition: Conceptual and Methodological Issues

Cognitive Bases of Musical Communication

Cognitive Dissonance: Progress on a Pivotal Theory in Social Psychology

Conceptualization and Measurement of Organism–Environment Interaction

Converging Operations in the Study of Visual Selective Attention

Creative Thought: An Investigation of Conceptual Structures and Processes

Developmental Psychoacoustics

Diversity in Work Teams: Research Paradigms for a Changing Workplace

Emotion and Culture: Empirical Studies of Mutual Influence

Emotion, Disclosure, and Health

Evolving Explanations of Development: Ecological Approaches to Organism–Environment Systems

Examining Lives in Context: Perspectives on the Ecology of Human Development

Global Prospects for Education: Development, Culture, and Schooling

Hostility, Coping, and Health

Measuring Patient Changes in Mood, Anxiety, and Personality Disorders: Toward a Core Battery

Occasion Setting: Associative Learning and Cognition in Animals

Organ Donation and Transplantation: Psychological and Behavioral Factors

Origins and Development of Schizophrenia: Advances in Experimental Psychopathology

The Perception of Structure

Perspectives on Socially Shared Cognition

Psychological Testing of Hispanics

Psychology of Women's Health: Progress and Challenges in Research and Application

Researching Community Psychology: Issues of Theory and Methods

The Rising Curve: Long-Term Gains in IQ and Related Measures

Sexism and Stereotypes in Modern Society: The Gender Science of Janet Taylor Spence

Sleep and Cognition

Sleep Onset: Normal and Abnormal Processes

Stereotype Accuracy: Toward Appreciating Group Differences

Stereotyped Movements: Brain and Behavior Relationships

Studying Lives Through Time: Personality and Development

The Suggestibility of Children's Recollections: Implications for Eyewitness Testimony
Taste, Experience, and Feeding: Development and Learning
Temperament: Individual Differences at the Interface of Biology and Behavior
Through the Looking Glass: Issues of Psychological Well-Being in Captive Nonhuman Primates
Uniting Psychology and Biology: Integrative Perspectives on Human Development
Viewing Psychology as a Whole: The Integrative Science of William N. Dember

APA Decade of Behavior Volumes

Acculturation: Advances in Theory, Measurement, and Applied Research
Aging and Cognition: Research Methodologies and Empirical Advances
Animal Research and Human Health: Advancing Human Welfare Through Behavioral Science
Behavior Genetics Principles: Perspectives in Development, Personality, and Psychopathology
Categorization Inside and Outside the Laboratory: Essays in Honor of Douglas L. Medin
Chaos and Its Influence on Children's Development: An Ecological Perspective
Child Development and Social Policy: Knowledge for Action
Children's Peer Relations: From Development to Intervention
Commemorating Brown: The Social Psychology of Racism and Discrimination
Computational Modeling of Behavior in Organizations: The Third Scientific Discipline
Couples Coping With Stress: Emerging Perspectives on Dyadic Coping
Developing Individuality in the Human Brain: A Tribute to Michael I. Posner
Emerging Adults in America: Coming of Age in the 21st Century
Experimental Cognitive Psychology and Its Applications
Family Psychology: Science-Based Interventions
Inhibition and Cognition
Medical Illness and Positive Life Change: Can Crisis Lead to Personal Transformation?
Memory Consolidation: Essays in Honor of James L. McGaugh
Models of Intelligence: International Perspectives
The Nature of Remembering: Essays in Honor of Robert G. Crowder
New Methods for the Analysis of Change
On the Consequences of Meaning Selection: Perspectives on Resolving Lexical Ambiguity
Participatory Community Research: Theories and Methods in Action
Personality Psychology in the Workplace

Perspectivism in Social Psychology: The Yin and Yang of Scientific Progress
Primate Perspectives on Behavior and Cognition
Principles of Experimental Psychopathology: Essays in Honor of Brendan A. Maher
Psychosocial Interventions for Cancer
Racial Identity in Context: The Legacy of Kenneth B. Clark
The Social Psychology of Group Identity and Social Conflict: Theory, Application, and Practice
Strengthening Couple Relationships for Optimal Child Development: Lessons From Research and Intervention
Strengthening Research Methodology: Psychological Measurement and Evaluation
Transcending Self-Interest: Psychological Explorations of the Quiet Ego
Unraveling the Complexities of Social Life: A Festschrift in Honor of Robert B. Zajonc
Visual Perception: The Influence of H. W. Leibowitz

This volume is dedicated to the personal memories and lasting theoretical insights of our friend, colleague, and mentor, Urie Bronfenbrenner. His thinking about human development has profoundly influenced so many students and colleagues in multiple areas of enquiry. We hope this volume will provide another vehicle through which Urie's ideas on the bioecology of human development can continue to flourish.

Contents

Contributors . . . . *xiii*

Foreword . . . . *xv*

Preface . . . . *xvii*

Part I. Foundations 1

   1. Chaos in Context 3
      Theodore D. Wachs and Gary W. Evans

   2. Chaos and the Diverging Fortunes of American Children:
     A Historical Perspective 15
      Daniel T. Lichter and Elaine Wethington

Part II. Chaos at the Microsystem Level 33

   3. Physical and Psychosocial Turmoil in the Home and
     Cognitive Development 35
      Brian P. Ackerman and Eleanor D. Brown

   4. The Dynamics of Family Chaos and Its Relation to
     Children's Socioemotional Well-Being 49
      Barbara H. Fiese and Marcia A. Winter

   5. Child-Care Chaos and Child Development 67
      Feyza Corapci

   6. Chaos Outside the Home: The School Environment 83
      Lorraine E. Maxwell

   7. Viewing Microsystem Chaos Through a Bronfenbrenner
     Bioecological Lens 97
      Theodore D. Wachs

   8. The Role of Temporal and Spatial Instability in
     Child Development 113
      Clyde Hertzman

Part III. Chaos at the Mesosystem Level 133

   9. From Home to Day Care: Chaos in the Family/Child-Care
     Mesosystem 135
      Robert H. Bradley

10. Disorder, Turbulence, and Resources in Children's Homes and Neighborhoods 155
Jeanne Brooks-Gunn, Anna D. Johnson, and Tama Leventhal

Part IV. Chaos at the Exosystem Level 171

11. Neighborhood Chaos and Children's Development: Questions and Contradictions 173
James R. Dunn, Nicole J. Schaefer-McDaniel, and Jason T. Ramsay

12. Parent Employment and Chaos in the Family 191
Rena Repetti and Shu-wen Wang

Part V. Chaos at the Macrosystem Level 209

13. Well-Being, Chaos, and Culture: Sustaining a Meaningful Daily Routine 211
Thomas S. Weisner

14. Chaos and the Macrosetting: The Role of Poverty and Socioeconomic Status 225
Gary W. Evans, John Eckenrode, and Lyscha A. Marcynyszyn

15. An Ecological Framework for the Refugee Experience: What Is the Impact on Child Development? 239
Stuart L. Lustig

Part VI. Conclusions 253

16. Dynamic Developmental Systems: Chaos and Order 255
Arnold Sameroff

Index 265
About the Editors 277

Contributors

Brian P. Ackerman, Professor, Department of Psychology, University of Delaware, Newark

Robert H. Bradley, Professor and Director, Family and Human Dynamics Research Institute, Arizona State University, Tempe

Jeanne Brooks-Gunn, Virginia and Leonard Marx Professor, Teachers College and the College of Physicians and Surgeons, Columbia University, New York, NY

Eleanor D. Brown, Assistant Professor, Department of Psychology, West Chester University, West Chester, PA

Feyza Corapci, Assistant Professor, Department of Psychology, Bogazici University, Istanbul, Turkey

James R. Dunn, Associate Professor, Department of Health, Aging and Society, McMaster University, Hamilton, Ontario, Canada; Scientist, Centre for Research on Inner City Health, St. Michael's Hospital, Toronto, Ontario, Canada

John Eckenrode, Director, Family Life Development Center; Professor, Department of Human Development, Cornell University, Ithaca, NY

Gary W. Evans, Elizabeth Lee Vincent Professor, Departments of Design and Environmental Analysis and of Human Development, Cornell University, Ithaca, NY

Barbara H. Fiese, Professor and Director, Family Resiliency Center, Department of Human and Community Development, University of Illinois at Urbana–Champaign

Clyde Hertzman, Chair, Human Early Learning Partnership, University of British Columbia, Vancouver, British Columbia, Canada

Anna D. Johnson, doctoral student, Department of Human Development, Teachers College, Columbia University, New York, NY

Tama Leventhal, Assistant Professor, Eliot-Pearson Department of Child Development, Tufts University, Medford, MA

Daniel T. Lichter, Ferris Family Professor, Departments of Policy Analysis and Management and of Sociology, Cornell University, Ithaca, NY

Stuart L. Lustig, Assistant Professor, Department of Psychiatry, University of California, San Francisco

Lyscha A. Marcynyszyn, Research Analyst, Casey Family Programs, Seattle, WA

Lorraine E. Maxwell, Associate Professor, Department of Design and Environmental Analysis, Cornell University, Ithaca, NY

Jason T. Ramsay, Postdoctoral Student, Centre for Urban Health, University of Toronto, Toronto, Ontario, Canada

Rena Repetti, Professor, Department of Psychology, University of California, Los Angeles

Arnold Sameroff, Professor, Department of Psychology, University of Michigan, Ann Arbor

Nicole J. Schaefer-McDaniel, postdoctoral student, Centre for Urban Health, University of Toronto, Toronto, Ontario, Canada

Theodore D. Wachs, Professor, Department of Psychology, Purdue University, West Lafayette, IN

Shu-wen Wang, doctoral student, Department of Psychology, University of California, Los Angeles

Thomas S. Weisner, Professor, Departments of Psychiatry (Semel Institute) and of Anthropology, University of California, Los Angeles

Elaine Wethington, Associate Professor, Departments of Human Development and of Sociology, Cornell University, Ithaca, NY

Marcia A. Winter, postdoctoral student, Department of Psychology, Syracuse University, Syracuse, NY

Foreword

In early 1988, the American Psychological Association (APA) Science Directorate began its sponsorship of what would become an exceptionally successful activity in support of psychological science—the APA Scientific Conferences program. This program has showcased some of the most important topics in psychological science and has provided a forum for collaboration among many leading figures in the field.

The program has inspired a series of books that have presented cutting-edge work in all areas of psychology. At the turn of the millennium, the series was renamed the Decade of Behavior Series to help advance the goals of this important initiative. The Decade of Behavior is a major interdisciplinary campaign designed to promote the contributions of the behavioral and social sciences to our most important societal challenges in the decade leading up to 2010. Although a key goal has been to inform the public about these scientific contributions, other activities have been designed to encourage and further collaboration among scientists. Hence, the series that was the "APA Science Series" has continued as the "Decade of Behavior Series." This represents one element in APA's efforts to promote the Decade of Behavior initiative as one of its endorsing organizations. For additional information about the Decade of Behavior, please visit http://www.decadeofbehavior.org.

Over the course of the past years, the Science Conference and Decade of Behavior Series has allowed psychological scientists to share and explore cutting-edge findings in psychology. The APA Science Directorate looks forward to continuing this successful program and to sponsoring other conferences and books in the years ahead. This series has been so successful that we have chosen to extend it to include books that, although they do not arise from conferences, report with the same high quality of scholarship on the latest research.

We are pleased that this important contribution to the literature was supported in part by the Decade of Behavior program. Congratulations to the editors and contributors of this volume on their sterling effort.

Steven J. Breckler, PhD
Executive Director for Science

Virginia E. Holt
Assistant Executive Director for Science

Preface

> Chaos is found in greatest abundance wherever order is being sought. It always defeats order, because it is better organized.
> —Terry Pratchett (1998, p. 4)[1]

> In the Beginning how the Heav'ns and Earth Rose out of Chaos
> —John Milton (1831, p. 3)[2]

The concept of *chaos* can easily be traced back to ancient Greek times, and refers to the primordial state from which all form and life emerged (Hesiod, trans. 1953).[3] Though the concept of chaos has a long history, it has not been a mainstream topic in the developmental sciences (e.g., a search of PsycINFO for literature published between 1900 and 1999 using the key words *chaos* and *development* yielded a total of 80 references). However, more recently there has been a growing interest in chaos as an influence on development (e.g., a similar search for literature published between 2000 and the present yielded 116 references). This increasing interest in chaos as a topic may reflect mounting evidence that chaos in the lives of children is inimical to healthy development. Further, trends also indicate that defining components of chaos, such as family instability, residential mobility, and living in noisy, crowded homes often lacking in routines and structure, along with deteriorated neighborhoods, are on the rise in the United States and throughout the world.

Unlike the heavens and the earth, this book did not arise out of chaos, but rather from an interdisciplinary conference at Cornell that we organized. Participants at this conference were drawn from such diverse fields as developmental and environmental psychology, psychiatry, geography, public health, sociology, and anthropology. The focus of this conference was on the role of chaotic environments in children's lives. Our interest in this topic can be traced back to the writings of and personal contacts with two of the most important developmental scientists over the past half century, Joachim Wohlwill and Urie Bronfenbrenner. Wohlwill's delineation of the physical environment and his emphasizing that too much stimulation may be as problematic for development as too little stimulation provided much of the conceptual framework for subsequent research on chaos and development. Bronfenbrenner's bioecological model provided a theoretical structure for developmental scientists to address questions such as the place of chaos in the overall environment of the child and the potential mechanisms through which chaos affects development. Further, toward the end of his life, Bronfenbrenner became increasingly concerned with what he viewed as escalating chaos in the lives of children and their families.

---

[1]Pratchett, T. (1953). *Interesting times.* New York: Harper Torch.
[2]Milton, J. (1831). *Paradise lost: Book I.* Boston: L. Coffin.
[3]Hesiod. (1953). *Theogony* (N. O. Brown, Trans.). New York: Liberal Arts Press.

The primary objectives for our conference and this volume were twofold. First, following in the footsteps of Wohlwill, we wanted to synthesize what is currently known about chaos and child development. Second, following the path that Bronfenbrenner laid out, we wanted to use the bioecological model as a framework to discuss how other aspects of chaos, particularly those outside the immediate microsettings inhabited by children, might influence child development. It is our hope that the chapters in this volume will provide evidence, methods, questions, and issues that will facilitate further research and theory by the next generation of developmental scientists on the nature and impact of chaotic environments.

We are indebted to our colleagues who attended our conference for the quality and intensity of discussion and the ideas that came from these discussions, many of which are presented in this volume. We also wish to acknowledge the financial support of the American Psychological Association's Science Directorate; the Bronfenbrenner Life Course Center; the College of Human Ecology, Institute for the Social Sciences, Provost's Office, Family Life Development Center, Cornell Cooperative Extension, and Departments of Design and Environmental Analysis and of Human Development, all at Cornell University; and the Department of Psychological Sciences at Purdue University for the First Bronfenbrenner Conference on the Ecology of Human Development. We are particularly grateful to Carrie Chalmers, who expertly handled the organization of the conference. Thanks are also due to Maureen Adams and Edward Porter of the American Psychological Association for their patience while chapters were being written and their editorial help on chapter drafts.

This volume has inspired a biennial book series: The APA Bronfenbrenner Series on the Ecology of Human Development. Each volume will be based on a conference held at Cornell University. Thanks are due to the editorial advisory committee for this new series: Stephen Ceci, Gary W. Evans, Daniel Lichter, Karl Pillemer, Valerie Reyna, and Elaine Wethington.

Part I

Foundations

1

Chaos in Context

Theodore D. Wachs and Gary W. Evans

The focus of this volume is on how chaotic environmental settings influence human development from infancy through adolescence. In addition to providing up-to-date reviews on relations between chaotic environmental settings and development, this volume also deals with a number of critical issues such as defining a chaotic environment, the dimensions of chaos, and the mechanisms through which chaotic environmental settings can act to influence development. In the various chapters in this volume, we draw on Bronfenbrenner's bioecological model of human development to provide a theoretical and organizational framework to address these issues.

Chaos and Development in Historical Context

The concept that environment plays a necessary role in children's development can be traced at least as far back as the writings of Plato: "The beginning in every task is the chief thing, especially for any creature that is young and tender. For it is then that it is best molded and takes the impression that one wishes to stamp upon it" (Plato, trans. 1937, p. 177). In spite of this long history, however, systematic research on the role of environment in development began only after the middle of the 20th century (Hunt, 1979). Further, much of the initial research on environment and development was based on infrahuman studies of animals living under conditions of low stimulation or children growing up in unstimulating conditions such as orphanages (Wachs & Gruen, 1982). It is not surprising that a primary conclusion from this research was what Wohlwill (1974) has called the "more the merrier hypothesis," namely, that the relation of stimulation to development was linear. In a linear model slower development occurs when environmental stimulation is low, and as stimulation increases, development is increasingly enhanced. One implication of a linear model was that the rate of development for economically disadvantaged children was lower because such children were "stimulus deprived" and required interventions that increased environmental enrichment (Wachs & Gruen, 1982).

Though a large body of both infrahuman and human research supported such a linear model (Hunt, 1979; Wachs & Gruen, 1982), not all evidence was consistent with this model (Wohlwill & Heft, 1987). Initial evidence from both observational studies of children in their natural environments (Deutsch, 1964;

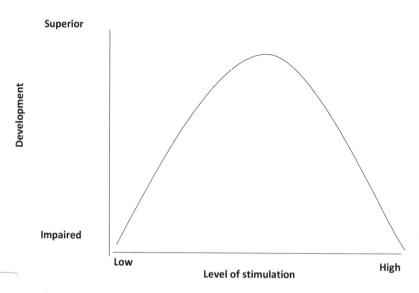

Figure 1.1. Curvilinear (optimal stimulation) model relating level of stimulation to development.

Heft, 1979; Klaus & Gray, 1968; Wachs, Uzgiris, & Hunt, 1971) and early intervention studies (Bronfenbrenner, 1974; White & Held, 1966) suggested that the relation of stimulation to development may be curvilinear in nature, with either too much or too little stimulation inhibiting development. These early studies led to what has been called the *optimal stimulation hypothesis* (Uzgiris, 1977), namely, that the relation of stimulation to development is best described by an inverted U curve. As shown in Figure 1.1, development is maximized when an individual encounters or is exposed to an optimal level of stimulation (top point of the inverted U curve), with development becoming increasingly inhibited as level of stimulation deviates from the optimal point in either direction. Researchers have increasingly used the term *environmental chaos* to describe the nature and consequences of overly high levels of stimulation (right-hand portion of the inverted U curve). As indicated later, we believe overstimulation from factors such as noise or crowding is a critical component of chaos but other environmental and temporal factors contribute to chaotic environments as well.

Supporting the validity of a nonlinear relation between stimulation and development, a growing body of research has found negative relations between measures of cognitive and socioemotional development and exposure of infants and children to ambient noise, crowding, and a lack of structure in their environments (for recent reviews, see Evans, 2006; Wachs & Corapci, 2003; see also chaps. 3, 4, 5, and 6, this volume).

What Is Chaos?

Though evidence supports the hypothesis that too much stimulation may be detrimental to development, at present chaos is not a well-developed construct within psychology. One of the primary objectives of this volume is to critically examine what is meant by chaos and the potential value of chaos as a concept within developmental science. At this time, the term *chaos* is typically used to describe environments that are characterized by high levels of noise, crowding, and instability, as well as a lack of temporal and physical structuring (few regularities, routines, or rituals; nothing has its time or place; Fiese, 2006; Wachs, 1989).[1]

Though it is a useful starting point for studying relations between environmental stimulation and development, this currently used definition of chaos is problematic in a variety of ways. For example, in light of evidence that higher levels of environmental chaos, as defined earlier, are more likely to occur in low-income families (see chaps. 3 and 14), it could be argued that chaos is nothing more than a proxy term for low family socioeconomic status (SES). Although chaos and low SES clearly covary, chaos cannot be reduced to low SES for three major reasons. First, if chaos were just a proxy for poverty, we would not expect to find a relation of chaos to development in middle-class samples. Yet the available evidence does show such a relation (e.g., Gottfried & Gottfried, 1984; Hygge, Evans, & Bullinger, 2002). Second, if chaos were just a marker for low income, then controlling for SES should wipe out the effects initially attributed to chaos. However, studies that statistically control for SES or income consistently show independent chaos–development associations over and above the impact of SES (e.g., Corapci & Wachs, 2002; Dumas et al., 2005; Evans, 2006; Hart, Petrill, Deckard, & Thompson, 2007; Petrill, Pike, Price, & Plomin, 2004; Pike, Iervolino, Eley, Price, & Plomin, 2006; Wachs, 1993; Wachs & Chan, 1986). Furthermore, several longitudinal studies have shown that changes in noise levels alter children's cognitive and physiological outcomes when no alterations in family SES have occurred (for a summary of these studies, see Evans, 2006). Third, as shown in chapter 14 of this volume, a number of mediational studies indicate that chaos may function as a mechanism through which low SES influences development. Thus, the overall pattern of evidence indicates that although chaos and low SES covary, the impact of chaos on development reflects something over and above the impact of low SES.

Although chaos as an influence on development is more than just low SES, multiple questions about the nature and definition of chaos still remain. For example, given that too much stimulation may be problematic for development, one might ask the critical question, What is meant by too much? This question is difficult to answer because the optimal stimulation point for a given individual will vary depending on individual characteristics (see chaps. 4 and 7) or characteristics of the individual's culture (see chap. 13).

Further, key dimensions of chaos can be operationalized in a variety of ways. For example, within the family, instability can be defined on the basis of

---

[1] It is important to distinguish our use of the term *chaos*, which represents specific characteristics of the child's environment, from use of the term in what has been called *dynamic systems theory* or *chaos theory*. This latter usage describes situations in which nonlinear solutions are applied to situations in which outcomes are unpredictable and there are sudden changes that cannot be modeled by the additive action of specific elements in the system (Barton, 1994). Chaotic environments may result in chaotic outcomes, as described by dynamic systems theory, but our focus in this volume is on chaotic environments.

random or nonsystematic changes in parental rearing styles, or by changes in family composition resulting from adult partner changes, or by residential instability that changes the proximity of family ties. Instability can also occur outside the family as seen in shifts in non–family members with whom the child interacts (e.g., midyear teacher change, residential relocation that uproots friendships). Nonfamily instability can also refer to changes that do not necessarily involve persons, as seen in changes in school curriculum or shifts in cultural customs.

In addition, the working definition of chaos used earlier does not exhaust other possible dimensions of this construct. A number of additional dimensions have been suggested by chapter authors in this volume. These potential dimensions include parental maladjustment (chap. 3); visual complexity (chap. 6); clutter and messiness (chap. 10); low supervision and monitoring (chap. 10); multiple caregivers and shuttling between locations (chap. 11); hurriedness and time pressure (chap. 11); cynicism and generalized mistrust in institutions (chap. 11); high workload, nonstandard working hours, and unstable employment (chap. 12); and high levels of fear, uncertainty, loss, or bewilderment, often linked with rapid economic or political upheaval or involuntary relocations (chap. 15). In addition to the conceptualization of chaos as an environmental characteristic, a case can be made that a richer understanding of the role of chaos in child development will occur if one also considers how individuals understand and interpret chaos, as reflected in states such as feeling hectic or that one's life is out of control (see chaps. 4 and 13). To what extent these alternative dimensions actually help to define chaos per se, versus being the result of existing chaos (e.g., maladjustment, mistrust, fear or bewilderment, sense of loss of control), is a critical question for future research and theorizing on this construct. As is discussed later and in a number of chapters in this volume, viewing chaos within the framework of Bronfenbrenner's bioecological theory may be one approach to dealing with issues involved in the definition of chaos and identification of the major dimensions defining chaos.

Why Does Chaos Influence Development?

Much of the initial research on chaos and development focused on noise and was based on the assumption that high noise levels resulted in the child's habituating to auditory input, thus depriving the child of an important source of cognitive input, namely, language (Cohen, Glass, & Singer, 1973; Deutsch, 1964). High noise levels can also interfere with information-processing skills, such as sensitivity to incidental information (Heft, 1979; see also chap. 8, this volume). More recent research also has documented how chaotic environments, as defined earlier, can adversely influence development by reducing the quality of parenting in such critical areas as parental responsiveness, parental involvement, promotion of children's exploration, linguistic or object stimulation of the child, and parental efficacy beliefs (Corapci & Wachs, 2002; Evans, Maxwell, & Hart, 1999; Matheny, Wachs, Ludwig, & Phillips, 1995; Wachs & Camli, 1991; see also chaps. 2, 4, and 12, this volume). However, additional mechanisms have been proposed as well. For example, the unpredictable, uncontrollable, and distracting nature of chaotic settings may interfere with the ability of children to acquire self-regulation skills such as inhibitory control of emotions and behaviors, or may compromise children's attention regulation or their development of a sense of mastery or self-efficacy (for more detailed discussion of these mechanisms, see chaps. 3, 4, 6, 7, and 14, this volume). Chaos may also disrupt the ability of families to access or profit from available community services (see chaps. 8 and 13, this volume).

In addition, physiological consequences may be associated with living in chaotic settings. In this volume, Ackerman and Brown (chap. 3) and Hertzman (chap. 8) propose that chaos adversely influences child development through dysregulation of physiological stress resulting from the relentless and uncontrollable demands that chaotic environments put on the developing child. Finally, several chapter authors allude to the potential for chaos to interfere with the ability to engage in tasks of daily living that provide ontological security as well as cultural and self-identity (see chaps. 11, 13, and 15). That is, without the ability to sustain meaningful daily activities and tasks in a regular manner, children and their families cannot acquire a sense of order, continuity, and purpose in life.

The issue of mechanisms is further complicated by questions of whether exposure to chaos is always harmful to children's development or whether the absence of chaos is necessarily healthy. In a classic early review, Wohlwill and Heft (1987) presented evidence that under specific conditions rearing in homes with high noise levels could be adaptive; a similar conclusion with regard to noise is also noted in chapter 2 of this volume. Similarly, there may be advantages to being reared in homes with a high density level (see review by Wachs and Corapci, 2003; also see chap. 13, this volume), as well as a possibility that under some circumstances discontinuity or instability in day-care environments can be beneficial to young children (see chap. 9, this volume). On the other hand, the absence of chaos does not necessarily have positive consequences. Overly regulated or highly structured situations grant little autonomy and are bereft of diversity (see chap. 11, this volume), and too much stability and sameness preclude opportunities to learn how to regulate external demands (see chap. 2, this volume).

Understanding the mechanisms through which environmental chaos can adversely impact development and the conditions under which exposure to environmental chaos does not have adverse outcomes is a second critical topic. As discussed later in this chapter, and in a number of chapters in this volume, viewing chaos within the framework of Bronfenbrenner's bioecological theory has the potential to substantially increase one's understanding of the mechanisms through which chaos can influence development and the conditions that influence the nature of chaos–development relations.

Bronfenbrenner's Bioecological Model

In this section we describe the main points of the bioecological model, indicate potentially useful linkages between the model and chaos, and describe how this model is used to organize this volume.

The Bioecological Model

The bioecological model of human development has four major components: process, person, context, and time. *Proximal processes* are the exchanges of energy between the developing organism and the persons, objects, and symbols in the immediate environment of the organism, and are the major driving force for development (Bronfenbrenner, 1999; Bronfenbrenner & Morris, 1998). Parental responsiveness and teacher scaffolding of learning opportunities are two examples of proximal processes. *Person* refers to characteristics of the organism, such as genetic composition, gender, or temperament. Person variables can either directly shape proximal processes or modify their impacts on development. An example of the former would be difficult temperament in an infant interfering with attentive, warm parenting (Rothbart & Bates, 2006). The moderation of a proximal process by a person variable is exemplified by high intelligence buffering some of the ill effects of a child living in a high-risk environment (Masten & Obradovic, 2006).

Probably the best known component of Bronfenbrenner's model, *context* refers to the multiple spheres of the social and physical environments that influence proximal processes, both directly and indirectly. For example, a higher day-care-provider-to-child ratio directly influences the quality of care. However, the sociocultural milieu in which the provider-to-child ratio varies can modify the relation of ratio to the quality of care. Bronfenbrenner's model delineates four developmentally salient levels of context. The *microsystem context*, wherein proximal processes occur, encompasses the immediate settings children inhabit, such as home, school, and neighborhood. The *mesosystem context* focuses on the capacity of one microsystem (e.g., home) to influence proximal processes in another microsystem (e.g., neighborhood). For example, proximal processes might affect the developing child differently in an impoverished versus a middle-class neighborhood. The next layer of context is termed the *exosystem*, which refers to microsystems that influence a child's development even though the child does not directly encounter these microsystems. For example, characteristics of a parent's place of work (e.g., complexity of work) might in turn affect parental views and expectancies for child self-directedness. The most distal contextual layer is termed the *macrosystem*. It incorporates large-scale cultural, political, economic, and natural forces that can shape both the quantity and quality of proximal processes, as well as the other levels of context (i.e., micro- to exosystem context) that surround the child. For example, income and race can dramatically influence risks and opportunities available to children, and harsh and unresponsive parenting is inversely related to parental levels of education (Evans, 2004; Repetti, Taylor, & Seeman, 2002).

The remaining major component of Bronfenbrenner's model is *time*. Various aspects of temporal exposure are included in this component such as chronological age, duration and continuity of exposure, and historical period. As children age, their range of contextual experiences expands, particularly at the micro- and mesosystem levels. In addition, *time* also can refer to the cumulative impact of proximal processes or context. For example, the impact of poverty on a child's development is different depending on whether poverty is chronic or transient (National Institute of Child Health and Human Development Early Childcare Study, 2005).

Chaos and the Bioecological Model

The process, person, context, and time aspects of the bioecological model can inform one's knowledge of the nature of chaos and chaos–development relations in multiple ways. In terms of process, chaos can directly affect the extent to which proximal processes will occur. For example, socially supportive relationships among adolescent peers are unlikely to be sustained under conditions of high residential mobility. Furthermore, for proximal processes to be effective, they must occur regularly, over extended periods, and become progressively more complex as the child matures (Bronfenbrenner & Evans, 2000). Person characteristics can directly influence chaos by shaping transactions with other persons and can also moderate the impact of chaos on the developing child (Wachs, 2000).

Context can directly and indirectly produce chaos. For example, with respect to mesosystem chaos, the detrimental effects of noise levels at school on children's reading acquisition are exacerbated by high levels of residential noise (Cohen, Evans, Stokols, & Krantz, 1986). In terms of exosystem chaos, consider the classroom setting. A teacher's caregiving burdens for a parent with Alzheimer's disease might readily manifest as fatigue, less sensitivity, and lack of patience with children in the classroom. War or major natural disasters that create refugees illustrate how a macrocontextual factor could precipitate chaos in children's immediate microsystems. With regard to time, there is evidence of cumulative deficits in cognitive performance (Hygge, Evans, & Bullinger, 2002) and social functioning (Evans & Lepore, 1993) with continued exposure to chaos. This cumulative effect may reflect the predictable and sustained nature of effective proximal processes being compromised by chronic chaos. For example, chronic chaotic living or working conditions can be tiring, often disrupting the sleep patterns of both children and their parents, or increasing fatigue, all of which have the potential to adversely influence the nature of parent–child or child–teacher relations (Brown & Low, 2008; Evans & Hygge, 2007).

The benefits of linking theory and research on chaos to the bioecological model are seen in multiple chapters in this volume. A number of chapters illustrate how viewing chaos within a bioecological framework may lead to more precise definitions of chaos. In various chapters authors note that although chaos is typically defined at the individual level, it may be important to look beyond the individual and consider what is meant by chaos at other contextual levels, such as the family, neighborhood, workplace, or community (see chaps. 3, 4, 8, 10, 12, and 15). For example, using the macrosystem as a frame of reference, Weisner (see chap. 13) argues that the key to understanding chaos is to see it in relation to the sustainability of daily routines for children and their families. Chaos occurs when actions or threats interfere with meaningful engagement of parents and children in everyday routines and activities that are viewed as appropriate in a given community. This definition of chaos leads to some additional candidates for inclusion within the chaos construct: chronic resource scarcity, unpredictability and inability to fit family routines into the resources available, continued conflict, violence and threat, and poor fit of routines with goals and values of the culture. In light of the nature of the macrosystem, Weisner also emphasizes the need to be cautious in defining whether a particular practice is chaotic without a proper understanding of the sociocultural context in which it unfolds. Thus, the use of multiple caretakers of children may or may not be chaotic, depending on how the organization of the community and the need for regular social interactions with family or others in the community is conceptualized.

Similarly, the benefits of viewing chaos within a bioecological framework are illustrated in a number of chapters dealing with the issue of identification of mechanisms through which chaos influences development. As noted earlier, interference with proximal processes is one hypothesized mechanism through which chaos can adversely influence developmental outcomes. Chaos may interfere with the maintenance of effective proximal processes that depend on predictable, sustained, and progressively more complex interactions between the developing child and his or her immediate environment. Elaborating on this theme, a number of chapter authors point toward chaos-influenced alterations in parenting, with particular emphasis on less responsive and more harsh parenting, as prime candidates for a mediating mechanism of chaos on child development (see chaps. 2, 4, 7, 12, and 14). Parents may also find it more difficult to allocate time to children as well as engage in important activities requiring effort, such as monitoring children or maintaining care regimens (e.g., medications), if they live or work under more chaotic conditions. In chapters 5, 6, and 9, respectively, Corapci, Maxwell, and Bradley each note parallel chaos-related processes among other caregivers, such as day-care providers or teachers. In addition, at a mesosystem level, active communication between parents and day-care providers may be compromised by chaos either in the home or in day care (see chap. 9).

With regard to the person dimension of the bioecological model, a number of chapters point out how critical parental characteristics can be compromised by chaos. For example, parental perceptions of their own level of self-efficacy may be diminished in situations of high chaos because of the unpredictable and uncontrollable nature of chaos (see chaps. 4, 7, and 15). Chaos can also adversely impact parental mental health, which in turn is bad for children (see chaps. 3 and 7). Further, an increasing body of evidence documents how the impact of chaos can be accentuated or attenuated as a function of individual characteristics (see chaps. 5, 7, and 8). These chapters illustrate that the mechanisms underlying chaos can be either main effect (chaos → parenting characteristics → development) or interactive in nature (Person × Chaos interactions).

Increased understanding of potential mechanisms can also result from integrating the context dimensions of the bioecological model with studies of chaos. At a mesosystem level Maxwell (chap. 6) raises the issue of whether constituent parts of chaos (e.g., noise, instability, lack of structure) are sufficient, or whether to be effective chaos requires the convergence of two or more components. From a bioecological perspective convergence may be built in, such that chaos at one level may normally ripple into other levels. For example, Repetti and Wang (see chap. 12) discuss the consequences of job loss at the level of the exosystem. Possible indirect effects of job loss include residential relocation, change in day-care or school settings, and rapid shifts in one or both parents' work schedules. The consequences of residential relocation are discussed by Evans, Eckenrode, and Marcynyszyn in chapter 14. At the interface between the microsystem and macrosystem, Lustig (see chap. 15) describes how the refugee experience invades family life, often rendering parents ineffectual in normal, expected functions such as being protectors and conveyers of cultural knowledge to children. Lustig further describes in some detail how the roles of parent and child are often reversed in refugees as children rapidly become more competent than their parents in negotiating the new environment. Lustig also points out that in addition to disempowering parents, the refugee experience often overly accelerates maturation in children because of the adultlike responsibilities they must assume at a young age. However, though these examples are compelling, Brooks-Gunn, Johnson, and Leventhal note in chapter 10 that it is not clear how much the various characteristics of chaos covary within and across levels. Research on the cross-context consequences of chaos offers a potentially important approach to understanding how chaos can influence development.

The Bioecological Model as the Organizational Structure of This Volume

In addition to serving as a rich conceptual model for thinking about how chaos might operate in the lives of children, the bioecological model provides a useful organizing heuristic for our presentation of research on chaos and children's development. The present volume is thus divided into six parts. Part I provides an overview consisting of the present chapter and a presentation of historical trends in chaos in children's lives (chap. 2). Given that nearly all of the empirical work to date on chaos and children has examined elements and processes within the microsystem, the chapters in Part II focus on microsystem chaos, reviewing what is known about chaos in the home (chaps. 3 and 4), child care (chap. 5), and school (chap. 6) in terms of cognitive and socioemotional outcomes in children. Part II concludes with two chapters that highlight aspects of microsystem chaos in need of further development. The first of these chapters discusses individual differences in vulnerability to chaos (chap. 7), and the second explores the temporal and spatial dimensions of chaos (chap. 8).

Part III of the volume explores relatively unknown territory, namely, the meaning and possible operation of chaos in other contextual layers beyond the microsystem. In the first chapter in this part, mesosystem chaos is explored in terms of how chaos in the home might transact with chaos in either child care (chap. 9) or neighborhoods (chap. 10) to produce effects on children. Part IV follows up with a discussion of the possible role of exosystem chaos using neighborhood (chap. 11) and work setting conditions (chap. 12) as examples of how chaos in a setting the child might not actually directly experience could spill over into the child's more immediate setting such as the home. Part V offers a discussion of how three macrocontextual variables could influence chaos among children: culture (chap. 13), poverty (chap. 14), and the refugee experience (chap. 15). The final chapter, and the only one in Part VI (chap. 16), identifies common themes throughout the volume, addresses key concepts, and proposes an agenda for future research on chaos and child development.

References

Barton, S. (1994). Chaos, self-organization and psychology. *American Psychologist, 49,* 5–14.
Bronfenbrenner, U. (1974). *Is early education effective?* (Publication No. OHD76-30025). Washington, DC: Department of Health, Education and Welfare.
Bronfenbrenner, U. (1999). Environments in developmental perspective: Theoretical and operational models. In S. Friedman & T. D. Wachs (Eds.), *Measuring environment across the life span: Emerging methods and concepts* (pp. 3–30). Washington, DC: American Psychological Association.
Bronfenbrenner, U., & Evans, G. (2000). Developmental science in the 21st century: Emerging theoretical models, research design, and empirical findings. *Social Development, 9,* 115–125.
Bronfenbrenner, U., & Morris, P. (1998). The ecology of developmental process. In W. Damon (Series Ed.) & R. Lerner (Vol. Ed.), *Handbook of child psychology, Vol. 1: Theoretical models of human development* (5th ed., pp. 992–1028). New York: Wiley.
Brown, E., & Low, C. (2008). *Chaotic living conditions and sleep problems predict children's responses to academic challenge.* Manuscript submitted for publication.
Cohen, S., Evans, G., Stokols, D., & Krantz, D. (1986). *Behavior, health and environmental stress.* New York: Plenum Press.
Cohen, S., Glass, D., & Singer, J. (1973). Apartment noise, auditory discrimination and reading ability in children. *Journal of Experimental Social Psychology, 9,* 407–422.
Corapci, F., & Wachs, T. D. (2002). Does parental mood or efficacy mediate the influence of environmental chaos upon parenting behavior? *Merrill-Palmer Quarterly, 48,* 182–201.
Deutsch, C. (1964). Auditory discrimination and learning. *Merrill-Palmer Quarterly, 10,* 276–296.
Dumas, J., Nissley, J., Nordstrom, A., Smith, E., Prinz, R., & Levine, D. (2005). Home chaos: Sociodemographic, parenting, interactional and child correlates. *Journal of Clinical Child and Adolescent Psychology, 34,* 93–104.
Evans, G. W. (2004). The environment of childhood poverty. *American Psychologist, 59,* 77–92.
Evans, G. W. (2006). Child development and the physical environment. *Annual Review of Psychology, 57,* 423–451.
Evans, G. W., & Hygge, S. (2007). Noise and performance in children and adults. In L. Luxon & D. Prasher (Eds.), *Noise and its effects* (pp. 549–566). London: Wiley.
Evans, G. W., & Lepore, S. J. (1993). Household crowding and social support: A quasi-experimental analysis. *Journal of Personality and Social Psychology, 65,* 308–316.
Evans, G. W., Maxwell, L. E., & Hart, B. (1999). Parental language and verbal responsiveness to children in crowded homes. *Developmental Psychology, 35,* 1020–1024.
Fiese, B. (2006). *Family routines and rituals.* New Haven, CT: Yale University Press.
Gottfried, A. W., & Gottfried, A. E. (1984). Home environment and cognitive development in young children of middle socio-economic status families. In A. Gottfried (Ed.), *Home environment and early cognitive development* (pp. 57–116). New York: Academic Press.
Hart, S., Petrill, S., Deckard, K., & Thompson, L. (2007). SES and CHAOS as environmental mediators of cognitive ability: A longitudinal genetic analysis. *Intelligence, 35,* 233–242.
Heft, H. (1979). Background and focal environmental conditions of the home and attention in young children. *Journal of Applied Social Psychology, 9,* 47–69.
Hunt, J. M. (1979). Psychological development: Early experience. *Annual Review of Psychology, 30,* 103–144.
Hygge, S., Evans, G. W., & Bullinger, M. (2002). A prospective study of some effects of aircraft noise on cognitive performance in school children. *Psychological Science, 13,* 469–474.
Klaus, R., & Gray, S. (1968). The early training project for disadvantaged children. *Monographs of the Society for Research in Child Development, 33*(4, Serial No. 120).
Masten, A., & Obradovic, J. (2006). Competence and resilience in development. In B. Lester, A. Masten, & B. McEwen (Eds.), *Annals of the New York Academy of Sciences: Vol. 1094. Resilience in children* (pp. 13–27). New York: New York Academy of Sciences.
Matheny, A., Wachs, T. D., Ludwig, J., & Phillips, K. (1995). Bringing order out of chaos: Psychometric characteristics of the Louisville Chaos Scale. *Journal of Applied Developmental Psychology, 16,* 429–444.
National Institute of Child Health and Human Development Early Childcare Study. (2005). Duration and developmental timing of poverty and children's cognitive and social development from birth through third grade. *Child Development, 76,* 795–810.

Petrill, S., Pike, A., Price, T., & Plomin, R. (2004). Chaos in the home and socioeconomic status are associated with cognitive development in early childhood: Environmental mediators identified in a genetic design. *Intelligence, 32*, 445–460.

Pike, A., Iervolino, A., Eley, T., Price, T., & Plomin, R. (2006). Environmental risk and young children's cognitive and behavioral development. *International Journal of Behavioral Development, 30*, 55–66.

Plato. (1937). *The republic* (Vol. 1). Cambridge, MA: Harvard University Press.

Repetti, R. L., Taylor, S. E., & Seeman, T. E. (2002). Risky families: Family social environments and the mental and physical health of offspring. *Psychological Bulletin, 128*, 330–366.

Rothbart, M., & Bates, J. (2006). Temperament. In N. Eisenberg (Ed), *Handbook of child psychology: Vol. 3. Social, emotional, and personality development* (6th ed., pp. 99–166). Hoboken, NJ: Wiley.

Uzgiris, I. (1977). Plasticity and structure: The role of experience in infancy. In I. Uzgiris & F. Weizmann (Eds.), *The structuring of experience* (pp. 89–114). New York: Plenum Press.

Wachs, T. D. (1989). The nature of the physical micro-environment: An expanded classification system. *Merrill-Palmer Quarterly, 35*, 399–420.

Wachs, T. D. (1993). Nature of relations between the physical and social microenvironment of the two year old child. *Early Development and Parenting, 2*, 81–87.

Wachs, T. D. (2000). *Necessary but not sufficient: The respective roles of single and multiple influences on individual development.* Washington, DC: American Psychological Association.

Wachs, T. D., & Camli, O. (1991). Do ecological or individual characteristics mediate the influence of the physical environment upon mother–infant transactions? *Journal of Environmental Psychology, 11*, 249–264.

Wachs, T. D., & Chan, A. (1986). Specificity of environmental action as seen in physical and social environmental correlates of three aspects of twelve-month infants' communication performance. *Child Development, 57*, 1464–1475.

Wachs, T. D., & Corapci, F. (2003). Environmental chaos, development and parenting across cultures. In C. Raeff & J. Benson (Eds.), *Social and cognitive development in the context of individual, social and cultural processes* (pp. 54–83). New York: Routledge.

Wachs, T. D., & Gruen, G. (1982). *Early experience and human development.* New York: Plenum Press.

Wachs, T. D., Uzgiris, I., & Hunt, J. M. (1971). Cognitive development in infants of different age levels and different environmental backgrounds; An exploratory study. *Merrill-Palmer Quarterly, 17*, 283–317.

White, B., & Held, R. (1966). Plasticity of sensorimotor development in the human infant. In J. Rosenblith & T. Allensmith (Eds.), *The causes of behavior* (2nd ed., pp. 60–71). Boston: Allyn & Bacon.

Wohlwill, J. (1974, October). *Environmental stimulation and the development of the child.* Paper presented to the Conference on Environment and Cognitive Development, Arad, Israel.

Wohlwill, J., & Heft, H. (1987). The physical environment and the development of the child. In I. Altman & D. Stokols (Eds.), *Handbook of environmental psychology* (pp. 281–328). New York: Wiley.

2

Chaos and the Diverging Fortunes of American Children: A Historical Perspective

Daniel T. Lichter and Elaine Wethington

Today's children will become tomorrow's political and civic leaders, entrepreneurs and workers, and spouses, parents, and caretakers. To many observers, America's future is threatened by increasing numbers of children at risk of family disruption, school dropout, drug abuse, delinquency, and teen pregnancy. Perhaps nostalgically, they cling to the belief that the chaos is something new in children's lives, that unprecedented social and economic changes are buffeting the life course experiences of America's children. Indeed, the past is often viewed in overly sentimental ways—strong family and kinship ties, and stable neighborhoods knitted together by shared ethnicity, religion, language, and supportive community support networks.

To be sure, nostalgic portrayals of the past often reflect past realities. But the truth is that children in the past faced chaos—chronic and persistent instability—in myriad ways that undoubtedly affected their healthy development and transitions to productive adult roles. As we argue in this chapter, chaos in the early 21st century is manifested in much different ways from the past: Chaos at the macrosystem level or structural level has been increasingly replaced over the past century by chaos at the microsystem level (i.e., in children's family environments). However, despite growing scholarly interest in chaos in children's developmental environments, it remains an empirical question whether the typical child today is worse off than children in the past or whether recent social changes threaten more children with chaotic conditions. We argue that America's children may be on ever more divergent tracks as they progress toward adulthood. Focusing on the "typical" or "average" child may obscure growing disparities.

Children's Lives in Historical Context

One of the first explicit statements of American national goals for children and youth was provided in 1930, at the onset of the Great Depression, under a Re-

---

This chapter was supported in part by the Bronfenbrenner Life Course Center at Cornell University. The authors acknowledge the helpful comments of Gary W. Evans, Clyde Hertzman, and Myra Sabir.

publican administration. The Children's Charter, produced by President Hoover's White House Conference on Child Health and Protection, laid out in precise language the national aims for the children of America. It pledged itself to 19 specific aims for the children of America, including some that clearly acknowledged what was perceived then as chaos in children's lives (White House Conference on Child Health and Protection, 1931):

- For every child a dwelling place safe, sanitary, and wholesome, with reasonable provision for privacy, free from conditions which tend to thwart this development; and a home environment harmonious and enriching;
- For every child a community which recognizes and plans for his needs, protects him again physical dangers, moral hazards, and disease; provides him with safe and wholesome places for play and recreation; and makes provision for his cultural and social needs;
- For every child the right to grow up in a family with an adequate standard of living, and the security of stable income as the surest safeguard against social handicaps; and
- For every child a protection against labor that stunts growth, either physical or mental, that limits education, that deprives children of the right of comradeship, of play, and of joy.

Concerns about the well-being of American children reflected the massive macroscale social and economic upheavals at the time. The early 20th century was a period of rapid social change in America, and in the lives of most children. In 1900, for example, the United States had a population of 76 million (U.S. Census Bureau, n.d.f). On October 17, 2006, the Census Bureau reported that the population had topped 300 million. The century also witnessed a sizeable demographic shift from rural to urban areas. In 1900, the large majority of the population lived in rural areas. Between 1900 and 1930 alone, the urban population more than doubled and the number of urban places increased from 1,743 to 3,183 (U.S. Census Bureau, n.d.f). Today, over 75% of the U.S. population lives in urban areas.

America's rapid growth and urbanization went hand-in-hand with a litany of other social and economic dislocations (e.g., two World Wars, the 1919 flu epidemic, massive immigration from Eastern and Southern Europe, the "Great Migration" of Blacks out of the rural South, the Great Depression, the Dust Bowl migration and other White migration from Appalachia to urban areas). It is easy to forget that the early 20th century was a period of great economic uncertainty, social unrest, and rootlessness. Rapid geographic mobility unleashed literally tens of thousands of new in-migrants from the traditional mechanisms of social control: family, community, and religion. In fact, the early Chicago School of Sociology (e.g., Wirth, 1938) sought to understand urbanization and increasing geographic mobility as direct causes of the modern social problems of anomie, crime, poverty, and homelessness. Chicago was a city of immigrants that increased from roughly 10,000 in 1860 to over 2 million in 1910. The city was a natural laboratory for the study of rapid social change—of chaos.

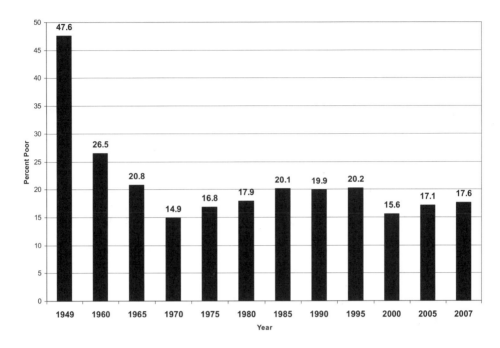

Figure 2.1. Child poverty rates in the United States, 1949–2007, selected years. Data from DeNavas-Walt, Proctor, and Smith (2008).

Then, as now, poverty and chaos are inextricably linked in children's lives (Evans, Connella, Marcynszyn, Gentile, & Salpekar, 2005). If one uses the current definition of poverty (i.e., based on inflation-adjusted income thresholds), the poverty rate for children in 1949—10 years after the Great Depression ended—was 49.7% (Danziger & Danziger, 1993), and it was almost certainly higher during the depth of the Great Depression (Ross, Danziger, & Smolensky, 1987). Child poverty rates dropped rapidly over the ensuing 2 decades, until 1970 (see Figure 2.1). Since then, progress has stalled, fluctuating from year to year between 15% and 20%. In 2007, for example, nearly 1 in 6 (17.6%), or 12.8 million, children lived in poverty (DeNavas-Walt, Proctor, & Smith, 2008). If poverty is a root cause of chaos in children's lives, then there is much less chaos today than in the first half of the 20th century.

For example, poverty is often linked to transience and geographic mobility. A large literature in fact suggests that geographically mobile children suffer from upheavals in friendship and familial support networks (Hagan, MacMillan, & Wheaton, 1996). School progression also is adversely affected by the movement from school to school. America is a highly mobile society, and its changing economy requires a highly mobile labor force. But even a cursory examination of the data suggests that geographical mobility rates even in the late 1940s were little different from the mobility rates of today (U.S. Census Bureau, n.d.e). In 1948, 20.8% of the population changed residences annually. Most moves (13.6%) were of short distance, usually within counties. In 2005, 86.1% of the population was immobile, and a smaller percentage of movers (5.3% vs. 6.4%)

moved long distances (i.e., between counties). These figures are perhaps surprising in light of the conventional wisdom, but unsurprising if considered in the context of rapid urbanization, unprecedented economic dislocations, and high rates of unemployment in early to mid-20th-century America.

At the turn of the 20th century, the death rate from maternal mortality was very high (i.e., nearly 900 maternal deaths per 100,000 live births; Loudon, 1992). Many children grew up without mothers. The children lucky enough to be living with their mothers benefited from full-time maternal care; most mothers did not work outside the home, except for African Americans and some Southern and Eastern European immigrant groups. But the homemaker role was much different at the time. Mothers had more children to care for and lacked access to labor-saving devices and food products. Refrigeration was a luxury. Rural electrification was not completed in many remote parts of America until the 1950s or 1960s. Mothers received little help from fathers, who typically worked long hours and were less engaged emotionally with their children (or, for that matter, with their wives). Female siblings were frequently family caregivers. Older male and female siblings often cut short their own education to help support large families (Sassler, 2000). Workplace injuries and death rates were very high by contemporary standards; occupational safety codes were not enacted until much later and only after considerable pressure from unions. Then, unlike today, single mothers were more likely to be widowed than divorced or never-married.

In rural areas, children were an economic asset and source of cheap labor, especially on farms. Young boys often worked alongside older male relatives in mining or in lumber camps. Child labor laws were not introduced until the 1930s. Big cities were congested, polluted, and unsafe—a breeding ground for infectious disease. Children often paid a big price. Infant and child mortality rates were extraordinarily high in the early 20th century. Most improvements in longevity over the 20th century resulted from mortality declines from infectious diseases among infants and children. Indeed, achievements of public health have been spectacular (Centers for Disease Control and Prevention, 1999). These included improvements in food safety and distribution, nutrition, public sanitation, clean water, and the development and delivery of vaccinations for infectious diseases (e.g., smallpox, measles, and diphtheria) that preyed heavily on infants and children.

Nothing in the preceding discussion is especially original or unknown. Our point is straightforward: Few Americans—even the most destitute—would want to return to the "good ol' days." And probably even fewer believe that the day-to-day circumstances of children and youth in the first half of the 20th century were superior to today's. Historian Stephanie Coontz (1992) made this point clearly in *The Way We Never Were*, exploding the nostalgia myths surrounding changing family, gender, and parenting relationships. To be sure, our intent is not to minimize the very real concerns about (possibly growing) chaos today, but neither should we view children's circumstances in the absence of the broader historical context. Our discussion is simply a reminder of the sweeping large-scale improvements in children's macrolevel environments (and lives) over the past half-century or more. Viewed in this broad historical light, chaos in children's lives is arguably on the wane. Compared with children a century ago, children today are living more stable, safer, and healthier lives along many different dimensions. As we outline in the next section, our concerns about the changing quality of children's lives largely reflect acceleration in family instability. Changes in family structure combined with deteriorating economic circumstances are promoting chaos in the lives of a substantial proportion of American children.

Childhood in Recent Historical Perspective

From Macro to Micro Chaos

In the past, chaos in children's lives reflected large-scale social and economic upheavals in American society. Home was often viewed as a safe haven for children (at least for native-born Whites) in a harsh and chaotic world. Today, much of the world may be a safer place for children, who benefit from better health and a government safety net, but the benefits may be offset by increasing chaos brought about by marital instability and lack of regularity in the home. Moreover, chaos in the home may have less to do with economic shocks or other society-wide economic transformations than with the growing freedom of choice—both good and bad—that parents have concerning marriage, childbearing, and work, and how their choices affect their children's lives. To be sure, parental choices in this regard are constrained by economic, social, and cultural factors. But sweeping cultural shifts in the meaning of marriage and family life often place adult personal satisfaction and happiness above concerns for the well-being of the community, larger kinship network, and even children (Bumpass, 1990). What is good for parents (e.g., leaving a bad marriage) may not be good for their children. Individualism increasingly trumps communalism—and children have suffered as a consequence.

Interestingly enough, our view—that increasing chaos is linked to changing family structure—is apparently at odds with recent studies in the developmental sciences. For example, Coldwell, Pike, and Dunn (2006) claimed that measures of family structure or demographics are correlated only weakly or not at all with measures of chaos in the home. Evans, Eckenrode, and Marcynyszyn (see chap. 14, this volume) document in some detail the data on poverty and chaos, noting the wide variation in findings from study to study, which seemingly reflects income heterogeneity in sampling designs. Evans et al. (2005) also suggested that "levels of chaos are accelerating and pushing beyond the confines of poverty into middle- and upper-income families" (p. 564). The implication, of course, is that chaos in the home is increasing independently of rapid family changes—chaos increasingly affects all American children. In our view, this assertion remains an empirical question.

In fact, direct longitudinal measures of chaos in the home are difficult to find in the literature. Most studies are cross-sectional or follow a survey panel, where any changes over time are confounded with individual development and parental age. For example, chaos has been measured using the Confusion, Hubbub, and Order Scale (Matheny, Wachs, Ludwig, & Phillips, 1995; Petrill, Pike, Tom, & Plomin, 2004). Parents respond to six statements on a 5-point Likert scale (1 = *definitely untrue*, 5 = *definitely true*): (a) "It's a real zoo in our home"; (b) "The atmosphere in our house is calm"; (c) "The children have a regular bedtime routine (e.g., same bedtime each night, a bath before bed, reading a story, saying prayers)"; (d) "You can't hear yourself think in our home"; (e) "We are usually able to stay on top of things"; and (f) "There is usually a television turned on somewhere in our home." Evans et al. (2005) summarized the developmental implications of family chaos succinctly:

> Frenetic activity, lack of structure, and unpredictability, in conjunction with intense background stimulation, take their toll by depriving the developing organism of the kinds of well-structured, predictable, and sustained exchanges of energy with the persons, objects, and symbols in the immediate environment critical to fostering and sustaining healthy development. (p. 560)

In light of this definition, it is perhaps unsurprising that much of the recent chaos literature in human development and psychology emphasizes the deleterious effects associated with excessive crowding, ambient noise, and other stressful neighborhood conditions (e.g., crime; see Evans, 2005). But even a cursory look at the demographic evidence arguably suggests overall declines rather than increases in chaos along these dimensions. For example, American homes have become less crowded. U.S. family size declined rapidly over the Baby Boom and Baby Bust cycles from the 1950s to 1970s. On average, household size declined from 3.33 to 2.57 persons between 1960 and 2006 (U.S. Census Bureau, n.d.b). The percentage of large households with seven or more people declined from 5.4% of all households in 1960 to only 1.2% in 2006 (U.S. Census Bureau, n.d.a). Homeownership rates are also up, rising from 63% to 69% from 1965 to 2006 (U.S. Census Bureau, n.d.c). If one assumes that owner-occupied housing is more spacious and therefore less crowded than renter-occupied housing, then the historical rise in homeownership rates augurs well for America's children. In fact, only 3% of all housing units averaged more than one occupant per room in 2007 (U.S. Census Bureau, n.d.d).

The past 40 years also have been marked by the centrifugal drift of densely settled urban populations to the suburbs, which, on average, are cleaner, quieter, and safer. Over one half of all Americans now reside in the suburbs or exurbs of metropolitan areas. Between 1993 and 2005 alone, the number and rate of serious crimes were cut in half (Bureau of Justice Statistics, 2008). Juvenile arrest rates (ages 0–17) followed a similar downtrend over the same period (Snyder & Sickmund, 2006). On the surface, it is not obvious that chaos, at least as typically measured in the developmental sciences, is a growing phenomenon among children, if measured along these macro trends.

Recent sociological studies have highlighted large class disparities in parenting styles that bear directly on children's development. In *Unequal Childhoods*, Annette Lareau (2003) emphasized the frenzied middle-class obsessions with children's lives—overregulation and overscheduling—rather than the "naturalistic" orientations found among many lower income families. One optimistic view is that working-class children create their own environments rather than have them created by parents and caretakers. Are less regulated environments more chaotic, or do some children learn by these experiences while becoming more self-directed, independent, or resilient? The answers are not obvious, at least to us.

The developmental sciences literature also assumes that chaos is increasing in children's lives, but the evidence is usually indirect rather than based on actual assessments of chaos over time. For example, a typical finding is that chaos is associated with poverty (Evans et al., 2005). On average, the home environments of poor children may be dirtier, less well-kept, and less routinized. The effects of family economic stressors on children's healthy socioemotional and cognitive development are presumably mediated by chaos in the home environment (Coldwell et al., 2006; Hart, Petrill, Keckard, & Thompson, 2007). The implication is clear: Societal levels of chaos therefore presumably fluctuate with changing U.S. poverty rates. But the fact is that poverty rates among U.S. children have been relatively stable over the past 3 decades after undergoing unusually rapid long-term declines over the 1940-to-1970 period. The inflation-adjusted income of children in families at the bottom of the income distribution has also increased over the past decade (Lichter, Qian, & Crowley, 2007). The key question seems clear: Have disadvantaged children or other subpopulations of children (immigrants or minorities) been left behind or excluded from mainstream society?

Perhaps more significantly, whether poverty per se has direct causal effects on children's emotional and cognitive well-being has been a question of considerable recent debate. The inferential problems are well known: Individual traits (e.g., IQ, ability, character traits) that lead directly to low income (and to other associated stressors, such as nonmarital childbearing and divorce) may also be linked to poor parenting and bad childhood outcomes. Mayer (1999), in fact, suggested that poverty or low income is largely unrelated to children's well-being, once other observed and unobserved individual characteristics (e.g., related to selection) are controlled. This situation raises a clear substantive question: Is home chaos a distinct construct, or, instead, does it simply serve as a proxy for other risk factors (including low socioeconomic status and family instability)?

If poverty has demonstrably causal effects and poverty rates provide indirect evidence of growing (or fluctuating) chaos in U.S. children's lives, then it also is important to evaluate whether the statistical relationship between poverty and chaos has changed over time. This is not a small issue. For example, the percentages of poor families and their children who have access to material commodities that are typically associated with a middle-class lifestyle (e.g., air conditioners, televisions, cars) are increasing. Compared with the early 1960s, children today have a larger federal safety net. The dollar value of cash public assistance may have eroded over the past 2 decades, but other "invisible" forms of support have increased, such as the State Children's Health Insurance Program, Social Security and Medicaid, and housing vouchers (Currie, 2006). Our point is not to minimize the cruel reality of poverty in children's lives but simply to acknowledge that the circumstances of low-income children today are much different from the circumstances pre-1960. As a result, poverty—as an indirect indicator of chaos—may misrepresent trends or fluctuations in the chaos in children's lives.

(Percent distribution)

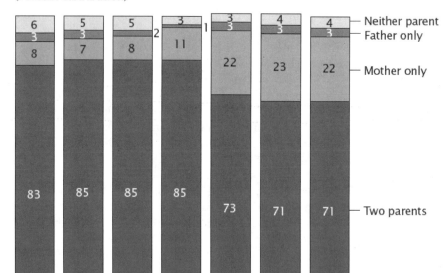

Figure 2.2. Historical living arrangements of children: Selected years: 1880 to 2001 (percent distribution). Reprinted from *Living Arrangements of Children: 2001* (Current Population Reports, No. P70-104; p. 14), by R. M. Kreider and J. Fields, 2005, Washington, DC: U.S. Census Bureau. In the public domain.

Family Change and Chaos

In the sociological and economic literature, children's socioemotional and cognitive development is often viewed in terms of parental investments in time and money, which in turn lead to better outcomes among children (for discussion, see Foster, 2002). Chaos is an unfamiliar concept. The emphasis instead is on changing family structure, which is linked directly with economic resources and parenting (e.g., warmth, monitoring) and with the quality of the home environment rather than on more proximal "causes" of divergent developmental outcomes among children linked to disorganization or chaos in the home. A crude cut on recent trends in children's living arrangements can be gleaned from data on the share of children living with two parents (see Figure 2.2). Living with a single parent is presumed to be associated with more chaos in the home. From 1880 to 1970, about 85% of America's children lived with both parents. Between 1970 and 1980, this percentage dipped to 70%, where it remains. Over 20% of children today reside with a single mother, which places them at risk of high poverty and other threats to well-being (Kreider & Fields, 2005).

Of course, prevalence measures of children's changing living arrangements mask the complexity of recent family trends (e.g., remarriages, new blended families). Crude divorce rates accelerated after 1970 and then leveled off after 1990. Of marriages begun from 1950 to 1954, 14% ended within 10 years; 30 years later, of those marrying between 1980 and 1984, 31% divorced within 10 years (Bramlett & Mosher, 2002). Today, more than 1 million children per year experience the divorce of their parents. Divorce and remarriage of children's parents have been associated with higher levels of sexual activity for daughters in adolescence, poor relationship choices, and depressive symptomatology during young adulthood. Although the data are controversial and analyses are often methodologically weak on this point, increasing instability during childhood in Western developed nations may be associated with rising rates of major depressive disorder in younger birth cohorts (Cross-National Collaborative Group, 1992). Throughout the world, major depressive disorder is one of the leading causes of work disability and years of healthy life lost (Lopez, Mathers, Ezzati, Jamison, & Murray, 2006).

Increasing instability of children's lives also is apparently reflected in high rates of cohabitation and remarriage, which expose children to two coresidential caretakers and economic providers but also introduce new complexities that are unique to stepfamilies (e.g., stepparents, stepsiblings, half-siblings, relationships with noncustodial parents). Remarriage and cohabitation have been linked to increased conflict and ambivalence between parents and children and decreased intergenerational solidarity and support (Eggebeen, 2005; Wethington & Dush, 2007). For children, serial cohabitation and remarriage of parents are linked to economic instability, neighborhood residential mobility, changes in schools and friends, severed emotional ties to adults (e.g., the parent's cohabiting partners), and the reorganization of family processes and rituals (for examples, see Guzzo & Furstenberg, 2007; Lichter & Qian, 2004).

Chaos is also reflected in increasing shares of children born to unmarried mothers. In fact, recent statistics from the National Center for Health Statistics indicate that over 38% of all births are to unmarried women (Hamilton, Martin, & Ventura, 2007). Of these, roughly one half are to cohabiting mothers, whose unions are typically short-lived (Lichter, Qian, & Mellott, 2006). Only about one half of these unions culminate in marriage, but this figure is much lower among poor cohabiting women. Recent estimates suggest that roughly 25% to 35% of American children can expect to live with cohabiting parents during childhood (Graefe & Lichter, 1999; Heuveline & Timberlake, 2004). The apparently growing instability of cohabiting unions may lead to elevated levels of family chaos in children's lives.

Rapid changes in American family life since the 1960s have many putative causes, including the historical rise in maternal employment and women's growing economic independence from men. The rise in women's status may have introduced an additional element of chaos and complexity into children's lives. From 1970 to 2000, labor force participation rates of mothers increased from 47% to 73% (see Figure 2.3). For mothers of very young children, the percentages increased from 34% to 61%. Whether mothers' increasing employment rate represents a new source of chaos in children's lives or has a net negative effect on children's healthy development is the subject of a large and contentious literature (see Brooks-Gunn, Han, & Waldfogel, 2002). It is not surprising that the recent debate over welfare reform has often centered on whether "work

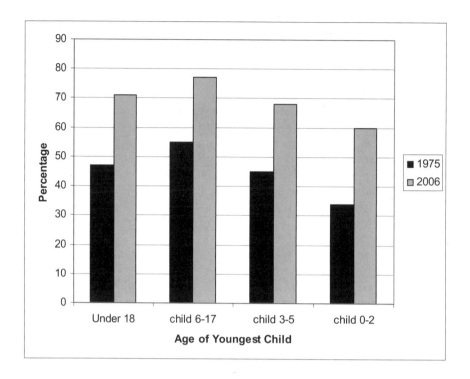

Figure 2.3. Labor force participation rates among mothers. Adapted from *Charting the U.S. Labor Market in 2006* (Chart 6-3), by Bureau of Labor Statistics, 2007, Washington, DC: U.S. Department of Labor. In the public domain.

first" programs ultimately hurt or help poor children (Morris, Duncan, & Clark-Kauffman, 2005).

The putative effects of maternal employment depend on many factors: hours worked, type of work, wage rates, availability of high-quality child care, or whether work is voluntary or forced by economic exigencies, among many other contingencies. On the one hand, working mothers (or parents, more generally) are less available to monitor their children's behaviors on a day-to-day basis or steer their children on a positive developmental trajectory. The typical 8-to-5 work schedule is increasingly being replaced with shift work, contingent work, and weekend schedules that can disrupt family time and routines (Han, 2005). The possible negative impacts appear to be exacerbated by poverty or lack of access to alternative sources of quality child care. Under these circumstances, maternal work can be an additional family stressor and source of marital conflict that places young children at risk.

On the other hand, for some families, maternal employment introduces a degree of routinization or regularity in children's lives. Working mothers must follow a clearly defined daily work schedule that must be synchronized with family routines. Growing children also may benefit from a working parent who provides a role model, additional income to ensure access to higher education, and a connection to positive social and organizational networks in the larger community. Research suggests that connections to positive social networks and effective educational institutions have clear developmental benefits for children who grow up in middle-class households (Davies & Guppy, 1997; Hatcher, 1998).

The long-standing concern that maternal work takes time away from children is being revisited. Bianchi (2000) suggested, for example, that parents today spend more time with their children than they did in the past, despite working more hours. Indeed, the time mothers spend with their children has apparently not been seriously affected by increases in work hours. Though employed mothers may spend less time with their children than do nonemployed mothers, it is also clear that working mothers arrange activities to maximize their time with their children. Fathers are more likely than in the past to contribute to homework and child care, and noncustodial fathers today are more involved with their children than in the past along many dimensions (e.g., economic support and visitation). These are positive developments.

Finally, arguments about growing chaos in the lives of children—especially chaos created by economic and family changes—must be viewed in the context of growing economic, family, and cultural diversity. Overall rates of poverty, single parenthood, divorce, and other risk factors potentially mask growing cultural and economic disparities in American society. Poverty rates have remained relatively stable over the past 25 years (e.g., within 5–7 percentage points), but the poor have fallen further behind the middle-class and affluent Americans (Lichter et al., 2007). The income gap between the "average" and poor children has widened. The British refer to this gap as *social exclusion*, which is characterized by an inability of the poor to fully participate in society. The evidence overwhelmingly indicates that income inequality has increased in the past 2 decades (Neckerman & Torche, 2007).

As we showed in this section, rising inequality may be ushering in a new period of growing chaos amid stability in America, a society of winners and losers. We also suggest the possibility—even likelihood—of childhood trajectories that will increasingly diverge by class, race, and nativity in America's multicultural society.

The Divergent Destinies of Children

Two Tracks?

In her 2004 presidential address to the Population Association of America, Sara McLanahan (2004) argued that the second demographic transition has led to two different trajectories in the lives of women and their children. On one trajectory are women who have delayed marriage and childbearing, invested in their own education and occupational skills, and entered well-paying jobs. Their children have parallel trajectories: social and economic stability and healthy development. On another trajectory are women who have children early, often out of wedlock, and face a life of economic and social instability—chaos—that creates persistent and chronic obstacles to providing resources for their children. McLanahan (2004) suggested that inequality has contributed to rapidly diverging trajectories among America's children. To prove her point, she documented a growing gap over 1960 to 2000 between low- and high-educated mothers, whom she defined by their location in the bottom and top quintiles in the education distribution, respectively. Over this period, low- and high-educated mothers diverged significantly on median age at childbearing, single motherhood, and employment rates. Women at the top of the education distribution were far more likely than other women to delay childbearing, avoid out-of-wedlock childbearing, and work outside the home.

McLanahan (2004) focused on women's experiences, but the clear implication of her thesis is that America's children may be increasingly divided into "haves" and "have-nots" (or even between "haves" and "have mores"). Our preceding discussion emphasized family change as an increasingly important axis of economic differentiation (and, by extension, chaos) in children's lives. To make this point, Table 2.1 provides poverty rates in 2006 for children living in different types of families and having parents with different work patterns. It is no surprise that children living in families supported by single mothers—especially nonworking mothers—show a marked disadvantage. Almost 75% of these children were poor in 2006. The poverty rate is less than 1% among children whose married parents worked full-time. Parental work and marriage clearly go hand-in-hand in shaping children's economic trajectories. Other recent studies show that college-educated women—those with good job prospects—are both more likely to marry and less likely to divorce than are other women. Lichter and Qian (2004) showed that the relationship between education and marriage reversed from a negative to a positive relationship over the 1980-to-2000 period. Highly educated women are not eschewing marriage.

In some respects, the circumstances at birth for disadvantaged children have probably improved overall over the past 3 decades. For example, the share of all births to teenagers declined from 17.6% to 11.8% between 1970 and 2000, and the share of all nonmarital births to teenage mothers declined from 50% to 28% (Lichter & Wooton, 2005). Most children, whether from married-couple families or not, are born to women rather than adolescents or teenagers. In fact, over one third of nonmarital births are born to women over age 25. Any negative implications for child development may be offset by positive changes in demographic characteristics of nonmarried mothers. For example, the share of all births to college-educated mothers increased threefold from 1970 to 2000, from 8.6% to 24.7%. Conversely, the share of children born to high school dropouts declined from 30.8 to 21.7 over the same period (Lichter & Wooton, 2005). Though these are potentially positive developments, they also reinforce our argument that some segments of American children, albeit perhaps a smaller proportion in relative terms, may be falling behind developmentally. They may be on a more chaotic trajectory while also falling behind other children on the road to adult success.

Chaos and Inequality

Whether winners and losers are increasingly differentiated by race and nativity also requires attention. Over 40% of American children today are racial mi-

Table 2.1. Poverty Rates Among Children Related to the Householder, by Family Structure and Work Status, 2006

	Worked last week	Full-time, full-year
Married couples		
Both worked	2.7	0.7
Husband only worked	14.4	9.9
Wife only worked	23.3	7.6
Single mothers		
Worked	31.1	17.0
Did not work	73.9	n/a

Note. n/a = not available. Data from U.S. Census Bureau (2007).

norities or immigrants, whose day-to-day living circumstances often differ substantially from those of the "typical" or modal child—native-born, White, middle-class children with two working married parents. Current estimates also indicate that one of every five children in America is the child of immigrants. For them, poverty rates are nearly twice the national average (Lichter et al., 2007). New immigrant families, mostly from Latin America and Asia, are poorer and less skilled than they were in the past. They often live in impoverished and highly congested urban ethnic enclaves, residentially segregated from native-born non-Hispanic Whites and cut off from mainstream society and employment opportunities. Minority and immigrant children are especially likely candidates for experiencing chaos in the home. At the same time, the meaning of chaos or of alternative expressions of instability may vary substantially across cultural and racial or ethnic groups. Furthermore, for the refugee population—the most economically disadvantaged immigrant population—the chaos they experience may pale in comparison to the oppression and abject poverty many of them faced in their origin counties. Commonly used definitions or measures seem to be insensitive to cultural variations in the way chaos is expressed across different cultural groups.

In the current era, a fundamental reason for differentiation between children at the top of the economic distribution and those at the bottom is that the former have parents who are "winners" in the job market. The winners can afford to purchase stability—less chaos—for their children on a number of important dimensions. They are more likely to be married, and they have a lower risk of divorce. For the low-income population, marriage is increasingly viewed as a luxury that only the affluent can afford (Burstein, 2007). With growing income inequality, middle-class and affluent children may be increasingly separating themselves—physically, socially, and culturally—from other children, especially minority and immigrant children, who reside disproportionately at the bottom of the income distribution. Their children are more likely to live in stable, low-crime neighborhoods, go to good schools, connect to effective social institutions, and live in relatively stable residential communities that provide good public and private support services. The affluent, in particular, may interact only with their own kind in private schools and (gated) communities (Lee & Marlay, 2007), cut off from "other" children who are culturally different and often poor. The experiences of poor, unmarried, minority, or immigrant mothers and their children are much different and are arguably diverging from the experiences of the "typical" child in America.

To be sure, chaos is also a social construction—a matter of definition. A real-life illustration makes our point. Early in 2007, *The New York Times* ran two stories that illustrate how income shapes the divergent social trajectories of America's young people (Barry, 2007; Rimer, 2007). One story profiled highly accomplished, suburban high school girls, the children of educated, middle-class working parents (Rimer, 2007). The girls were at the top of their high school classes academically and scored highly on standardized tests. The article focused on the uncertainty about whether their excellent grades, extracurricular activities, and test scores were sufficient to guarantee admission into an Ivy League university or another prestigious private college. This was a worrisome issue for parents. Admission to an elite college was a critical turning point in the life course, with long-term implications for success. The anxieties of girls and parents were portrayed sympathetically—the future was at stake.

Another *New York Times* article featured a related but very different story (Barry, 2007). This one focused on a teenage girl soon to graduate from a small-town high school in the Southwest. As the oldest child in an economically struggling family, she too had ambitions to go to college after graduating at the top of her class. But without family savings and with younger siblings still at home, these goals would have to wait. She chose instead to enlist in the Army. After high school graduation, she would report for basic training, and then (everyone was certain) she would be on her way to Iraq. GI benefits would take care of college expenses later at the local community college. The article did not convey family stress or uncertainty, but instead the girl's family and friends were cast as united in support of her choice—a rational one under the circumstances. There was no sense of either entitlement or bitterness.

These two stories make contrasting points. On the one hand, they illustrate our general point that middle-class children and lower income children routinely follow diverse life course trajectories, with clear implications for upward mobility and the putative openness of America's system of social stratification. On the other hand, they illustrate that seemingly more chaotic life circumstances can nevertheless produce mature, admirable, courageous, and prosocial young people. Averages mask great heterogeneity in the experiences (and outcomes) of today's young people. It is interesting that comments from readers, published later, focused less on the fascinating juxtaposition of divergent life stories than on the negative impacts of the intense competition among middle-class children (and their parents) for too few prestigious college slots.

Looking Backward and Forward

In this chapter, we have reviewed research on chaos and on the question of whether it has increased in the lives of children over the past century. The answer clearly depends on how chaos is defined—at the macrosystem or microsystem level. As we have argued here, many more children 100 years ago than today suffered from financial and social instability. Children were more likely to be threatened by ill health and even death. Life was hard and children suffered in myriad ways. From a macrolevel perspective, children are arguably better off overall today than they have been in the past. We have argued that chaos in children's lives today is more likely linked to changes in the micro environment, especially to changing family structure and processes that directly affect parental inputs in time and money.

To this end, we have reviewed evidence that chaos in children's lives is associated with the confluence of cultural and economic changes that have reshaped American family life through rising nonmarital fertility, the "retreat from marriage," cohabitation, and divorce. These changes have been played out in the growing share of children living in disrupted or nonintact families during critical developmental periods. Family changes relate in one way or another to the undermining of family solidarity and traditional strengths (for nurturing and socialization and other forms of family social capital) in an increasingly autonomized and individualized world. Strong kinship ties are irreplaceable connections to important social networks and to the wider community of opportunity. At the same time, we fully recognize that any claims that chaos is increasing (or decreasing) over time depend on some consensus about what chaos is, how it is measured, and what historical time frame should be used (long- versus short-term).

The question about whether chaos has increased may be less germane than whether the risk of chaos differs across divergent populations in the increasingly multiracial, multicultural society of the United States. We have argued that the emphasis should be on "chaos amid stability"—on how growing inequality and diversity are possibly steering children along different developmental trajectories. Indeed, American children may be on increasingly divergent trajectories, with varying exposure to many different forms of chaos in their lives. Chaos is a useful concept for understanding developmental threats during childhood, but our review highlights the need for more contextualized empirical applications of the chaos perspective. Glen Elder's (1974) well-known volume *Children of the Great Depression* offers a reminder of the extraordinary success of this birth cohort, despite the threats (and chaos) they faced during childhood. The lesson seems clear: Contextualized applications may highlight many compensating factors that contribute to positive development or resilience among children.

Our historical perspective is meant to stimulate additional research on at risk populations that may be diverging from the cultural and economic mainstream. An important but often neglected dimension of chaos research is the political and policy context in which prevention and intervention programs are developed to mitigate the developmental impact of family instability. A fundamental-cause perspective on intervention and prevention (Link & Phelan, 1995) suggests that individually based interventions are less likely to be effective than more broadly based societal interventions. The past decade ushered in a political era characterized by a resurgence of ideologies that emphasize laissez-faire capitalism and privatization, personal responsibility rather than "big government," and cutbacks in entitlements and the social safety net. This may change under the new Democratic administration, especially as it responds to the financial crisis and the economic recession. Our concern is that the gap in living conditions (and of chaos) between the poorest and richest children will widen—that more children will be put at risk and that the great progress of the 20th century will evaporate.

References

Barry, D. (2007, March 4). A teenage soldier's goodbyes on the road to over there. *The New York Times.* Available at http://select.nytimes.com/2007/03/04/us/04land.html?_r=1&scp=1&sq=&st=nyt

Bianchi, S. M. (2000). Maternal employment and time with children: Dramatic change or surprising continuity? *Demography, 37,* 401–414.

Bramlett, M. D., & Mosher, W. D. (2002). Cohabitation, marriage, divorce, and remarriage in the United States (Vital and Health Statistics, Series 23, No. 22). Hyattsville, MD: National Center for Health Statistics.

Brooks-Gunn, J., Han, W. J., & Waldfogel, J. (2002). Maternal employment and child cognitive outcomes in the first three years of life: The NICHD study of early child care. *Child Development, 73,* 1052–1072.

Bumpass, L. L. (1990). What's happening to the family: Interactions between demographic and institutional change. *Demography, 27,* 483–498.

Bureau of Justice Statistics. (2008). *National crime victimization survey violent crime trends, 1973-2005.* Retrieved January 7, 2008, from http://www.ojp.usdoj.gov/bjs/glance/tables/viortrdtab.htm

Bureau of Labor Statistics. (2007). *Charting the U.S. labor market in 2006.* Washington, DC: U.S. Department of Labor.

Burstein, N. R. (2007). Economic influences on marriage and divorce. *Journal of Policy Analysis and Management, 26,* 387–429.

Centers for Disease Control and Prevention. (1999, April 2). Ten great public health achievements: United States, 1900-1999. *Morbidity and Mortality Weekly Report, 48,* 241–243.

Coldwell, J., Pike, A., & Dunn, J. (2006). Household chaos: Links with parenting and child behaviour. *Journal of Child Psychology and Psychiatry, 47,* 1116–1122.

Coontz, S. (1992). *The way we never were: American families and the nostalgia trap.* New York: Basic Books.

Cross-National Collaborative Group. (1992). The changing rate of major depression. *JAMA, 268,* 3098–3105.

Currie, J. M. (2006). *The invisible safety net: Protecting the nation's poor children and families.* Princeton, NJ: Princeton University Press.

Danziger, S. K., & Danziger, S. (1993). Child poverty and public policy: Toward a comprehensive antipoverty agenda. *Daedalus, 122*(1), 57–84.

Davies, S., & Guppy, N. (1997). Fields of study, college selectivity, and student inequalities in higher education. *Social Forces, 75,* 1417–1438.

DeNavas-Walt, C., Proctor, B. D., & Smith, J. (2008). *Income, poverty, and health insurance coverage in the United States: 2007* (Current Population Reports, No. P60-235). Washington, DC: U.S. Census Bureau.

Eggebeen, D. J. (2005). Cohabitation and exchanges of support. *Social Forces, 83,* 1097–1110.

Elder, G. H., Jr. (1974). *Children of the Great Depression: Social change in life experience.* Chicago: University of Chicago Press.

Evans, G. W. (2005). Child development and the physical environment. *Annual Review of Psychology, 57,* 423–451.

Evans, G. W., Connella, C., Marcynszyn, L. A., Gentile, L., & Salpekar, N. (2005). The role of chaos in poverty and children's socioemotional adjustment. *Psychological Science, 16,* 560–565.

Foster, E. M. (2002). How economists think about family resources and child development. *Child Development, 73,* 1904–1914.

Graefe, D. R., & Lichter, D. T. (1999). Life course transitions of American children: Parental cohabitation, marriage, and single motherhood. *Demography, 36,* 205–217.

Guzzo, K. B., & Furstenberg, F. F. (2007). Multipartnered fertility among American men. *Demography, 44,* 583–601.

Hagan, J., MacMillan, R., & Wheaton, B. (1996). New kid in town: Social capital and the life course effects of family migration on children. *American Sociological Review, 61*, 368–385.

Hamilton, B. E., Martin, J. A., & Ventura, S. J. (2007). *Births: Preliminary data for 2006.* Retrieved December 9, 2007, from http://www.cdc.gov/nchs/data/nvsr/nvsr56/nvsr56_07.pdf

Han, W. J. (2005). Maternal nonstandard work schedules and child cognitive outcomes. *Child Development, 76*, 137–154.

Hart, S. A., Petrill, S. A., Keckard, K. D., & Thompson, L. A. (2007). SES and CHAOS as environmental mediators of cognitive ability: A longitudinal genetic analysis. *Intelligence, 35*, 233–242.

Hatcher, R. (1998). Class differentiation in education: Rational choice? *British Journal of Sociology of Education, 19*, 5–23.

Heuveline, P., & Timberlake, J. M. (2004). The role of cohabitation in family formation: The United States in comparative perspective. *Journal of Marriage and Family, 66*, 1214–1230.

Kreider, R. M., & Fields, J. (2005, July). *Living arrangements of children: 2001* (Current Population Reports, No. P70-104). Washington, DC: U.S. Census Bureau. Available at http://www.census.gov/prod/2005pubs/p70-104.pdf

Lareau, A. (2003). *Unequal childhoods: Class, race, and family life.* Berkeley: University of California Press.

Lee, B. A., & Marlay, M. (2007). The right side of the tracks: Affluent neighborhoods in the metropolitan United States. *Social Science Quarterly, 88*, 766–789.

Lichter, D. T., & Qian, Z. (2004). *Marriage and family in a multiracial society* (The American People: Census 2000 Series). New York: Russell Sage Foundation and Population Reference Bureau.

Lichter, D. T., Qian, Z., & Crowley, M. L. (2007). Poverty and economic polarization among America's minority and immigrant children. In R. Crane & T. Heaton (Eds.), *Handbook of families and poverty: Interdisciplinary perspectives* (pp. 119–143). New York: Sage.

Lichter, D. T., Qian, Z., & Mellott, L. (2006). Marriage or dissolution? Union transitions among poor cohabiting women. *Demography, 43*, 223–240.

Lichter, D. T., & Wooton, J. (2005). The concentration of reproduction in low-fertility societies: The case of the United States. In A. Booth & A. C. Crouter (Eds.), *The new population problem: Why families in developed countries are shrinking and what it means* (pp. 213–224). Mahwah, NJ: Erlbaum.

Link, B. G., & Phelan, J. C. (1995). Social conditions as fundamental causes of disease. *Journal of Health and Social Behavior, 36*(Extra issue), 80–94.

Lopez, A. D., Mathers, C. D., Ezzati, M., Jamison, D. T., & Murray, C. J. L. (Eds.). (2006). *The global burden of disease and risk factors.* Oxford, England: The World Bank.

Loudon, I. (1992). *Death in childbirth: An international study of maternal care and maternal mortality, 1800-1950.* New York: Oxford University Press.

Matheny, A. P., Wachs, T. D., Ludwig, J. L., & Phillips, K. (1995). Bringing order out of chaos: Psychometric characteristics of the Confusion, Hubbub, and Order Scale. *Journal of Applied Developmental Psychology, 16*, 429–444.

Mayer, S. (1999). *What money can't buy: Family income and children's life chances.* Cambridge, MA: Harvard University Press.

McLanahan, S. (2004). Diverging destinies: How children are faring under the second demographic transition. *Demography, 41*, 607–627.

Morris, P., Duncan, G. J., & Clark-Kauffman, E. (2005). Child well-being in an era of welfare reform: The sensitivity of transitions in development to policy change. *Developmental Psychology, 41*, 919–932.

Neckerman, K. M., & Torche, F. (2007). Inequality: Causes and consequences. *Annual Review of Sociology, 33*, 335–357.

Petrill, S. A., Pike, A., Tom, P., & Plomin, R. (2004). Chaos in the home and socioeconomic status are associated with cognitive development in early childhood: Environmental mediators identified in a genetic design. *Intelligence, 32*, 445–460.

Rimer, S. (2007, April 1). For girls, it's be yourself, and be perfect, too. *The New York Times.* Available at http://www.nytimes.com/2007/04/01/education/01girls.html?scp=1&sq=&st=nyt

Ross, C., Danziger, S., & Smolensky, E. (1987). The level and trend of poverty in the United States, 1939–1979. *Demography, 24*, 587–600.

Sassler, S. (2000). Learning to be an "American lady"? Ethnic variation in daughters' pursuits in the early 1900s. *Gender & Society, 14*, 184–209.

Snyder, H., & Sickmund, M. (2006). *Juvenile offenders and victims: 2006 National Report.* Washington, DC: U.S. Department of Justice, National Center for Juvenile Justice, Office of Juvenile Justice and Delinquency Prevention.

U.S. Census Bureau. (n.d.a). *FM-3. Average number of own children under 18 per family, by type of family: 1955 to present.* Retrieved January 7, 2008, from http://www.census.gov/population/socdemo/hh-fam/fm3.csv

U.S. Census Bureau. (n.d.b). *HH-4. Households by size: 1960 to present.* Retrieved January 7, 2008, from http://www.census.gov/population/socdemo/hh-fam/hh4.csv

U.S. Census Bureau. (n.d.c). *Housing vacancies and homeownership (CPS/HVS).* Retrieved January 7, 2008, from http://www.census.gov/hhes/www/housing/hvs/historic/histt14.html

U.S. Census Bureau. (n.d.d). Percent of occupied housing with 1.01 or more occupants per room. Retrieved April 26, 2009, from http://factfinder.census.gov/servlet/GRTTable?_bm=y&_geo_id=01000US&-_box_head_nbr=R2509&-ds_name=ACS_2007_1YR_G00_&-redoLog=false&-format=US-30&-mt_name=ACS_2005_EST_G00_R2509_US30

U.S. Census Bureau. (n.d.e). *Table A-1. Annual geographic mobility rates, by type of movement: 1947–2006.* Retrieved January 7, 2008, from http://www.census.gov/population/socdemo/migration/tab-a-1.pdf

U.S. Census Bureau. (n.d.f). *Table 4. Population: 1790 to 1990.* Retrieved January 7, 2008, from http://www.census.gov/population/censusdata/table-4.pdf

U.S. Census Bureau (2007). *Current Population Survey: 2007 Annual Social and Economic Supplement.* Retrieved April 27, 2009, from http://pubdb3.census.gov/macro/032007/pov/new21_100_01.htm

Wethington, E., & Dush, C. M. K. (2007). Parents' assessments of relationship quality with offspring in midlife: Intact and blended families. In T. J. Owens & J. J. Suitor (Eds.), *Advances in life course research: Interpersonal relations across the life course* (Vol. 12, pp. 123–152). New York: Elsevier.

White House Conference on Child Health and Protection. (1931). *The Children's Charter [government document].* Retrieved April 26, 2009, from http://chnm.gmu.edu/cyh/primary-sources/124

Wirth, L. (1938). Urbanism as a way of life. *American Journal of Sociology, 44*, 1–24.

Part II

Chaos at the Microsystem Level

3

Physical and Psychosocial Turmoil in the Home and Cognitive Development

Brian P. Ackerman and Eleanor D. Brown

Many developmental researchers use Bronfenbrenner's (Bronfenbrenner & Evans, 2000) bioecological model as a research frame for understanding progressive transactions between the child and the immediate environment. The extraordinary influence of the bioecological perspective is apparent in the abundant research on the multiple contexts of child development represented by family, neighborhood, school, and cultural variables and their nested interrelations. Other core features include the focus on mediating and proximal processes linking contextual factors and child outcomes, with a strong recent emphasis on stress physiology and self-regulation, and the search for demographic and child moderators of those links. However, units of analysis beyond the dyad have received less attention until recently. Much is known about interpersonal processes in the family, for instance, but not much is known about family-level variables and their relations to child outcomes. Chaos centers on family-level variables concerning background stimulation, social and physical disorganization, and a lack of predictability and structure in the home (for detailed discussion of the chaos construct, see chaps. 1 and 16, this volume).

In this chapter, we discuss developmental research describing chaotic aspects of the home environment in early and middle childhood in relation to child cognitive outcomes. Following the lead of Evans (2003; Evans, Kim, Ting, Tesher, & Shannis, 2007), we distinguish between aspects of the physical setting, including noise, crowding, and household routines, and psychosocial aspects that contribute to family turmoil. The psychosocial aspects include instability in caregiver intimate relationships and separations from caregivers, residential moves, parent maladjustment, and perhaps dynamic income changes for disadvantaged families.

Our discussion is divided into four sections. First, we discuss preliminary theoretical considerations as a way of establishing an interpretive frame. In the second section we describe empirical evidence concerning the relations between physical and psychosocial chaos at a family level and child outcomes. The third and fourth sections focus on possible mechanisms that mediate the relations and possible moderators of the relations. These sections highlight missing links that might contribute to a research agenda.

Theoretical Considerations

Five theoretical issues frame interpretation of evidence relating chaotic home environments to children's cognitive processes. We discuss these issues with the goal of providing insights into design elements and limitations of studies. A second goal is to delimit the set of studies that concern these relations. The empirical base is huge for individual variables that may contribute to environmental chaos (Evans, 2006), for instance, but small for chaos defined as an aggregate variable. But aggregates may be most appropriate for representing chaos at a family level. Similarly, different aspects of environmental adversity tend to relate to specific developmental outcomes. Physical chaos relates most directly to cognitive development, whereas psychosocial turmoil relates primarily to socioemotional processes and psychological distress. This specificity limits the relevance of studies about psychosocial chaos for our review.

Family-Level Variables

Family-level variables influence all aspects of family functioning, including individual behavior and adaptation and interpersonal processes. High levels of background noise, for instance, or unpredictable and irregular family mealtimes may impact the ability of every family member to plan and control a daily agenda. The implication is that the explanatory value of family chaos as a construct requires specification of pathways of potential influence on child outcomes. The contribution may be direct or mediated through family and parenting variables, such as coercive parenting or hostile marital conflict. Coldwell, Pike, and Dunn (2006) provided a good recent example by distinguishing unique effects of family chaos and interpersonal processes in the home.

Income Poverty

A second important issue concerns the theoretical and empirical distinction between environmental chaos in the home and income poverty. The issue is complex because income poverty also is a family-level variable. The distinction thus rests on the argument that chaos cuts across income levels. There is good evidence that it does (Evans, Maxwell, & Hart, 1999; Gottfried & Gottfried, 1984; Wachs, 1979).

But to what extent? Abundant evidence shows that chaos varies with family income and that disadvantaged homes are more likely to suffer from high levels of chaos than are the homes of more advantaged families (Evans, 2004; Evans, Gonnella, Marcynyszyn, Gentile, & Salpekar, 2005). A good case also can be made that extremely poor families are by definition chaotic, as perhaps revealed in environmental correlates of the scramble to find adequate daily resources for the family. The issue then is the extent to which representations of environmental chaos add generally to the explanation of child cognitive outcomes beyond family income and strong correlates concerning maternal education and markers of cognitive competence.

It is possible to phrase the issue the other way as well, which may work to the advantage of family chaos as an explanatory variable. A common problem with studies showing relations between income poverty and child cognitive outcomes is the failure to distinguish between family income per se and environmental cofactors. When poverty is treated as an undifferentiated variable, relations to outcomes could well reflect unexamined aspects of environmental chaos (Duncan & Brooks-Gunn, 1997). Similarly, when there are robust correlations between chaos variables and family income, controlling for family income in exploring unique effects for chaos variables is likely to provide an underestimate of chaos effects.

Representation

A third issue concerns the appropriate level of representation for chaotic environments, as individual variables or as a cumulative aggregate. Good reasons to use aggregate representations are that individual risk factors usually cluster together in families, and individual variables usually explain only a small percentage of the unique variance in child outcomes (Sameroff, Seifer, & Bartko, 1997) and potentially describe only an aspect of a potentially chaotic environment rather than the environment as a whole. The argument is that individual elements such as noise or crowding or residential moves may not be experienced as disorganizing or disruptive if other aspects of the family system are stable, predictable, nondebilitating, and not particularly stressful. The sense of family life as hectic, disorganized, and unpredictable (i.e., chaotic), and perhaps out of control, seems like a summary judgment emerging from the disruptive effects of multiple variables. In this respect, a cumulative aggregate composed of multiple chaotic elements seems most appropriate for representing the environment as a whole.

Researchers have focused on various individual variables such as environmental noise (Evans, Bullinger, & Hygge, 1998; Evans, Hygge, & Bullinger, 1995) and crowding (Evans, Lepore, Shejwal, & Palsane, 1998; Evans et al., 1999; Wachs & Gruen, 1982) or have combined variables in a single scale, such as CHAOS (Matheny, Wachs, Ludwig, & Phillips, 1995), or in a cumulative risk index (Evans, 2003; Evans & Kim, 2007; Evans et al., 2007). *CHAOS* refers to environmental confusion, hubbub, and order, with items focusing on background noise, crowding, and family routines. Several recent studies have related the scale to parenting (Coldwell et al., 2006; Corapci & Wachs, 2002) and to aspects of children's attentional control (Dumas et al., 2005; Valiente, Lemery-Chalfant, & Reiser, 2007). Evans has gone further in combining these aspects of the physical environment with variables representing psychosocial turmoil in a cumulative risk index. The advantage is that an index coding multiple environmental stressors may more adequately represent the child's experience of the home environment.

In general, a strong argument can be made that high levels of cumulative risk generate and reflect environmental chaos in the home by definition. Cumulative representations, however, have difficulties that urge specification of relations to environmental chaos. One difficulty is that most cumulative risk indexes in the literature contain an amalgam of variables, with some clearly not family-level variables (e.g., parent–child interactions) and some relating only vaguely and inferentially to physical or psychosocial chaos in the home. Another difficulty is that variables may differ substantially in contributing to family disorganization and unpredictability, though a cumulative representation weights all variables equally. A third and related difficulty is that physical and psychosocial variables generally seem to relate differently to child outcomes.

Cognitive Development

The fourth issue concerns conceptualization of cognitive development and outcomes. Because many processes develop, it is difficult to draw general conclusions about relations between environmental chaos and cognitive development. Another difficulty is that some cognitive products seem derivative of others and are influenced by multiple factors. Early literacy skills, intelligence, and behavioral regulation (e.g., attention) in the preschool years, for example, predict academic skill acquisition and achievement in school (Duncan et al., 2007). But academic achievement may also reflect motivational variables and more social variables that foster engagement (Blair, 2002) and that foster convergence of academic and behavioral problems in school. These variables suggest the possibility of a more social pathway linking chaotic home environments and later cognitive outcomes.

Directions of Effects

Chaos effects assume a social causation perspective in which the effects of environmental causes cannot be reduced to or explained by social selection processes focusing on person variables. An advantage of conceptualizing chaos as a family-level and aggregate variable, in this regard, is that chaos is somewhat distal to the child and cannot be reduced easily to child effects. This advantage, however, does not protect the explanatory value of environmental chaos from caregiver characteristics that select and shape environments, or from genetic explanations. Valiente et al. (2007) provided a good recent example of the issue in showing relations between the CHAOS Scale, parents' effortful control, and children's effortful control. The issue is whether environmental chaos undermined parents' ability to regulate their own behavior or was a result of that lack of regulation.

This problem of eliminating social selection explanations of chaos effects is largest for studies with concurrent measures of environmental, caregiver, and child variables. The problem is reduced in prospective longitudinal designs and in designs focusing on explaining change in caregiver or child behavior (or both). Only a few studies about physical or psychosocial chaos in the home, however, are prospective or attempt to explain change in child behavior.

Cognitive Outcomes

With these theoretical issues in mind, in this section we describe evidence relating chaotic home environments to cognitive outcomes in early and middle childhood. We distinguish between direct and indirect effects.

Physical Chaos

In a recent review, Evans (2006) comprehensively described the many studies over several decades linking noise, crowding, and housing quality as individual variables to cognitive functioning and development in children. The studies frequently controlled for socioeconomic status (SES) but focused mostly on concurrent relations among environmental and cognitive variables. The studies showed both direct effects on markers of children's psychological stress and indirect effects through such mediators as parental responsiveness and the home learning environment. We highlight findings and issues rather than repeat Evans's review.

Evans's studies (Evans, Bullinger, & Hygge, 1998; Evans et al., 1995) provide good illustrations of direct effects found in examining relations between chronic noise exposure for third and fourth graders and neuroendocrine and cardiovascular markers of psychological stress. The manipulation of noise exposure contrasted children living in noisy (near the Munich International Airport) with those in quiet urban neighborhoods. Demographic characteristics of the neighborhoods (SES) and households were controlled. Evans et al. (1995) found relations to attention, memory, and reading, and both studies found relations to quality-of-life indicators. The relation to reading deficits is a common finding for both noise and crowding (Evans, 2006; Evans & Lepore, 1993; Wachs & Gruen, 1982).

Studies of the effects of residential density (i.e., crowding) provide good illustrations of indirect effects, with effects on language development and other cognitive processes mediated by parent–child interactions. Parents in crowded homes, for instance, talk less (Wachs, 1979; Wachs & Camli, 1991) and in less complex ways to their children (Evans et al., 1999), and generally are less responsive (Evans et al., 1999) and more negative (Evans, Lepore, et al., 1998). Crowding also relates generally to aspects of the home learning environment as reflected in the Home Observation for Measurement of the Environment (HOME; Caldwell & Bradley, 1984), and relates through these aspects to cognitive development across the preschool period (Bradley & Caldwell, 1984; Gottfried & Gottfried, 1984). Residential density (and noise), for instance, may discourage caregiver efforts to read to children.

Evans et al. (1999) provided a particularly good example in showing relations between residential density, parental verbal responsiveness to their children, and the complexity of parent speech to their children in the University of Kansas Language Acquisition Project. The example is particularly apt because findings from this project have been seminal in showing relations among SES, the home verbal environment, and measures of child cognitive competence (Hart & Risley, 1995). Evans et al. (1999) showed, however, that the effect for the parent responsiveness variable was independent of SES.

Fewer studies have focused on the relations between disorganization, confusion, and unpredictability in home routines and child cognitive developments per se in the context of controls for family income and child effects. Widespread use of the HOME inventory is a good candidate in this regard, because this measure of the household environment consistently explains variance in measures of child cognitive competence. Only one scale, however, represents chaotic aspects of the environment per se, focusing on the organization of the physical and temporal environment. Bradley and Caldwell (1984) showed zero-order relations between the scale and infant cognitive competence, but the unique effects are questionable in the context of other scales and controls.

A few other studies with SES controls also show relations between household routines and structure and both cognitive development (Petrill, Pike, Price, & Plomin, 2004) and academic achievement (Brody & Flor, 1997), but most studies focus on relations to infant temperament and socioemotional development (Fiese et al., 2002). Indeed, recent studies using the CHAOS measure (Matheny et al., 1995) related household routines and disorganization mostly to parenting interactions and child problem behaviors (Coldwell et al., 2006; Corapci & Wachs, 2002; Dumas et al., 2005; Evans et al., 2005; Valiente et al., 2007), though child attentional processes are outcomes as well (Dumas et al., 2005).

Other stressful aspects of the family environment also may have an impact on household routines and predictable functioning, which may not necessarily reflect turmoil, which we describe next. Hsueh and Yoshikawa (2007), for instance, prospectively related nonstandard and variable work schedules to parent and teacher reports about school achievement for low-income families participating in the New Hope Project. The New Hope Project is a work-based antipoverty program in Milwaukee, Wisconsin. The results of this study suggest the need to cast a broader net in conceptualizing variables that may impact family routines affecting cognitive outcomes.

Psychosocial Chaos

Aspects of psychosocial turmoil that may stress all aspects of family functioning include caregiver relationship instability, parent maladjustment (i.e., incarceration, substance abuse, psychiatric morbidity), residential instability (i.e., moves), separations from primary caregivers and unstable foster care arrangements, nonstandard and variable employment schedules, and dynamic income changes (i.e., for disadvantaged families). These aspects as individual variables or as cumulative aggregates may affect child behavior indirectly through mediators such as caregiver mood, marital discord, and parenting processes (Conger et al., 2002) and perhaps directly through a variety of proximal person-based processes (Ackerman, Brown, & Izard, 2003).

Although it is reasonable to conceive of these aspects of family turmoil as family-level variables, interpreting relations between turmoil and child cognitive outcomes is problematic in three ways. First, the aspects covary with family income, which means that effects associated with environmental chaos and with poverty must be distinguished. Second, relations to confusion, disorganization, and unpredictability in the environment (i.e., chaos) usually are implied but not shown directly. Third, few studies show direct relations to cognitive outcomes in the context of controls for family income. In contrast, psychosocial turmoil shows strong selective relations to problem behaviors (Ackerman, Brown, & Izard, 2004).

Despite these difficulties, some evidence suggests that several family-level variables relate directly to children's cognitive outcomes. Residential mobility, for instance, predicts children's academic achievements independent of SES (Adam, 2004), though perhaps not cognitive competence (Stoneman, Brody, Churchill, & Winn, 1999). Similarly, changes in family income may destabilize family functioning in various ways for disadvantaged families. These changes may be frequent, given the relative instability of unskilled and low-wage jobs. Although it is theoretically important to distinguish environmental chaos from income poverty per se, dynamic poverty represents an exception because of the disruptive influence on families. These dynamic income changes also are associated with the cognitive outcomes of preschool children (Yeung, Linver, & Brooks-Gunn, 2002), perhaps mediated through parenting practices and school involvement (Dearing, Kreider, Simpkins, & Weiss, 2006).

Child Mechanisms

We have described parenting mediators that relate aspects of chaotic home environments to child cognitive outcomes. This section describes child mechanisms that potentially function as mediators. These mechanisms could reflect the direct effects of environmental chaos or the indirect effects of parenting interactions that educate attention and encourage the development of inhibitory control. We are cautious here because the requirements of a mediational model for cognitive outcomes such as literacy and academic competencies have not been met.

Behavioral Regulation

Behavioral regulation is an ambiguous construct that includes a variety of processes, is a core component of self-regulation and overlaps substantially with effortful control (focused on emotion regulation) and temperament constructs, and is closely associated with developments in executive functioning in the late preschool period and middle childhood (Blair, 2002; Blair & Razza, 2007). The construct is useful theoretically because it is a core component of school readiness and the ability to take advantage of learning opportunities. The construct also is useful because the development of regulatory processes and the processes per se seem likely to be disrupted by unpredictable, disorganized, and stressful environments.

McClelland et al. (2007) defined behavioral regulation as centering on inhibitory control, attention, and working memory. These processes represent core aspects of the readiness to learn and cooperate in school (Blair, 2002), and accordingly predict a variety of cognitive achievements involving literacy and mathematics skills (Blair & Razza, 2007; McClelland et al., 2007). These processes also relate to chaotic aspects of the home environment. Evans and English (2002), for instance, related a multiple risk index representing physical and psychosocial stressors to inhibition in 8- to 10-year-olds in the form of a delayed gratification task. Evans (2003; Evans et al., 2005) also has replicated the relation to measures of inhibition in several studies. Others have directly linked chaotic aspects of the physical environment to attention focusing (Dumas et al., 2005) and attentional persistence (Evans et al., 2005; Evans, Lepore, et al., 1998; Wachs & Gruen, 1982), and relations between more general cumulative risk indexes and attentional function are well documented (Barocas et al., 1991).

Stress Physiology and Reactivity

The assumption underlying most of the research on environmental chaos in the home is that it affects parenting processes and child characteristics through stress mechanisms. The mechanisms include physiological responses focusing on cardiovascular and neuroendocrine functions and individual differences in stress-reactivity and stress-sensitization. Several studies by Evans (2003, 2004; Evans & Kim, 2007; Evans et al., 2007) have led the way in documenting this assumption by relating aspects of environmental chaos in cumulative risk indexes directly to children's physiological responses. Relating these biomarkers to children's social or cognitive functioning, however, remains a challenge, and few studies provide evidence that markers of stress physiology mediate between environmental stressors and children's behavior.

Recent studies by El-Sheikh and colleagues (Buckhalt, El-Sheikh, & Keller, 2007; El-Sheikh, Buckhalt, Keller, Cummings, & Acebo, 2007; El-Sheikh, Buckhalt, Mize, & Acebo, 2006) posit children's sleep disruption as an intervening variable between an environmental stressor (e.g., marital conflict) and academic achievement and a variety of cognitive functions. The argument is that sleep reflects biological regulation that is sensitive to arousal and that affects aspects of executive functions. The research provides interesting leads in conceptualizing concrete ways in which environmental chaos may affect children's habits and health-related functioning and impact on cognitive processes.

Motivational Processes

Academic success requires persistence in the face of challenge. One of the factors that may inhibit the ability of children in highly stressful and chaotic home environments from taking advantage of school opportunities is a helpless or hopeless approach to academic challenge and failure, in which the children give up early. This approach may contribute to the failure to close the achievement gap over the elementary years between children who are disadvantaged and those who are not (Entwisle, Alexander, & Olson, 2004). The approach may reflect a mixture of socioemotional and cognitive processes reflecting negative emotionality, difficulty in tolerating frustration, and attentional control processes that are shaped by experiences in chaotic home environments. The unpredictability and disorganization in such environments may not reward persistent effort in the face of challenge.

An emerging body of evidence documents links between environmental stress and a helpless or hopeless approach to challenge that are independent of markers of family SES. A seminal study by Wachs (1987) showed relations between physical chaos in the home and infant mastery motivation. In more recent studies, Evans (Evans, 2003; Evans & English, 2002; Evans et al., 2005; Evans, Lepore, et al., 1998) has shown links between physical chaos and psychosocial turmoil and children's approach to challenge and feelings of self-worth.

Moderators

Our discussion has focused on main effects for chaos variables. On both theoretical and empirical grounds, however, there are good reasons to look for moderators of the effects and interactions among chaos variables and both child and sociodemographic variables (for a thorough discussion of this point, see chap. 7). A key theoretical issue is that the effects of variables that work through stress mechanisms are likely to be contingent on differences in stress reactivity and sensitization among parents and children. Thus, the organismic specificity of stress effects is likely to be the rule rather than the exception (Wachs, 1987, 2000). The empirical issue is that such interactions help explain relatively small (main) effect sizes for most chaos variables in relation to children's cognitive outcomes.

Only a few studies examine or show moderation effects in relation to children's cognitive processes, in contrast to the robust literature showing interactions between environmental adversity and child temperament variables in relation to socioemotional outcomes. Work by Boyce and his colleagues (Boyce, 2006; Boyce & Ellis, 2005; Quas, Bauer, & Boyce, 2004) illustrates the issue by showing that children vary in stress reactivity and that the variation interacts with aspects of environmental adversity and support in predicting child cognitive and social adjustment. In explicit opposition to main effect models of stress responses, Boyce argued that stress reactivity reflects increased biological sensitivity to context, which may be disadvantageous for children in challenging and stressful situations but advantageous in supportive and protected contexts. The specific relation to cognitive processes is that the stress response systems centrally involve the hippocampus, which in turn is centrally involved in memory and learning. Quas et al. (2004) provided support for the interaction model by showing that the event memory of highly reactive children was better in a supportive than in a nonsupportive recall context. Viewed the other way, this study also is interesting in suggesting that supportive dyadic relations could protect vulnerable children against the effects of environmental stress.

Other studies provide leads about potential moderators of the relations between chaotic environments and cognitive processes. Likely moderators include differences in stress reactivity, sex, health status, and age. Because chaos works through stress mechanisms, potential moderators concern differential sensitivity or vulnerability to stress. Caspi et al. (2003) provided evidence, for example, for a genetic vulnerability to environmental stress involving alleles for serotonin transporters that seem to increase the likelihood of depressive and anxious behavior. Something similar could help explain diversity in the vulnerability of children's cognitive processes to chaotic home environments. Similarly, developmental theory argues in general that the developmental processes of males and children with health issues are particularly vulnerable to environmental stressors.

Child age is likely to function as a moderator of chaos effects in the same way that age moderates relations between income poverty and different aspects of cognitive development. The review by Duncan and Brooks-Gunn (2000), for instance, provides strong evidence that family poverty in the preschool years relates more strongly than later poverty to indices of cognitive ability (e.g., intelligence, verbal ability). Early experiences in this regard clearly have primacy in building and constraining cognitive competence. In contrast, later family adversity may have specific effects on cognitive skill deployment and acquisition. Effects associated with chaotic home environments could follow a similar pattern for aspects of cognitive competence.

Sociodemographic moderators may include poverty and the chronicity of environmental adversity. Though studies focusing on household chaos now routinely control SES as a way of distinguishing different aspects of environmental adversity, few explore interactions of chaos and poverty status. There are good reasons to do so, however, given evidence that dynamic income changes matter more for economically disadvantaged preschoolers than for other children in relation to school readiness and linguistic competence (Dearing, McCartney, & Taylor, 2006). Income changes represent a marker of environmental stress. Similarly, stress sensitivity may represent a cumulative effect of chronic stress exposure as opposed to more episodic exposure (Evans & English, 2002; Evans & Kim, 2007).

Summary

Conceptualizing relations between chaotic home environments and child cognitive processes and outcomes is easy at a practical and common-sense level. Everyone has experienced cognitive difficulties, perhaps in noisy situations or in situations of psychosocial stress at home, and it is easy to imagine how noise, crowding, unpredictable home routines and agendas, and social turmoil in the home might disrupt children's attention, learning, and cognitive skill acquisition.

The theoretical and empirical cases are harder to make in individual studies, however, because chaos in the home environment must be distinguished from and yet related to other aspects of family adversity. Destructive marital conflict and harsh parenting are strong and clear predictors of a variety of child adjustment problems, for instance, which means that isolating effects for chaos requires a distinction between more individual and dyadic processes and more family-level processes. Disruption in family routines, for example, could be a mechanism relating marital conflict to child adjustment or it could be a correlate of other mechanisms. Chaos also must be distinguished from income poverty, with which it is correlated and which functions at a family level, but the distinction poses problems because controls for family income often vitiate relations between family-level variables describing turmoil and instability and child cognitive outcomes. Similarly, interpretation of chaos effects often is vulnerable to third-variable issues that press for a social selection rather than a social causation perspective. Poorly educated mothers with low verbal skills may be a common factor, for instance, linking a chaotic home environment and child cognitive difficulties.

With these theoretical caveats in mind, we reviewed several strong studies linking physical chaos in the home environment and a variety of child cognitive processes, often mediated through parenting processes. Evans (2006) provided a comprehensive review in this regard. Many of the studies are well-controlled, as researchers are fully aware of the interpretive issues. Also, abundant evidence relates psychosocial chaos to children's behavioral adjustment, but relations to cognitive outcomes may be explained fully or in part by family income.

Because chaos is a family-level variable, a key challenge is specifying mechanisms that mediate relations to cognitive and behavioral adjustment. Assuming that chaos constitutes environmental stress, we focused on aspects of behavioral regulation and stress reactivity as proximal mediators. Here good evidence links regulation and reactivity to school readiness and academic skills, and good evidence also links environmental chaos and stress to regulation and stress reactivity. But few studies show direct links between environmental chaos per se, child behavioral regulation, and cognitive outcomes. The same is true for moderators of relations to cognitive outcomes. There are good reasons to posit moderation by child and social variables, but only a few studies have provided direct evidence to date. Working out mechanisms and moderators of chaos effects constitutes a robust research agenda.

References

Ackerman, B. P., Brown, E. D., & Izard, C. E. (2003). Continuity and change in levels of externalizing behavior in school of children from economically disadvantaged families. *Child Development, 74*, 694–709.

Ackerman, B. P., Brown, E. D., & Izard, C. E. (2004). The relations between contextual risk, earned income, and the school adjustment of children from economically disadvantaged families. *Developmental Psychology, 40*, 204–216.

Adam, E. K. (2004). Beyond quality: Parental and residential stability and children's achievement. *Current Directions in Psychological Science, 13*, 210–213.

Barocas, R., Seifer, R., Sameroff, A. J., Andrews, T. A., Croft, R. T., & Ostrow, E. (1991). Social and interpersonal determinants of developmental risk. *Developmental Psychology, 27*, 479–488.

Blair, C. (2002). School readiness: Integrating cognition and emotion in a neurobiological conceptualization of child functioning at school entry. *American Psychologist, 57*, 111–127.

Blair, C., & Razza, R. P. (2007). Relating effortful control, executive function, and false belief understanding to emerging math and literacy ability in kindergarten. *Child Development, 78*, 647–663.

Boyce, W. T. (2006). Symphonic causation and the origins of childhood psychopathology. In D. Cicchetti & D. J. Cohen (Eds.), *Developmental psychopathology: Vol. 2. Developmental neuroscience* (2nd ed., pp. 797–817). Hoboken, NJ: Wiley.

Boyce, W. T., & Ellis, B. J. (2005). Biological sensitivity to context: I. An evolutionary-developmental theory of the origins and functions of stress sensitivity. *Development and Psychopathology, 17*, 271–301.

Bradley, R. H., & Caldwell, B. M. (1984). 174 children: A study of the relationship between home environment and cognitive development during the first 5 years. In A. W. Gottfried (Ed.), *Home environment and early cognitive development* (pp. 5–56). Orlando, FL: Academic Press.

Brody, G. H., & Flor, D. L. (1997). Maternal psychological functioning, family processes, and child adjustment in rural, single-parent, African American families. *Developmental Psychology, 33*, 1000–1011.

Bronfenbrenner, U., & Evans, G. W. (2000). Developmental science in the 21st century: Emerging theoretical models, research designs, and empirical findings. *Social Development, 9*, 115–125.

Buckhalt, J. A., El-Sheikh, M., & Keller, P. (2007). Children's sleep and cognitive functioning: Race and socioeconomic status as moderators of effects. *Child Development, 78*, 213–231.

Caldwell, B., & Bradley, R. H. (1984). *Home Observation for Measurement of the Environment.* Little Rock: University of Arkansas at Little Rock.

Caspi, A., Sugden, K., Moffitt, T. E., Taylor, A., Craig, I. W., Harrington, H., et al. (2003, July 18). Influence of life stress on depression: Moderation by a polymorphism in the 5-HTT gene. *Science, 301*, 386–389.

Coldwell, J., Pike, A., & Dunn, J. (2006). Household chaos—links with parenting and child behaviour. *Journal of Child Psychology and Psychiatry, 47*, 1116–1122.

Conger, R. D., Wallace, L. E., Sun, Y., Simons, R. L., McLoyd, V. C., & Brody, G. H. (2002). Economic pressure in African American families: A replication and extension of the family stress model. *Developmental Psychology, 38*, 179–193.

Corapci, F., & Wachs, T. D. (2002). Does parental mood or efficacy mediate the influence of environmental chaos upon parenting behavior? *Merrill-Palmer Quarterly, 48*, 182–201.

Dearing, E., Kreider, H., Simpkins, S., & Weiss, H. B. (2006). Family involvement in school and low-income children's literacy: Longitudinal associations between and within families. *Journal of Educational Psychology, 98*, 653–664.

Dearing, E., McCartney, K., & Taylor, B. A. (2006). Within-child associations between family income and externalizing and internalizing problems. *Developmental Psychology, 42*, 237–252.

Dumas, J. E., Nissley, J., Nordstrom, A., Smith, E. P., Prinz, R. J., & Levine, D. W. (2005). Home chaos: Sociodemographic, parenting, interactional, and child correlates. *Journal of Clinical Child and Adolescent Psychology, 34*, 93–104.

Duncan, G. J., & Brooks-Gunn, J. (1997). Income effects across the life span: Integration and interpretation. In G. J. Duncan & J. Brooks-Gunn (Eds.), *Consequences of growing up poor* (pp. 596–610). New York: Russell Sage Foundation.

Duncan, G. J., & Brooks-Gunn, J. (2000). Family poverty, welfare reform, and child development. *Child Development, 71*, 188–196.

Duncan, G. J., Dowsett, C. J., Claessens, A., Magnuson, K., Huston, A. C., Klebanov, P., et al. (2007). School readiness and later achievement. *Developmental Psychology, 43*, 1428–1446.

El-Sheikh, M., Buckhalt, J. A., Keller, P. S., Cummings, E. M., & Acebo, C. (2007). Child emotional insecurity and academic achievement: The role of sleep disruptions. *Journal of Family Psychology, 21*, 29–38.

El-Sheikh, M., Buckhalt, J. A., Mize, J., & Acebo, C. (2006). Marital conflict and disruption of children's sleep. *Child Development, 77*, 31–43.

Entwisle, D., Alexander, K. L., & Olson, L. S. (2004). The first-grade transition I life course perspective. In J. T. Mortimer & M. J. Shanahan (Eds.), *Handbook of the life course* (pp. 229–250). New York: Kluwer Academic/Plenum Publishers.

Evans, G. W. (2003). A multimethodological analysis of cumulative risk and allostatic load among rural children. *Developmental Psychology, 39*, 924–933.

Evans, G. W. (2004). The environment of childhood poverty. *American Psychologist, 59*, 77–92.

Evans, G. W. (2006). Child development and the physical environment. *Annual Review of Psychology, 57*, 423–451.

Evans, G. W., Bullinger, M., & Hygge, S. (1998). Chronic noise exposure and physiological response: A prospective study of children living under environmental stress. *Psychological Science, 9*, 75–77.

Evans, G. W., & English, K. (2002). The environment of poverty: Multiple stressor exposure, psychophysiological stress, and socioemotional adjustment. *Child Development, 73*, 1238–1248.

Evans, G. W., Gonnella, C., Marcynyszyn, L. A., Gentile, L., & Salpekar, N. (2005). The role of chaos in poverty and children's socioemotional adjustment. *Psychological Science, 16*, 560–565.

Evans, G. W., Hygge, S., & Bullinger, M. (1995). Chronic noise and psychological stress. *Psychological Science, 6*, 333–338.

Evans, G. W., & Kim, P. (2007). Childhood poverty and health: Cumulative risk exposure and stress dysregulation. *Psychological Science, 18*, 953–957.

Evans, G. W., Kim, P., Ting, A. H., Tesher, H. B., & Shannis, D. (2007). Cumulative risk, maternal responsiveness, and allostatic load among young adolescents. *Developmental Psychology, 43*, 341–351.

Evans, G. W., & Lepore, S. J. (1993). Nonauditory effects of noise on children. *Children's Environment, 10*, 31–51.

Evans, G. W., Lepore, S., Shejwal, B. R., & Palsane, M. N. (1998). Chronic residential crowding and children's well being: An ecological perspective. *Child Development, 69*, 1514–1523.

Evans, G. W., Maxwell, L., & Hart, B. (1999). Parental language and verbal responsiveness to children in crowded homes. *Developmental Psychology, 35*, 1020–1023.

Fiese, B. H., Tomcho, T. J., Douglas, M., Josephs, K., Poltrock, S., & Baker, T. (2002). A review of 50 years of research on naturally occurring family routines and rituals: Cause for celebration? *Journal of Family Psychology, 16*, 381–390.

Gottfried, A. W., & Gottfried, A. E. (1984). Home environment and cognitive development in young children of middle-socioeconomic-status families. In A. W. Gottfried (Ed.), *Home environment and early cognitive development* (pp. 57–115). Orlando, FL: Academic Press.

Hart, B., & Risley, T. R. (1995). *Meaningful differences in the everyday experiences of young American children*. Baltimore: Paul H. Brookes.

Hsueh, J., & Yoshikawa, H. (2007). Working nonstandard schedules and variable shifts in low-income families: Associations with parental psychological well-being, family functioning, and child well-being. *Developmental Psychology, 43*, 620–632.

Matheny, A. P., Wachs, T. D., Ludwig, J. L., & Phillips, K. (1995). Bringing order out of chaos: Psychometric characteristics of the Confusion, Hubbub, and Order scale. *Journal of Applied Developmental Psychology, 16*, 429–444.

McClelland, M. M., Cameron, C. E., Connor, C. M., Farris, C. L., Jewkes, A. M., & Morrison, F. J. (2007). Links between behavioral regulation and preschooler's literacy, vocabulary, and math skills. *Developmental Psychology, 43*, 947–959.

Petrill, S. A., Pike, A., Price, T., & Plomin, R. (2004). Chaos in the home and socioeconomic status are associated with cognitive development in early childhood: Environmental mediators identified in a genetic design. *Intelligence, 32*, 445–460.

Quas, J. A., Bauer, A., & Boyce, W. T. (2004). Physiological reactivity, social support, and memory in early childhood. *Child Development, 75*, 797–814.

Sameroff, A. J., Seifer, R., & Bartko, T. (1997). Environmental perspective on adaptation during childhood and adolescence. In S. S. Luthar, J. A. Burack, D. Cicchetti, & J. R. Weisz (Eds.), *Developmental psychopathology* (pp. 507–526). Cambridge, England: Cambridge University Press.

Stoneman, Z., Brody, G. H., Churchill, S. L., & Winn, L. L. (1999). Effects of residential instability on Head Start children and their relationships with older siblings: Influences of child emotionality and conflict between family caregivers. *Child Development, 70*, 1246–1262.

Valiente, C., Lemery-Chalfant, K., & Reiser, M. (2007). Pathways to problem behaviors: Chaotic homes, parent and child effortful control, and parenting. *Social Development, 16*, 249–267.

Wachs, T. D. (1979). Proximal experience and early cognitive-intellectual development: The physical environment. *Merrill-Palmer Quarterly, 25*, 3–41.

Wachs, T. D. (1987). Specificity of environmental action as manifest in environmental correlates of infant's master motivation. *Developmental Psychology, 23*, 782–790.

Wachs, T. D. (2000). *Necessary but not sufficient: The respective roles of single and multiple influences on individual development*. Washington, DC: American Psychological Association.

Wachs, T. D., & Camli, O. (1991). Do ecological or individual characteristics mediate the influence of the physical environment upon maternal behavior. *Journal of Environmental Psychology, 11*, 249–264.

Wachs, T. D., & Gruen, G. (1982). *Early experience and human development*. New York: Plenum Press.

Yeung, W. J., Linver, M. R., & Brooks-Gunn, J. (2002). How money matters for young children's development: Parental investment and family processes. *Child Development, 73*, 1861–1879.

4

The Dynamics of Family Chaos and Its Relation to Children's Socioemotional Well-Being

Barbara H. Fiese and Marcia A. Winter

A discussion of chaos effects on child socioemotional well-being necessarily draws attention to family-level processes. *Environmental chaos* typically refers to disruptions in multiple domains, including sensory overload, physical crowding, and routine family life (Evans, Gonnella, Marcynyszyn, Gentile, & Salpekar, 2005; Matheny, Wachs, Ludwig, & Philips, 1995). When routines are diminished in frequency and family life is disorganized, questions about how the group works together collectively to promote and sustain healthy development are raised. This systems-level question extends beyond parenting practices and family structure. Though there is considerable theoretical support for how systems, as a whole, can support or derail development, systematic investigations of family process as a whole are less prominent (Fiese & Spagnola, 2006). Parent–child relationships, parental warmth, and attachment relationships certainly are essential to the health and well-being of children. However, these qualities are not equivalent to (although they may be related to) family-level processes. From a systems perspective, family-level processes are intermediary to dyadic processes and larger institutional effects on child development such as neighborhoods and schools. In this regard, they may form a nexus in studying chaos and child socioemotional well-being and are essential ingredients in Bronfenbrenner's ecological model (Bronfenbrenner, 1979; see also chap. 1, this volume). Take, for example, the fact that neighborhoods characterized by chronic noise and crowding are associated with more behavior problems in children. Similarly, crowding and elevated noise levels are also associated with poorer parent–child relationships. An apparent missing feature from a socioecological perspective is how the family as a whole, organizing itself in the service of child health and well-being, responds to chaotic environments or itself is the progenitor of chaotic processes. In this chapter we address not only how environmental chaos is associated with socioemotional problems in children but also how the family as a whole, as an organizing influence, responds to chaos in the environment. We do this by focusing largely on chaos conceptualized as a family-wide construct, particularly in studies of family routines.

Preparation of this manuscript was supported, in part, by a grant to the first author from the National Institute of Mental Health (#51771).

We structure our observations as follows: (a) an overview of the potential multidimensional nature of family chaos, (b) a consideration of the daily dynamic nature of chaos in family settings, and (c) a brief discussion of potential mediators and moderators of the link between family risk and child socioemotional functioning.

Multidimensional Nature of Family Chaos

Bronfenbrenner and Evans (2000) offered several perspectives on chaos in family life, sometimes describing it as "frenetic activity, lack of structure, unpredictability in everyday activities, [and] high levels of ambient stimulation" (p. 121), which assumedly results, at least in part, from more hours spent away from home, families being less likely to share meals together, domestic time becoming compressed, and family members characterizing their home life as "hectic, unstructured, unpredictable, and, at times, simply out of control" (Evans et al., 2005, p. 560). Embedded in these descriptors are two dimensions of family chaos: the disruption of daily activities and the associated felt experience (hectic, out of control). These two aspects of disruption—the disturbance of the event itself and the subsequent interpretation or representation of the event as a whole—may have differential sources and pathways of effect on child socioemotional functioning. Thus, it is useful to consider how chaos operates both at the event level and at the level of interpreting and making meaning of circumstances. In addition to different aspects of disruption, chaos also operates on several different levels in family life. In the ensuing discussion, we focus on three diverse but interrelated levels: the construction of family time, the frequency and disruption of family activities, and the meaning created out of disrupted or irregular activities. We discuss each of these in turn.

Construction of Family Time

Bronfenbrenner's person, process, context, time model clearly anchors time as a regulatory force of development (Bronfenbrenner & Morris, 1998). Children are proximally influenced by parenting through consistent exposure to supportive interactions or through direct contact, and thus such features as sensitivity and warmth are essential. Parents also produce and allocate resources that indirectly influence development by virtue of their effects on access to materials and protection from harm (e.g., shelter, neighborhood location, food, educational materials). Whereas parenting may influence development at the proximal level through responsiveness and sensitivity, at the level of the family, time and allocation of time are valuable for development. Time is used not only to support child-care activities but to organize collective family activities in the interest of creating what is euphemistically referred to as "family time."

It may be helpful to consider how different family activities evolved over time as a source of protection for the family unit as a whole. Collective family gatherings such as mealtimes and communal family celebrations are activities that evolved during the mid-1800s. During the Victorian era, individuals rarely ate with family members. The family dining room as a separate space in the home did not appear until the late 1800s. The family meal as a time set aside for family communication with the expectation of attendance did not develop until the mid-19th century (Caplow, Bahr, Chandwick, Hill, & Williamson, 1982). Even religious holidays such as Christmas were considered events to celebrate with the larger community rather than with the family until the late 19th century (Gillis, 1996). Family historians speculate that following the Industrial Revolution, families reorganized their daily lives in accordance with shifting work patterns outside the home. During the day, family members went to work or school, and on their return home they participated in the family dinner, homework routines, and bedtime routines. Thus, in many respects the routines that are identified as missing in the chaotic lives of contemporary families were those initially created to protect the family being pulled apart from outside economic forces. Today's home is no longer a retreat from the social world or a place where routines are organized around the needs of children (Hareven, 1985); instead, the impoverishment of economic resources often disrupts daily activities as energy is focused outside the home, resulting in little organized time for children's needs (Roy, Tubbs, & Burton, 2004). These activities are embedded in a socioeconomic context. While some families were developing special family time in the parlor, other families were pulled apart because of service responsibilities in the kitchen, nursery, or fields (Cinotto, 2006). Thus, the structure of daily activities and routines in the home is in part dictated by outside economic forces as well as the intentions of caregivers. Lichter and Wethington (see chap. 2, this volume) suggested the existence of increasingly class-related divisions in contemporary America, with those at the bottom facing escalating chaos whereas well-educated families purchase better and more stable living conditions for their children.

The link between economic resources and allocation of family time is seen in parents' perceptions about the time they spend with their children in relation to leisure activities. Bianchi and Raley (2005) examined the time-use responses to the 1975–1976 and 2000 National Survey of Parents. They found that despite the increase in maternal employment over the 25-year time span, on average mothers' overall time with children has remained at 1975 levels. However, what has changed is the way that they spend time with their children. Parents (both mothers and fathers) report a greater proportion of time multitasking while caring for their children (presumably activities such as shopping, going to the park, visiting friends). These analyses are consistent with Hofferth and Sandberg's (2001) report on children's activities, which found that a substantial amount of household work time with children was spent shopping. What these large-scale reports suggest is that families generally organize time to meet multiple demands in single-activity settings. Mealtimes are not just for being fed but also to catch up on the day's events, go over homework, arrange transportation schedules, and assign household chores. The organization of daily activities overlaps such that homework routines, reading time, and dinnertime tend to co-occur regardless of income and race (Serpell, Sonnenschein, Baker, & Ganapathy, 2002). Thus, if one activity is disrupted, order is affected in other arenas of the child's life, such as communicating with school, completing homework, and engaging in peer activities.

Family time can be directly measurable as the frequency of shared activities, but it also includes the pressures experienced by parents in organizing the day and balancing obligations. A consistent thread throughout national surveys and qualitative interviews is parents' and children's desire to spend more time with each other (Bianchi & Raley, 2005; Fagan, 2003) while at the same time feeling at a loss as to how to capture family time. This desire is most poignant when economic resources are strained and parents struggle to carve family time into a 24-hour time frame that is consumed by juggling public transportation schedules, making child-care arrangements, and working shift jobs. As part of the Three-City Study, Roy, Tubbs, and Burton (2004) described how women in poverty make difficult decisions to preserve family time:

> For poor mothers with fewer resources, the shifts from picking up children to dinner preparation, or from waking, feeding, and grooming children to dropping them off on the way to work, were more than simply hectic. . . . This hectic pace and persistent demand of overlapping time obligations led to feelings that time was beyond her control and that daily routines were driven by others outside family.

Other parents reported juggling their routines through serial meals so that at least one parent could eat with at least one child on a regular basis, resulting in meals being served over several hours to accommodate everyone in the household (Tubbs, Roy, & Burton, 2005). These staggered schedules and extended days do not come without cost, as the emotional burden of orchestrated arrangements may ultimately express itself through compromised health of the caregiver (Adler et al., 1994) and portends how children ultimately engage with civic institutions (Lareau, 2003).

When one considers chaos in family life, one must consider not only how families structure their time together but also what these collective activities mean to them. "Family time" as a set-aside moment is created to promote close relationships (Daly, 2001). When there are demands to be in multiple places at the same time, schedules become disorganized or there is a pressure to create or participate in collective activities, a sense of urgency takes over and there is the potential for bonds to be broken. Whereas highly organized family environments can be characterized by direct and meaningful forms of communication that bring the group together to use time in a deliberate way, chaotic environments are more likely to be characterized by patterns of communication that are indirect, used to exclude others, and create meanings based on derision (Fiese, 2006). Thus, when we consider the empirical evidence linking chaotic family environments to children's socioemotional well-being, both the daily practice of collective activities as well as their personal meanings must be taken into account.

Family Routines and Child's Socioemotional and Physical Well-Being

It is reasonable to question whether the absolute frequency of family routine activities is associated with better socioemotional outcomes for children. Consistent findings link frequency of family mealtimes with positive outcomes, including reduced risk of alcohol problems, sexual risk taking, cigarette smoking, and mental health problems (Compan, Moreno, Ruiz, & Pascual, 2002; Eisenberg, Olson, Neumark-Sztainer, Story, & Bearinger, 2004). Frequency of mealtimes has been used as a proxy for family engagement and monitoring. For example, the 2006 National Center on Addiction and Substance Abuse (CASA) report on the importance of family dinners indicated that compared with parents who ate with their teenage children 5 or more nights a week, parents who shared dinners with their teens on 3 or fewer nights per week were 5 times more likely to have a poor relationship with their child, 1.5 times more likely to say they did not know their child's friends, and more than twice as likely not to know their child's teachers' names (CASA, 2007). Thus, frequent routine mealtimes provide an opportunity to engage and be involved as a family unit and monitor the whereabouts of others, and may ultimately help to prevent risky behaviors. The actual mechanisms of this effect have yet to be tested because these reports are correlational and it is plausible that third explanatory constructs, such as sensitive parenting, may account for much of the variance.

The mealtime frequency data also do not paint a complete picture as to whether this effect is linear or whether four meals a week would be "good enough" to ensure positive outcomes. Further, the data associated with socioeconomic status are incomplete. Bradley, Corwyn, McAdoo, and Coll (2001) reported both poverty and ethnicity effects in number of days that children eat a meal, with both mothers and fathers with poorer African American families reporting the lowest rate. However, as pointed out previously, these frequency counts could potentially miss "off-time" meals noted in qualitative reports (Tubbs, Roy, & Burton, 2005).

Chaos, however, is not just the sheer absence or reduced frequency of an activity but the conditions under which expectable activities are held. There is some evidence that disruptions in mealtime routines are associated with poorer outcomes for youth. When mealtimes are accompanied by the background noise of television, they are associated with a 5% increase in consumption of less healthy foods such as pizza, salty snacks, and soda, and a 5% decrease in consumption of fruits, vegetables, and juices (Coon, Goldberg, Rogers, & Tucker, 2001). Family routines tend to co-occur, and the smooth operation of one suggests a positive climate for another. For example, in a nationally representative sample of 5- to 11-year-olds, amount of time spent on family meals during the week was positively associated with amount of time children spent sleeping (Adam, Snell, & Pendry, 2007). In the same study, the amount of time spent in family meals was negatively related to conflict and positively associated with parental warmth (as detected on the Home Observation for Measurement of the Environment [HOME] Scale) and family rules (including bedtime and homework).

Family instability has the potential to disrupt daily routines and is clearly linked to poor socioemotional outcomes for children. For example, the number of residential moves that adolescents experience has been found to be positively associated with adjustment problems even while controlling for family demographic characteristics and quality of current environment (Adam & Chase-Landsdale, 2002). Ackerman, Kogos, Youngstrom, Schoff, and Izard (1999) demonstrated that family instability during preschool and first grade is related to teacher and parent report of child behavior problems. In fact, the first-grade effects held when controlling for behavior problems at preschool, indicating that the effects were not due to just persistence of behavior problems. The mechanisms linking to poorer socioemotional outcomes are multifold. Instability is not due to just economic strain, as evidenced by analyses presented by Ackerman et al. (1999) and Adam and Chase-Landsdale (2002) that accounted for income effects. Plausible mechanisms include threats to feelings of felt security (Cummings, Davies, & Campbell, 2000), loss of social connections (Pribesh & Downey, 1999), and stress reactivity (Ackerman et al.). More proximal variables such as disruptions in family routine practices are alluded to in the family instability literature (e.g., Adam & Chase-Landsdale, 2002) but not directly measured.

Direct Observation and Meaning of Family Routines

To truly capture chaos, one needs to know what is happening during the course of expectable routines. Direct observations of family mealtimes suggest that how roles are assigned, the relative ease with which the meal is carried out, how conversations are conducted, and how behavior is regulated are related to child mental and physical health. Mealtimes marked by unevenness in assigning tasks and roles, communication that degrades or excludes others, and harsh or lax forms of behavior control are associated with parental depression (Dickstein et al., 1998), child internalizing (Fiese, Foley, & Spagnola, 2006) and externalizing problem behaviors (Stark et al., 2000), and overweight conditions (Jacobs & Fiese, 2007). Even though these mealtimes last on average 18 minutes, the family dynamics of risk in this routine setting are characterized by awkwardness in initiating and carrying out the group activity, uneven communication among members, and erratic behavior control. Exceptions to this pattern are situations in which a preexisting condition in the child calls for overly controlling strategies and lengthened meals, as in the case of cystic fibrosis (Stark et al., 1997) or eating disorders (Neumark-Sztainer et al., 2003) where the dynamics of parental control and intrusiveness dominate the meal. It is interesting that the opposite pattern is seen in children who are overweight; their parents have a more laissez-faire style of mealtime interactions (Jacobs & Fiese, 2007; Johnson & Birch, 1994).

Children are keenly sensitive to disorganization in family life, and the representation of chaos is linked to poorer socioemotional outcomes. Dumas et al. (2005) reported that children raised in chaotic home environments have more difficulty responding to social cues and are more likely to have problem behaviors as reported by their parents and teachers. The compromised ability to read social cues is an understudied one but is consistent with Evans's (2001) report of the deleterious impact of exposure to high levels of ambient noise. We now turn to children's representations of family chaos to gain a better understanding of what it means to them and how these representations, in turn, are related to child well-being.

Children need to make sense of chaos, and they often do so by creating narrations or stories about what happens in their daily lives. Researchers have been able to tap into this process by examining child responses to structured story stems as well as analysis of personal accounts made by children in high-risk environments. Fivush, Hazzard, Sales, Sarfati, and Brown (2003) interviewed 5- to 12-year-old children living in inner-city Atlanta who had been exposed to high rates of violence. Children used more internal state language when talking about negative events, and the more stressful the event (e.g., witnessing a gunfight, loss of a family member), the more they used internal state language. Grych, Wachsmuth-Schlaefer, and Klockow (2002) also examined children's representations of self and family relationships using the MacArthur Story Stem Battery comparing responses made by children whose mothers had experienced domestic violence with those whose mothers had not. (For the purposes of our discussion it is important to note that Grych's sample included children who were living in domestic violence shelters and thus had moved from their residential homes.) These children saw their mothers as less nurturing (but not more aggressive), and their self-representations were less powerful and more obedient than their community peers. In addition, their narratives were less coherent overall. Similar findings are reported for children who have experienced maltreatment (Toth, Cicchetti, Macfie, Maughan, & Vanmeenen, 2000). Despite a considerable lack of specificity in linking children's representations to chaotic environments with this limited empirical base, there is the hint that unpredictable and disruptive family environments have the potential to set the stage for internal representations marked by poor self-worth and a sense of powerlessness. In support of this notion, research by Forman and Davies (2003) showed that greater family instability, measured via an index of unstable family events, was related to greater internalizing and externalizing symptoms in adolescents, in part by fostering appraisals of the family unit as not protective of the child and family and not promoting emotional security. Lustig (see chap. 15, this volume) describes in detail a similar process that often unfolds among refugee families. In research using a similar measure of family instability, Winter, Davies, and Cummings (2008) found that security in young, school-age children's representations (as assessed via a modified version of the MacArthur Story Stem Battery) hinged on the interplay between family instability and patterns of maternal communication pertaining to challenging events. It is interesting to note that these two studies both focused on samples of middle-class families that showed only modest levels of instability, further attesting to the power of chaos in predicting children's socioemotional outcomes.

Summary

Chaos operates on several levels in family life. The first is the construction of family time. How much time do families allocate to being together to accomplish the task of raising children? At its most basic level this includes feeding, clothing, and sheltering youth. However, national surveys indicate that parents want to spend more time with their children to provide care that extends beyond routine physical attention to cement close relational bonds. When family time becomes pressured and overwhelming, there is little order to the day, interactions are strained, and a sense of obligation builds. This is often the background to family chaos.

The second level to consider is the frequency and disruption of family activities. Routinization of activities relies on repetition, expectations for attendance, assignment of roles, and deliberate planning (Fiese et al., 2002). Under chaotic conditions such as moving frequently, spending 2 to 5 hours using public transportation for work, and complicated and shifting child-care needs, the essential elements of routines are threatened. Planning can be at the mercy of external forces (e.g., public transportation schedules, child-care hours, public clinic hours) rather than inside the family unit and is therefore rarely successful. Family presence and attendance at collective events becomes strained as there is a serial progression of individuals who come in and out of a setting rather than set aside time for the group as a whole.

The third level or aspect is the meaning created out of disrupted or irregular activities. Though the literature is relatively sparse, there is some consistency between parental and child accounts. The qualitative interviews of parents' experience in chaotic environments highlight feelings of burden, being overwhelmed, and often being exhausted with few emotional resources left at the end of the day (Roy et al., 2004). The draining of emotional resources in chaotic environments appears to be intertwined, in part, with instability of economic resources. Likewise, when parents are overwhelmed and burdened by medical regimens associated with care for children with chronic health conditions, routines are disrupted and chaos and anxiety predominate (Fiese & Everhart, 2006; Fiese & Wamboldt, 2003). This pileup of limited economic and emotional resources contributes to a chaotic climate characterized by irregular responsiveness to children's emotional and physical needs. Thus, the disruptions in the proximal caregiving environment are a reflection of the strains and pressures experienced in the work and economic environment more distal to the child as well as parental mental health often formative before the child's birth.

The child's understandings of this chaotic world can be accessed through narratives of daily life. Children's narratives drawn from samples of children experiencing exposure to moderate amounts of instability as well as extreme amounts of violence, residential instability, or both are characterized by internal representations of emotional insecurity of the self as having little control over the environment. For both children and adults, there are personal feelings of powerlessness and a sense of giving up. Where order and predictability can often give the illusion of control, family chaos is associated with representations of self and others as subservient, unworthy of trust, and feeling hassled.

To further appreciate how these levels may transact and evolve over time, it may be helpful to consider not only how chaos operates across extended periods but also how daily fluctuations may portend chaotic responses.

Cyclic Variations and Family Chaos

A traditional and reasonable way to consider family chaos effects on children's socioemotional functioning is through the accumulative exposure to highly disruptive and disorienting environments. Evans's documentation of the effects of airplane noise on children's socioemotional well-being is a case in point. Though a plane passing overhead can interrupt conversation, it is the chronic exposure to noise that is associated with poorer outcomes for children (Evans, 2001). A similar argument can be made for overcrowding in the home, where it is the accumulated effect of multiple members in the household that is associated with poorer outcomes, not just the effects of having a birthday sleepover with 10 close friends. Nonetheless, an accumulation necessarily consists of a series of regularly occurring events or circumstances that probably fluctuate in frequency and intensity through time. Therefore, an examination of cyclic variability and disruptions of daily routines has the potential to shed light on some of the dynamics of family chaos.

Ackerman et al. (1999) considered whether the persistence of contextual risk is a stronger predictor of behavior problems than intermittent or cyclic adversity. If one focuses only on their findings related to relationship and residential instability, a complex pattern emerges. It is not surprising that recent exposure to relationship instability was related to behavior problems for children in elementary school. Intermittent risk, however, was often just as toxic as recent risk in explaining child adjustment, particularly in regards to relationship instability. The cycling in and out of chaotic environments calls into question timing of measurement of the negative effects of chaos on children's socioemotional well-being. From a family stress perspective (Conger & Donnellan, 2007), it is plausible to expect concurrent stress and disruptions to affect parenting, which in turn will affect child adjustment. On the other hand, intermittent exposure to chaos may suggest alternative pathways to maladjustment including how individuals respond to stress in different environments (B. J. Ellis, Essex, & Boyce, 2005) as well as how routines may be sustained or serve a protective function under highly stressed environments (Brody & Flor, 1997).

In contrast to large-scale longitudinal studies, qualitative interview-based studies provide an in-depth, insider's view of how routines have a cyclic nature under high-risk child-raising conditions. In studies using interview-based data, the disruption of daily routines in households with an alcoholic member has received longstanding attention. Indeed, much of the work on family rituals has its roots in this literature (Wolin & Bennett, 1984; Wolin, Bennett, Noonan, & Teitlebaum, 1980). Wolin et al.'s (1980) early findings suggested that families who were able to maintain distinctive routines (particularly mealtime) in the context of parental alcoholism were less likely to pass down problematic drinking to their offspring.

A recent qualitative report identifies which routines may be most susceptible to disruption and what patterns are most closely associated with protecting youth from problem behaviors (Haugland, 2005). When fathers were actively drinking they disengaged from daily parenting routines such as helping to get children ready for school, attending dinner, and participating in bedtime routines. These roles were deliberately planned and upheld by the mothers in the study. Family meals were described as being altered in affective tone during and surrounding the times of heavy drinking and as being characterized by arguments, irritability, or silence. Discipline routines were also altered as fathers would frequently disengage: "Having no right to reprimand the children when I am drinking, I rather try to make things right when I am sober" (p. 231).

In contrast, other fathers would become harsher in their disciplinary style because of irritability related to drinking or hangovers. In the typology that was offered, chaotic families are identified as those with little distinctiveness in routines during both sober and active drinking periods. Little compensation was made by the mother in these families to take over the additional roles and responsibilities abdicated by the father. Major holidays and celebrations were disrupted, and children were exposed not only to paternal drinking but also to household violence. Though this qualitative study was small, the pattern of routine and ritual protection is consistent with previous reports using a more quantitative strategy (Bennett, Wolin, & Reiss, 1988; Fiese, 1993).

A look at cyclic variations calls attention to how even chaos itself is not a stable phenomenon; some families may move in and out of chaotic states before chaos becomes a stable pattern. The variation in health outcomes in high risk families also points to the protective mechanism of routines and rituals. The studies reviewed on parental alcoholism suggest that when families are able to maintain daily routines such as mealtimes, and holiday celebrations are not subsumed by the parent's drinking patterns, then a buffer is created for offspring. Other examples of where routines and rituals serve a protective function in high-risk conditions include the promotion of academic success and self-esteem in low-income rural African American boys (Brody & Flor, 1997; Brody, Flor, & Gibson, 1999), protection from learned helplessness under conditions of extreme poverty (Evans et al., 2005), and physiological reactivity in high-stress family environments (B. J. Ellis et al., 2005). Routines can also ease the strains and stresses associated with normative transitions such as the transition to parenthood (Fiese, Hooker, Kotary, & Schwagler, 1993) and transition to kindergarten (Wildenger, McIntyre, Fiese, & Eckert, 2008). An examination of the cyclic nature of chaos suggests that chaos is not always a stable phenomenon in family life and that routines and rituals may serve a protective function when chaos begins to escalate.

Another way to regard daily variation is to consider within-family variability of routines and how daily disruptions affect child health and well-being. One such area that we have examined is nighttime waking. When a child wakes in the middle of the night, there is the potential for the entire household to feel the effect. Disturbed sleep for children with chronic health conditions such as pediatric asthma is associated with disruptions in other daily activities, such as going to school and parents missing work (Diette et al., 2000). Emerging evidence suggests that family climate plays an important role in regulating children's sleep. Even in cases in which there is not an existing health risk, family climate variables such as marital conflict have been related to both quantity and quality of sleep (El-Sheikh, Buckhalt, Mize, & Acebo, 2006). In a recent telephone-based diary study conducted over 1 year, 500 observations were made of 47 families with a child with asthma (a risk condition for nighttime waking; Fiese, Winter, Sliwinski, & Anbar, 2007). Increased odds for waking in the middle of the night were associated with disruptions of bedtime routines, parents having to change plans, and parents feeling hassled by their children not listening to them within that given 24-hour period. The odds ratios were comparable with previous reports linking nighttime waking and children's allergic response to cats, cockroaches, and ragweed (Strunk, Sternberg, Bacharie, & Szefler, 2002).

This view of a daily (and nightly) disruption paints a picture of how chaos may begin to unfold. All parents feel hassled and have to change plans (Crnic & Acevedo, 1995), but when hassles are linked to disruptions in routines there are consequences for children's health and well-being.

These disruptions may affect health behaviors as well as contribute to feelings of burden and obligation. When families are able to minimize chaos by constructing a clear set of routines around health behaviors, their children are more likely to adhere to prescribed protocols (e.g., taking medications) and less likely to use emergent care (Fiese, Wamboldt, & Anbar, 2005). In terms of socioemotional functioning, the sense of burden of care and obligation in carrying out these routines is most closely related to both parent and child emotional well-being. Parents who perceive the daily routines necessary to keep their child healthy as a chore and a burden are more likely to have children who worry about their condition, experience somatic complaints unrelated to their condition, and feel that they cannot fully engage in out-of-school activities even when they are asymptomatic. This glimpse of disruption to one routine (taking medications) reinforces the multiple influences that chaos can have on children's health and well-being. Other daily routines also deserve further consideration for children's health and well-being. Increasing evidence points to the importance of good sleep hygiene and regularity of bedtimes for children and adolescents. Late bedtimes and abbreviated sleep cycles have been associated with increased risk for childhood obesity (Snell, Adam, & Duncan, 2007), internalizing problems (Hofferth & Sandberg, 2001), and cognitive functioning (Dahl & Lewin, 2002). Good sleep habits are established early in a child's life and there appears to be an overlap between regularity of bedtime routines and other daily routines such as mealtimes (Adam, Snell, & Pendry, 2007).

As in other events indicative of chaos, a single missed dose of medicine or one late bedtime does not indicate chaos in the household. The accumulation of poor health behaviors, disrupted routines, and added burden in a multiple-risk model will likely best explain outcomes for both parents and children (Evans, 2003; Everhart, Fiese, & Smyth, 2008). Indeed, measurement of environmental chaos very often assumes an aggregate of multiple risks (Ackerman et al., 1999; Evans & English, 2002).

For chaos to have a lasting effect on child well-being, it must be established over time. It is somewhat remarkable that the daily dynamics of family chaos is a relatively uncharted territory. Several key developmental questions have not been addressed. For example, in light of the documented powerful early effects of poverty on child development (Conger & Donnellan, 2007; Duncan & Brooks-Gunn, 1997), do the dynamics of disrupted family routines follow a complementary timetable? How long does it take for dreaded obligations, burden, and disrupted routines to have an effect on child well-being? Is it possible to interrupt the process at key family and developmental periods—transition to school, birth of the second child, moving to a new home? What is known about the stability of routine practices? It is known that even for new parents with relatively good economic resources, it takes a few years to stabilize family life after the birth of a child (Fiese et al., 1993) and new roles and rules need to be negotiated during adolescence (Furstenberg, Cook, Eccles, Elder, & Sameroff, 1999). Yet, relatively little is known about the daily dynamics of these choices and what makes for good practices under what conditions at which points in the family development life cycle. Without a clearer understanding of these developmental processes, it will be difficult to set an agenda for reducing chaos in children's lives.

Processes by Which Chaos Impacts Child Well-Being

One of the primary reasons for studying family chaos is to more clearly specify factors that are linked to socioemotional outcomes in the hopes that they may be ameliorated. It would be far easier, one suspects, to address lack of routinization than to address widespread poverty. As a first pass, researchers have considered chaos as a main effect in explaining children's socioemotional development and behavioral outcomes. In general, more organized households and less chaos are associated with fewer behavior problems (Coldwell, Pike, & Dunn, 2006; Dumas et al., 2005; Keltner, 1990; Valiente, Lemery-Chalfant, & Reiser, 2007). Others have extended this approach and considered whether chaos in and of itself contributed to child socioemotional functioning above and beyond parenting and process variables. For example, Coldwell, Pike, and Dunn (2006) demonstrated that chaos (as measured by the Confusion, Hubbub, and Order Scale [CHAOS]) predicts mothers' reports of child behavior problems above and beyond parenting measures. Similarly, Ackerman et al. (1999) have shown unique effects of instability on children's externalizing over other robust family process variables (e.g., family conflict).

However, it should be noted that though these zero-order correlations are often robust even in the context of other family variables, they also often tend to be modest in magnitude. Further, many of the studies that have examined the effects of chaos on children's lives have relied on middle-class families and did not examine in careful detail the stresses experienced at the most extreme levels of poverty. This suggests heterogeneity in outcomes even among children exposed to similar levels of chaos and gives rise to questions of when, or for whom, these relationships are strong or weak (i.e., moderator questions). Even with more robust main effect relationships, questions arise as to mechanisms by which chaos affects children. Just as chaos was identified in part as a mediating mechanism by which poverty would impact children, so are there mechanisms by which chaos impact children. Thus, though it is plausible to expect that more disorganized homes will be associated with more dysregulated behavior—either externalizing or internalizing—what is perhaps even more interesting is identifying those variables that may mediate or moderate the effects of chaotic home environments on child socioemotional well-being.

Drawing from process models of economically disadvantaged families (Conger & Donnellan, 2007), researchers have found a possible link between chaos and compromised parenting practices and harsh discipline. Two different patterns have been noted in the literature. One pattern suggests parental withdrawal and lack of responsiveness (Evans, 2001; Matheny et al., 1995; see also chap. 12, this volume). Another pattern recently reported by Valiente, Lemery-Chalfant, and Reiser (2007) linked low levels of family chaos with parents' positive reactions to children's effortful control, which in turn predicted externalizing problems. Similarly, the data of Forman and Davies (2003) fit a structural model in which greater instability led to increased parenting difficulties, which in turn increased adolescents' internalizing and externalizing in part by decreasing their appraisals of the family as secure. The various explanatory models that rely on dyadic parenting features offer a reminder that a full understanding of how parenting influences chaos requires taking into account how parenting itself is embedded in a larger system including socioeconomic factors, institutional supports and barriers, and cultural context. It is unlikely that parenting variables, though necessary, will be sufficient to explain child outcomes in the chaotic environments.

As previously indicated, chaos has also been considered as the mediating variable itself. Evans et al. (2005) explicitly took this approach in predicting learned helplessness from a model of income-to-needs ratio with chaos as the mediator. Though the overall effect size was small, the model was statistically significant. Brody and Flor (1997) presented a more complicated model for rural single-parent African American families. Overall there was an indirect relation between financial resources and child internalizing and externalizing problems mediated by family routines, mother–child harmony, and child self-regulation. For boys, a model could be fit whereby routines mediated the link between financial resources and internalizing problems. With the same sample and a composite indicator of family competence (routines, mother–child harmony, maternal involvement in school), family organization was one in a series of mediators between financial resources and child social and academic competence; other mediators included mothers' efficacy beliefs and child self-regulation (Brody, Flor, & Gibson, 1999). Consistent with the notion that chaos leads to feelings of being overwhelmed and feeling out of control, the inclusion of maternal efficacy beliefs in this model is important. There are also instances in which the mediating role of routines cannot be demonstrated. Seaton and Taylor (2003) reported links between routines and maternal optimism and adolescent academic self-concept. However, there were no significant relations between financial resources and routine.

Ackerman et al. (1999) authored a series of reports not only presenting compelling evidence as to the links between chaotic home environments as indexed by instability and child socioemotional outcomes but also suggesting moderating variables worthy of further investigation. Very briefly put, family instability was conceptualized as an aggregate of number of residences, number of intimate adult relationships, number of families with whom the child has lived, significant illnesses in the child's history, and negative life events. In a longitudinal analysis, family instability accounted for unique variance in caregiver and teacher reports of internalizing behaviors for children in the first grade (controlling for preschool problem behaviors). Other researchers have noted a link between chaotic environments and internalizing symptoms. Brody and Flor (1997) reported a link between family routines and internalizing symptoms but not externalizing symptoms for African American boys, Dumas et al. (2005) found relations between chaos and internalizing symptoms but only in an economically disadvantaged sample, and associations between internalizing symptoms and family routines have been reported (Fiese et al., 2006; Markson & Fiese, 2000). We mention this pattern because frequently attention is paid to externalizing problem behaviors and conduct disorders as a result of harsh and inconsistent parenting practices (Dishion & Stormshak, 2007). Chaotic home environments, by virtue of their unpredictable nature, disruption of daily activities, and lack of attention to positive emotions, may result in internalizing symptoms particularly during the elementary school years when most of these studies were conducted. Thus, moderators suggested by this research include race and ethnicity, parenting practices, economic disadvantage, and age or developmental status.

Other potential moderating variables that were suggested are biological reactivity and sensitivity, which may also attenuate the effects of family chaos. Unfortunately, the empirical literature is very scant in this regard. Ackerman et al. (1999) identified temperamental adaptability as a significant moderating variable for family instability in predicting parent report of child internalizing symptoms. Effortful control has also been suggested as a significant link between family chaos and child externalizing problems (Valiente et al., 2007). Other family processes offer an additional promising avenue of identifying moderating effects. Evans (2003) proposed that extensive exposure to chaotic environments results in a biological sensitivity or reactivity. It is also plausible that temperamental or biological sensitivities interact, or transact, with chaotic environments. Boyce's work suggests that level of routinization in the environment may distinguish children's biological sensitivities (D. A. Ellis et al., 2005). Consistent with his argument, however, is the fact that it is not the sheer amount of routinization in the environment but rather the interaction between social support, routines, and biology that predicts biological sensitivity. For the further advancement of the study of chaos in children's lives, a more complex model that incorporates temperament, biological sensitivity, and socioemotional adaptation seems warranted (see chap. 7, this volume, for further discussion on some of the candidate moderators of chaos on children's development).

Conclusion

Our intention with this chapter was not to provide an exhaustive account of evidenced links between chaos and socioemotional development but to focus more specifically on how the family unit as an organizing influence creates or responds to chaos in the environment. We did this by concentrating largely on chaos conceptualized as a family-wide construct, particularly in studies of family routines. From this standpoint, chaos in the family is multidimensional in nature and requires attention to how families find and maintain family time, how family activities are routinized and disrupted, and how families and members make meaning out of activities. In addition, we highlighted the value in attending to the frequently fluctuating but accumulating dynamic nature of chaos in the family setting. Last, we briefly discussed processes by which chaos has been linked to child socioemotional development.

In raising these foci and issues, we hoped to foster thinking about additional avenues of research pertaining to chaos and child development. The evidence is clear that chaos—whether it is conceptualized as we have done here or in other ways throughout this volume—consistently affects children's adaptation and behavioral outcomes negatively, even over and above other important contributors to child functioning. A salient challenge is to extend the rich existing research by (a) focusing on elucidating the effects of chaos in the context of other powerful individual and family processes, (b) attending to intervening mechanisms and moderating processes in the link between chaos and child socioemotional functioning, (c) considering chaos as an accumulation of events or circumstances that also fluctuate frequently and have distinct developmental courses, and (d) paying attention to the family unit as a whole. Previous work has been limited by using sample selections that either overrepresent low-income families, calling into question the inextricable link between poverty and chaos, or by relying on small self-selected samples in qualitative studies that call into question representativeness not only of the samples but also of the generalizability of the selected experiences. We noted how the timing and length of exposure may affect some children under some circumstances depending on characteristics of both children and resources available to the family. The ability of the family to reorganize itself on a daily, nightly, and weekly basis is one feature to consider in future research aimed at understanding the rhythms of chaos and its effects on children's lives. Fortunately, as our agendas broaden and deepen in focus, our data analytic (e.g., growth curve modeling) and measurement (e.g., biological stress reactivity) tools also increase in sophistication.

The challenge for the future will be how to engage families and corral resources before demands on their time pull them apart. As Bronfenbrenner (1979) noted, proximal processes are the engines of development, and preserving family time and routines is one resource that can be garnered in the service of reducing chaos in children's lives.

References

Ackerman, B. P., Kogos, J., Youngstrom, E., Schoff, K., & Izard, C. (1999). Family instability and the problem behaviors of children from economically disadvantaged families. *Developmental Psychology, 35*, 258–268.

Adam, E. K., & Chase-Landsdale, P. L. (2002). Home sweet home(s): Parental separations, residential moves and adjustment in low-income adolescent girls. *Developmental Psychology, 38*, 792–805.

Adam, E. K., Snell, E. K., & Pendry, P. (2007). Sleep timing and quantity in ecological and family context: A nationally representative time-diary study. *Journal of Family Psychology, 21*, 4–19.

Adler, N. E., Boyce, T., Chesney, M. A., Cohen, S., Folkman, S., Kahn, R. L., et al. (1994). Socioeconomic status and health: The challenge of the gradient. *American Psychologist, 49*, 15–24.

Bennett, L. A., Wolin, S. J., & Reiss, D. (1988). Deliberate family process: A strategy for protecting children of alcoholics. *British Journal of Addiction, 26*, 821–829.

Bianchi, S. M., & Raley, S. B. (2005). Time allocation in families. In S. M. Bianchi, L. M. Casper, & R. B. King (Eds.), *Work, family, health, and well-being* (pp. 21–48). Mahwah, NJ: Erlbaum.

Bradley, R. H., Corwyn, R. F., McAdoo, H. P., & Coll, C. G. (2001). The home environments of children in the United States Part I: Variations by age, ethnicity, and poverty status. *Child Development, 72*, 1844–1867.

Brody, G. H., & Flor, D. L. (1997). Maternal psychological functioning, family processes, and child adjustment in rural, single-parent, African American families. *Developmental Psychology, 33*, 1000–1011.

Brody, G. H., Flor, D. L., & Gibson, N. M. (1999). Linking maternal efficacy beliefs, developmental goals, parenting practices, and child competence in rural single-parent African American families. *Child Development, 70*, 1197–1208.

Bronfenbrenner, U. (1979). *The ecology of human development.* Cambridge, MA: Harvard University Press.

Bronfenbrenner, U., & Evans, G. W. (2000). Developmental science in the 21st century: Emerging questions, theoretical models, research designs, and empirical findings. *Social Development, 9,* 115–125.

Bronfenbrenner, U., & Morris, P. A. (1998). The ecology of developmental processes. In W. Damon & R. M. Lerner (Eds.), *Handbook of child psychology* (pp. 993–1028). Hoboken, NJ: Wiley.

Caplow, T., Bahr, H. M., Chandwick, B. A., Hill, R., & Williamson, M. H. (1982). *Middletown families: Fifty years of change and continuity.* Minneapolis: University of Minnesota Press.

CASA. (2007). *The importance of family dinners III.* New York: Columbia University.

Cinotto, S. (2006). "Everyone would be around the table": American family mealtimes in historical perspective, 1850-1960. *New Directions for Child and Adolescent Development, 111,* 17–34.

Coldwell, J., Pike, A., & Dunn, J. (2006). Household chaos—links with parenting and child behavior. *Journal of Child Psychology and Psychiatry, 47,* 1116–1122.

Compan, E., Moreno, J., Ruiz, M. T., & Pascual, E. (2002). Doing things together: Adolescent health and family rituals. *Journal of Epidemiology and Community Health, 56,* 89–94.

Conger, R. D., & Donnellan, M. B. (2007). An interactionist perspective on the socioeconomic context of human development. *Annual Review of Psychology, 58,* 175–199.

Coon, K. A., Goldberg, J., Rogers, B. L., & Tucker, K. L. (2001). Relationships between use of television during meals and children's food consumption patterns. *Pediatrics, 107,* 1–9.

Crnic, K. A., & Acevedo, M. (1995). Everyday stresses and parenting. In M. H. Bornstein (Ed.), *Handbook of parenting* (pp. 277–297). Mahwah, NJ: Erlbaum.

Cummings, E. M., Davies, P. T., & Campbell, S. B. (2000). *Developmental psychopathology and family process.* New York: Guilford Press.

Dahl, R. E., & Lewin, D. S. (2002). Pathways to adolescent health: Sleep regulation and behavior. *Journal of Adolescent Health, 31,* 175–184.

Daly, K. J. (2001). Deconstructing family time: From ideology to lived experience. *Journal of Marriage and Family, 63,* 283–294.

Dickstein, S., Seifer, R., Hayden, L. C., Schiller, M., Sameroff, A. J., Keitner, G. I., et al. (1998). Levels of family assessment: II. Impact of maternal psychopathology on family functioning. *Journal of Family Psychology, 12,* 23–40.

Diette, G. B., Markson, L., Skinner, E. A., Nguyen, T. T. H., Algatt-Bergstrom, P., & Wu, A. W. (2000). Nocturnal asthma in children affects school attendance, school performance, and parents' work schedule. *Archives of Pediatric and Adolescent Medicine, 154,* 923–928.

Dishion, T. J., & Stormshak, E. A. (2007). *Intervening in children's lives: An ecological, family-centered approach to mental health care.* Washington, DC: American Psychological Association.

Dumas, J. E., Nissley, J., Nordstrom, A., Smith, E. P., Prinz, R. J., & Levine, D. W. (2005). Home chaos: Sociodemographic, parenting, interactional, and child correlates. *Journal of Clinical Child and Adolescent Psychology, 34,* 93–104.

Duncan, G. J., & Brooks-Gunn, J. (1997). *Consequences of growing up poor.* New York: Russell Sage Foundation.

Eisenberg, M. E., Olson, R. E., Neumark-Sztainer, D., Story, M., & Bearinger, L. H. (2004). Correlations between family meals and psychosocial well-being among adolescents. *Archives of Pediatric and Adolescent Medicine, 158,* 792–796.

El-Sheikh, M., Buckhalt, J. A., Mize, J., & Acebo, C. (2006). Marital conflict and disruption of children's sleep. *Child Development, 77,* 31–43.

Ellis, B. J., Essex, M. J., & Boyce, W. T. (2005). Biological sensitivity to context: II. Empirical explorations of an evolutionary-developmental theory. *Development and Psychopathology, 17,* 303–328.

Ellis, D. A., Naar-King, S., Frey, M., Templin, T., Rowland, M., & Cakan, N. (2005). Multisystemic treatment of poorly controlled Type I Diabetes: Effects on medical resource utilization. *Journal of Pediatric Psychology, 30,* 656–666.

Evans, G. W. (2001). Environmental stress and health. In A. Baum, T. Revenson, & J. E. Singer (Eds.), *Handbook of health psychology* (pp. 365–385). Mahwah, NJ: LEA.

Evans, G. W. (2003). A multimethodological analysis of cumulative risk and allostatic load among rural children. *Developmental Psychology, 39,* 924–933.

Evans, G. W., & English, K. (2002). The environment of poverty: Multiple stressor exposure, psychophysiological stress, and socioemotional adjustment. *Child Development, 73*, 1238–1248.

Evans, G. W., Gonnella, C., Marcynyszyn, L. A., Gentile, L., & Salpekar, N. (2005). The role of chaos and poverty and children's socioemotional adjustment. *Psychological Science, 16*, 560–565.

Everhart, R. S., Fiese, B. H., & Smyth, J. S. (2008). A cumulative risk model for predicting caregiver quality of life in pediatric asthma. *Journal of Pediatric Psychology, 33*, 809–818.

Fagan, A. (2003). *Poll shows teens value ties, time with family.* Retrieved September 23, 2004, from http://www.washingtontimes.com/national/20030805-113337-717r.htm

Fiese, B. H. (1993). Family rituals in alcoholic and nonalcoholic households: Relation to adolescent health symptomatology and problematic drinking. *Family Relations, 42*, 187–192.

Fiese, B. H. (2006). *Family routines and rituals.* New Haven, CT: Yale University Press.

Fiese, B. H., & Everhart, R. S. (2006). Medical adherence and childhood chronic illness: Family daily management skills and emotional climate as emerging contributors. *Current Opinions in Pediatrics, 18*, 551–557.

Fiese, B. H., Foley, K. P., & Spagnola, M. (2006). Routine and ritual elements in family mealtimes: Contexts for child wellbeing and family identity. *New Directions for Child and Adolescent Development, 111*, 67–90.

Fiese, B. H., Hooker, K. A., Kotary, L., & Schwagler, J. (1993). Family rituals in the early stages of parenthood. *Journal of Marriage and the Family, 57*, 633–642.

Fiese, B. H., & Spagnola, M. (2006). The interior life of the family: Looking from the inside out and the outside in. In A. Masten (Ed.), *Multilevel dynamics in developmental psychopathology: Pathways to the future* (pp. 119–150). Mahwah, NJ: Erlbaum.

Fiese, B. H., Tomcho, T., Douglas, M., Josephs, K., Poltrock, S., & Baker, T. (2002). Fifty years of research on naturally occurring rituals: Cause for celebration? *Journal of Family Psychology, 16*, 381–390.

Fiese, B. H., & Wamboldt, F. S. (2003). Tales of pediatric asthma management: Family-based strategies related to medical adherence and health care utilization. *Journal of Pediatrics, 143*, 457–462.

Fiese, B. H., Wamboldt, F. S., & Anbar, R. D. (2005). Family asthma management routines: Connections to medical adherence and quality of life. *Journal of Pediatrics, 146*, 171–176.

Fiese, B. H., Winter, M. A., Sliwinski, M., & Anbar, R. D. (2007). Nighttime waking in children with asthma: An exploratory study of daily fluctuations in family climate. *Journal of Family Psychology, 21*, 95–103.

Fivush, R., Hazzard, A., Sales, J. M., Sarfati, D., & Brown, T. (2003). Creating coherence out of chaos? Children's narratives of emotionally positive and negative events. *Applied Cognitive Psychology, 17*, 1–19.

Forman, E. M., & Davies, P. T. (2003). Family instability and young adolescent maladjustment: The mediating effects of parenting quality and adolescent appraisals of family security. *Journal of Clinical Child and Adolescent Psychology, 32*, 94–105.

Furstenberg, F. F., Cook, T. D., Eccles, J., Elder, G. H., & Sameroff, A. J. (1999). *Managing to make it: Urban families and adolescent success.* Chicago: University of Chicago Press.

Gillis, J. R. (1996). Making time for family: The invention of family time(s) and the reinvention of family history. *Journal of Family History, 21*, 4–21.

Grych, J. H., Wachsmuth-Schlaefer, T., & Klockow, L. L. (2002). Interparental aggression and young children's representations of family relationships. *Journal of Family Psychology, 16*, 259–272.

Hareven, T. (1985). Historical changes in the family and the life course: Implications for child development. *Monographs of the Society for Research in Child Development, 50*(4–5), 8–23.

Haugland, B. S. M. (2005). Recurrent disruptions of rituals and routines in families with paternal alcohol use. *Family Relations, 54*, 225–241.

Hofferth, S. L., & Sandberg, J. F. (2001). How American children spend their time. *Journal of Marriage and Family, 63*, 295–308.

Jacobs, M. P., & Fiese, B. H. (2007). Family mealtime interactions and overweight children with asthma: Potential for compounded risks? *Journal of Pediatric Psychology, 32*, 64–68.

Johnson, S. L., & Birch, L. L. (1994). Parents' and children's adiposity and eating style. *Pediatrics, 94*, 653–661.

Keltner, B. (1990). Family characteristics of preschool social competence among Black children in a Head Start program. *Child Psychiatry and Human Development, 21*, 95–108.
Lareau, A. (2003). *Unequal childhoods: Class, race, and family life.* Berkeley: University of California Press.
Markson, S., & Fiese, B. H. (2000). Family rituals as a protective factor against anxiety for children with asthma. *Journal of Pediatric Psychology, 25*, 471–479.
Matheny, A., Wachs, T. D., Ludwig, J., & Philips, K. (1995). Bringing order out of chaos. *Journal of Applied Developmental Psychology, 16*, 429–444.
Neumark-Sztainer, D., Wall, M. M., Hannan, P. J., Story, M., Croll, J., & Perry, C. (2003). Correlates of fruit and vegetable intake among adolescents: Findings from Project EAT. *Preventive Medicine, 37*, 198–208.
Pribesh, S., & Downey, D. B. (1999). Why are residential and school moves associated with poor school performance? *Demography, 36*, 521–534.
Roy, K. M., Tubbs, C. Y., & Burton, L. M. (2004). Don't have no time: Daily rhythms and the organization of time for low-income families. *Family Relations, 53*, 168–178.
Seaton, E. K., & Taylor, R. D. (2003). Exploring familial processes in urban, low-income African American families. *Journal of Family Issues, 24*, 627–644.
Serpell, R., Sonnenschein, S., Baker, L., & Ganapathy, H. (2002). Intimate culture of families in the early socialization of literacy. *Journal of Family Psychology, 16*, 391–405.
Snell, E. K., Adam, E. K., & Duncan, G. J. (2007). Sleep and body mass index and overweight status of children and adolescents. *Child Development, 78*, 309–323.
Stark, L. J., Jelalian, E., Powers, S. W., Mulvihill, M. M., Opipari, L. C., Bowen, A., et al. (2000). Parent and child mealtime behavior in families of children with cystic fibrosis. *Journal of Pediatrics, 136*, 195–200.
Stark, L. J., Mulvihill, M. M., Jelalian, E., Bowen, A., Powers, S. W., Tao, S., et al. (1997). Descriptive analysis of eating behavior in school-age children with cystic fibrosis and healthy control children. *Pediatrics, 99*, 665–671.
Strunk, R. C., Sternberg, A. L., Bacharie, L. B., & Szefler, S. J. (2002). Nocturnal awakening caused by asthma in children with mild-to-moderate asthma in the Childhood Asthma Management Program. *Journal of Allergy and Clinical Immunology, 110*, 395–403.
Toth, S. L., Cicchetti, D., Macfie, J., Maughan, A., & Vanmeenen, K. (2000). Narrative representations of caregivers and self in maltreated preschoolers. *Attachment and Human Development, 2*, 271–305.
Tubbs, C. Y., Roy, K. M., & Burton, L. M. (2005). Family ties: Constructing family time in low-income families. *Family Process, 44*, 77–91.
Valiente, C., Lemery-Chalfant, K., & Reiser, M. (2007). Pathways to problem behaviors: Chaotic homes, parent and child effortful control, and parenting. *Social Development, 16*, 249–267.
Wildenger, L. K., McIntyre, L. L., Fiese, B. H., & Eckert, T. (2008). Children's daily routines during kindergarten transition. *Early Childhood Education Journal, 36*, 69–74.
Winter, M. A., Davies, P. T., & Cummings, E. M. (2008). *Children's security in the context of family instability and maternal communications.* Unpublished manuscript, University of Rochester, Rochester, NY.
Wolin, S. J., & Bennett, L. A. (1984). Family rituals. *Family Process, 23*, 401–420.
Wolin, S. J., Bennett, L. A., Noonan, D. L., & Teitlebaum, M. A. (1980). Disrupted family rituals: A factor in generational transmission of alcoholism. *Journal of Studies of Alcohol, 41*, 199–214.

5

Child-Care Chaos and Child Development

Feyza Corapci

Child-care quality is an important, contemporary, and highly sensitive issue. Many child-care experts and researchers view the goals of child-care services as the provision of nurturant care as well as cognitive and social stimulation based on developmentally appropriate practices and beliefs (Lamb, 1998). However, there is concern that child-care programs fall short of keeping children safe and promoting growth and competence (Brauner, Gordic, & Zigler, 2004; Gallagher, Rooney, & Campbell, 1999). This chapter addresses environmental chaos in nonparental care or child-care settings. Though a number of researchers (e.g., Brauner et al., 2004) have noted the "child-care crisis" in the United States, the problem also exists around the world, especially in most of the developing countries (Montie & Olmsted, 2001). Therefore, research pertinent to environmental chaos conducted outside the United States has been integrated into this chapter wherever possible.

The focus on the environmental conditions of child care is of key interest to researchers, clinicians, and policymakers for a number of reasons. First, as social, economic, and demographic changes take place worldwide, especially with increasing maternal employment outside the home, rising numbers of children experience extensive child care early on. As a consequence, child care has turned into another central microsystem that provides complementary opportunities for children's motor, cognitive, and socioemotional development in addition to those provided at home (Lamb, 1998; Shonkoff & Phillips, 2000). The second and more alarming reason concerns the wide variation in child-care quality in middle-class and working-poor communities (Adams, Tout, & Zaslow, 2007; Fuller, Kagan, Loeb, & Chang, 2004; National Institute of Child Health and Human Development Early Child Care Research Network [NICHD ECCRN], 2003). The wide variation in quality and overall lack of access to affordable and high-quality child care are even more worrisome in light of longitudinal research documenting enduring negative effects of poor-quality child care on cognitive and social development throughout the elementary school years (Belsky et al., 2007; Burchinal, Peisner-Feinberg, Pianta, & Howes, 2002; NICHD ECCRN, 2006). There is also growing awareness that low-income families' access to quality child care is a particularly critical issue given the evidence that extensive amount of time spent in poor-quality child care is associated with poorer language competence and increases in externalizing problems among children from disadvantaged families (Burchinal, Peisner-Feinberg, Bryant, & Clifford, 2000), whereas high-quality child care acts as a protective effect fostering the social and emotional adjustment of children at high risk (Love et al., 2003; Votruba-Drzal, Levine Coley, & Chase-Lansdale, 2004). Finally, poor child-care quality during the early years partly explains the effects of early full-time maternal employment on children's poor cognitive skills (Brooks-Gunn, Han, & Waldfogel, 2002).

This chapter is organized around four issues. First, the constructs of chaos and quality in child care are delineated. Second, relevant research on how chaotic child-care settings impact children's developmental outcomes as well as proximal processes is summarized. Third, available data on variables that moderate the influence of chaos are reviewed. Finally, major gaps in knowledge about the nature and impact of environmental chaos are identified, and a future research agenda is offered.

Nonparental Child Care

Nonparental child care is offered by different arrangements and auspices. *Center-based* settings include day-care centers or preschools (typically affiliated with a public or private agency such as a religious organization, corporation, or community center), public school-based programs, and funded programs such as Head Start. *Home-based* care can be in the form of informal arrangements with relatives or friends in the child's home or in the form of highly structured, formal arrangements in the home of a care provider (i.e., family child-care homes). Family child-care home, especially for infants and toddlers, appeals to parents because of flexible schedules offered by these arrangements, familiarity with the caregiver, and proximity to the child's home (Burchinal, Howes, & Kontos, 2002; Early & Burchinal, 2001). The NICHD Study of Early Child Care and Youth Development (SECCYD) is a multisite, prospective longitudinal study with a sample of over 1,100 children from nine states in the United States. This study has documented that "no one care type is uniformly better than another" given the variability in quality across settings (NICHD ECCRN, 2004, p. 226).

Definition and Measurement of Chaos and Child-Care Quality

Before we review the literature on the role of environmental chaos on developmental outcomes and proximal processes, it is necessary to define and elucidate the constructs of chaos and child-care quality. Do chaos and high child-care quality constitute the two ends of a continuum, or do these constructs represent two independent but complementary features of the child-care environment?

The construct of *environmental chaos* refers to high levels of crowding, environmental traffic, and ambient background noise, as well as a lack of temporal and structural regularity such that there are few regularities or routines, little is scheduled, or nothing has its place (Matheny, Wachs, Ludwig, & Phillips, 1995; discussion on the conceptualization of chaos is also presented in chap. 1, this volume). One source of information about chaos in child-care settings comes from the Life in Early Childhood Programs (LECP) Scale (Kontos & Wachs, 2000). The 16-item LECP Scale is a teacher report of classroom chaos based on both previous research on environmental chaos and the Confusion, Hubbub, and Order Scale (CHAOS; Matheny et al., 1995), which was developed for use in the home. The items pertain to child-care crowding (e.g., "There are too many children in the classroom given the amount of space"), environmental traffic (e.g., "There are many adults in and out of my classroom during the day"), the overall noise level (e.g., "There is often a fuss going on in our classroom"), and the degree of control of the structural and temporal organization in the classroom (e.g., "Interruptions make it difficult to keep a schedule in the classroom"). The scale's reliability and validity have been demonstrated in recent research (Wachs, Gurkas, & Kontos, 2004).

NICHD SECCYD, the largest child-care study to date, has focused on multiple components of the child-care context, including chaos (NICHD ECCRN, 2000). Using the Profile for Early Childhood Programs (Abbott-Shim & Sibley, 1987) and three items from Wachs (1991), researchers found that child-care chaos reflected a crowded, cluttered, and unsafe physical environment as well as a lack of developmentally appropriate toys and learning materials.

Child-care quality is a multidimensional construct including characteristics of the physical context, known as *structural quality*, and characteristics of the caregiver–child interaction, known as *process quality* (Lamb, 1998; NICHD ECCRN, 2000; Vandell, 2004). High structural quality is evident when the child-care setting has a small group size (i.e., low number of children taken care of by a caregiver or team of caregivers), low child–adult ratio, and high staff qualifications such as education level (Howes, 1997; NICHD ECCRN, 2000; Scarr, Eisenberg, & Deater-Deckard, 1994; Vandell, 2004). Other structural quality indicators such as teacher wages and staff turnover are affected not only by government regulations but also by center policies and economic climate (Phillipsen, Burchinal, Howes, & Cryer, 1997; Scarr et al., 1994; Whitebook, Howes, & Phillips, 1998).

High process quality is evident when children have affectionate, sensitive, responsive, and stimulating caregivers who engage them in age-appropriate learning and play activities as well as in positive interactions with peers (Lamb, 1998; Vandell, 2004). Assessment of process quality involves conducting direct observations of focal children as they interact with their teachers and peers (NICHD ECCRN, 1997) or conducting direct observations of the classroom features (i.e., furnishing, arrangement of space, schedule, health and safety, materials and activities, interactions among all children and adults) by using rating scales such as the Early Childhood Environment Rating Scale—Revised (ECERS–R) developed by Harms, Clifford, and Cryer (1998).

Some evidence suggests that poor quality and chaos are distinct but complementary constructs of the child-care environment. First, Wachs et al. (2004) revealed that each construct contributed unique predictive variance in children's situational compliance, even after statistically controlling for the other measure. Committed compliance was predicted only by child-care quality, whereas passive noncompliance was predicted only by child-care chaos. Second, the NICHD study of early child care also found that the inclusion of the chaos measure described earlier significantly improved the prediction of positive caregiving, over and above the other quality indicators at 15, 24, and 36 months, both in home- and center-based care settings (NICHD ECCRN, 2000). Finally, a low total ECERS–R score might reflect an underprivileged economic climate such that the classroom is deprived of various play materials including blocks, gross motor equipment, and other expensive materials but does not necessarily imply a chaotic child-care environment.

Taken together, the inclusion of both types of environmental measures can advance understanding of how different components of day-care context contribute to developmental outcomes. Without a doubt, certain components of environmental chaos such as large group sizes and high child–adult ratios overlap with the indices of poor structural quality. Yet other components of environmental chaos are not typically emphasized as part of the structural or process quality. These components include instability in care, such as variability in the caregivers and program routine, high levels of ambient noise, environmental traffic, and a sense of upheaval. In the next section, these environmental features are examined in relation to child developmental outcomes and proximal processes between caregivers and children.

Crowding

In the sections that follow, we review research on crowding in center- and home-based child care in relation to child development outcomes and caregiving quality. Studies on crowding conducted outside the United States have also been integrated into this discussion to examine the influence of environmental chaos on children's development across cultures.

Group Size and Child–Adult Ratios in Center- and Home-Based Child Care

Crowding in child care is measured by large group sizes and child–adult ratios. Given young children's greater need for closer monitoring and sensitive care, the National Association for the Education of Young Children recommended a child–adult ratio of 4:1 for infants with a group size of 8, 5:1 for toddlers with a group size of 10, and 8:1 for preschoolers with a group size of 16 (Committee on Early Childhood, Adoption, and Dependent Care, 2005). In child-care settings serving mostly middle-income, European American families, the average child–adult ratios were 2.6:1, 3.3:1, and 4.6:1 at 15, 24, and 36 months of age, respectively (NICHD ECCRN, 2000). However, other researchers have found considerable variability in child–adult ratios in center-based settings, reflecting large differences in licensing and regulatory policies across the states (Peisner-Feinberg & Burchinal, 1997; Phillips, Mekos, Scarr, McCartney, & Abbott-Shim, 2000; Phillipsen et al., 1997).

Recent studies have shown that the average child–adult ratios and group sizes in centers serving children from disadvantaged families were within the ranges suggested by professional organizations, reflecting stricter state regulation and targeted subsidies in some states (Adams et al., 2007; Clifford et al.,

2005; Li-Grining & Coley, 2006; LoCasale-Crouch et al., 2007; Loeb, Fuller, Kagan, & Carrol, 2004; Pianta et al., 2005). Finally, in contrast to the center-based settings, family child-care homes have smaller group sizes and child–adult ratios, in both middle-class and low-income communities (Fuller et al., 2004; Li-Grining & Coley, 2006; NICHD ECCRN, 2000).

Crowding, Child Developmental Outcomes, and Proximal Processes

A growing body of evidence indicates that crowding in child care is associated with poor language skills, hostile and aggressive behavior, aimless wandering in the classroom, fewer positive social behaviors, less school readiness, and high rates of communicable illnesses even after controlling for family characteristics such as maternal education and family income (Bradley & Vandell, 2007; Lamb, 1998; Love et al., 2003). There is also evidence for higher levels of cortisol at day care compared with levels at home, suggesting children's attempts to cope with demands in the day-care context (Legendre, 2003; Sims, Guilfoyle, & Parry, 2006; Vermeer & Van IJzendoorn, 2006). Furthermore, a low adult–child ratio was found to be a predictor of poor attachment security in infants, when data from the NICHD study of early child care, which lacks the poorest quality child-care settings, were combined with data from the Haifa Study of Early Child Care to yield a broader range of child-care quality in the sample (Love et al., 2003). It is important to note that available reviews have revealed smaller child-care effect sizes compared with family effect sizes. A number of researchers nevertheless contend that even the modest effect sizes can be meaningful for policy implications given the large number of children attending child care from early on (NICHD ECCRN, 2006).

Research also indicates that crowding acts to influence child outcomes by undermining the quality of caregiver–child interactions (NICHD ECCRN, 2002). For instance, child-care providers in crowded, poor-quality child care tend to be less responsive, less involved, less likely to demonstrate how something works, and less vocally stimulating compared with their counterparts in less crowded child-care settings (E. J. De Schipper, Riksen-Walvaren, & Geurts, 2006; Lamb, 1998; Montie, 2001; Vandell, 2004). A recent meta-analysis (Ahnert, Pinquart, & Lamb, 2006) also documented that crowding attenuates the association between day-care providers' responsiveness and children's attachment security to day-care providers. In other words, children formed a secure attachment relation to their sensitive child-care providers only in the context of small group sizes.

A few recent studies have unexpectedly suggested that "better structural features of quality appear to support and facilitate, but do not assure higher process quality" (Adams et al., 2007, p. 15). With the exception of the Early Head Start and Head Start programs, minimal process quality was observed in some child-care settings in compliance with high child–adult ratios, especially in classrooms with children from disadvantaged families (Adams et al., 2007; Li-Grining & Coley, 2006; Love et al., 2003; Pianta et al., 2005). Examination of child-care conditions in disadvantaged communities is particularly important because about half of children younger than 6 years of age in low-income families experience nonparental child care following the welfare-to-work program

(Adams et al., 2007). As an indicator of chaos, it may be especially important to differentiate between the number of children in the classroom and the number of children with disruptive behavior problems given the higher prevalence of behavioral and emotional problems among poor children (Duncan & Brooks-Gunn, 1999). The presence of even a few children with severe behavioral problems is likely to disrupt regular classroom routine and compromise teachers' responsive and sensitive caregiving over time, resulting in growing chaos, even in classrooms with favorable child–adult ratios.

Research findings on the relation between crowding and caregiver–child interactions in family child-care homes have also been somewhat inconsistent. According to the Family Child Care and Relative Care Study (Kontos, Howes, Shinn, & Galinsky, 1995) and the Vancouver Day Care Project (Goelman & Pence, 1987, cited in Burchinal, Howes, & Kontos, 2002), there was higher process quality in crowded family child-care settings. In both studies, day-care providers responsible for more children had higher levels of education and training than did those responsible for fewer children. After statistical controls for day-care provider background variables, less responsive and less sensitive care toward infants was observed in crowded family child-care homes (Clarke-Stewart, Vandell, Burchinal, O'Brien, & McCartney, 2002; NICHD ECCRN, 1996).

In contrast to these findings, secondary analyses using data from two other large-scale studies of family child-care homes did not reveal a link between crowding and proximal processes (Burchinal, Peisner-Feinberg, et al., 2002). Group sizes in these licensed family child-care homes averaged around six children, with a range from 1 to 13 children. Burchinal, Peisner-Feinberg, et al. (2002) argued that caregiver education and training, rather than the child–adult ratio, appears to act as a powerful predictor of interaction quality in home-based care settings where group sizes are small to moderate. Other researchers have also identified inadequate health and safety provisions, prolonged exposure to television, caregivers' poor training, and lack of stimulation in family child-care homes as factors for increased chaos and poor quality, especially in unregulated settings (Adams et al., 2007; Clarke-Stewart et al., 2002; Kontos et al., 1995).

Crowding in Child-Care Settings Outside the United States

Cross-national investigations have revealed that early child-care settings for preschool-age children in non-Western, developing countries are fairly crowded, partly reflecting the challenge to use the limited resources in these countries in the service of early child-care expansion (Montie, 2001). For example, Chinese and Hong Kong kindergartens had the largest median group sizes (29–35 children) followed by Thailand's educational programs and Indonesian rural kindergartens with median child–adult ratios of 31:1, 29:1, and 17:1, respectively.

Research on the impact of child-care crowding on child development from different countries has yielded results inconsistent with research conducted in the United States (Montie, Xiang, & Schweinhart, 2006). Rather than the child–adult ratio, teachers' education level, the implementation of free-choice activities in class, and the variety in equipment and materials predicted children's language and cognitive competence at age 7 across 10 diverse countries (e.g., Finland, Greece, Hong Kong, Indonesia, Poland, Spain). Findings on the link between crowding and proximal processes have also been mixed. Child-care research conducted in Europe revealed that different patterns of structural quality indicators predicted process quality in center-based child-care settings compared with the United States (Cryer, Tietze, Burchinal, Leal, & Palacios, 1999; E. J. De Schipper, Riksen-Walraven, & Geurts, 2007; Gevers Deynoot-Schaub & Riksen-Walraven, 2005). For example, a low child–adult ratio was a predictor of process quality only in the sample from Germany, whereas physical space and teacher education were strong predictors of process quality in the samples from Portugal and Spain, respectively. In addition, data from a study done in China revealed that high child-care quality, defined as the presence of qualified staff and favorable child–adult ratio, was related to better process quality (Rao, Koong, Kwong, & Wong, 2003). This differential pattern of relations between the structural and process quality indices may be due to different child-care policies in different countries or to methodological issues such as the restricted range in the child–adult ratios and low statistical power to detect the modest associations between the structural and process quality indices (Cryer et al., 1999).

An intriguing finding on the role of cultural attitudes toward care and education comes from a recent study conducted in the United States and Korea (Clarke-Stewart, Lee, Allhusen, Kim, & McDowell, 2006). This cross-national comparison study illustrated that compared with teachers in the United States, Korean teachers engaged in less verbal and playful interactions with children even when child–adult ratios in the classrooms were controlled statistically. The study also noted that Korean classrooms, where teachers are responsible for almost twice as many children as in the United States, primarily contained materials for math and music to engage children in activities that require concentration and sustained attention. Fewer physical resources such as varied materials to use for gross motor activities and free play were observed. Clarke-Stewart et al. (2006) argued that culturally valued activities may compensate for the unavailability of one-on-one interaction with teachers in classrooms where socialization goals emphasize group membership, hard work, and discipline. More research is certainly needed on potential culturally unique moderators to determine whether the influence of environmental chaos on children's development and interactions with caregivers are truly similar across cultures (Wachs & Corapci, 2003).

Instability and Lack of Temporal and Structural Regularities

Three independent aspects of instability in child care have been identified as indicators of environmental chaos: (a) daily instability, (b) arrangement multiplicity, and (c) long-term instability (J. C. De Schipper, Tavecchio, Van IJzendoorn, & Linting, 2003; Tran & Weinraub, 2006).

Daily Instability

Daily instability refers to variability in the number of caregivers in charge of the group. The term also includes instability in the child's peer group and unpredictability in the daily program schedule. Children's daily instability experiences in a Dutch child-care setting were assessed via teacher report in a sample of 186 children, ages 6 to 30 months (J. C. De Schipper et al., 2003). Children who experienced less daily stability with caregivers and peers were perceived by their teachers as more stressed and more apprehensive in child care after controlling for child gender, age, and temperament. In a follow-up study about one and a half years later, when the children were 26 to 50 months old, daily instability did not predict behavioral observations of child noncompliance or the quality of peer interactions in child care. Rather, more daily stability in care unexpectedly predicted poorer quality of observed caregiver behavior (i.e., more detachment and flatness of affect in caregivers) after controlling for mother's daily stress, child's temperament, and staff turnover rate. According to J. C. De Schipper et al. (2003), the unexpected relation detected in this study is more understandable when daily stability facilitates social engagement among the day-care caregivers, who consequently orient more toward each other and less toward the children in the group. Future research is warranted to replicate the link between daily stability and caregiving quality.

The notion of "flexible" child care was recently introduced with the increases in nonstandard working hours of parents (J. C. De Schipper et al., 2003). *Flexible care* refers to the extent to which a child experiences evening care (i.e., after 5:30 p.m.), has evening meals in child care, attends child care a varying number of days each week, and experiences the absence of parents' fixed time schedule for dropping off or picking up the child. Flexible child-care use as reported by mothers was related to less stability in day-care caregivers and peer contacts during the course of a day and a less predictable daily routine as well as higher rates of child noncompliance toward caregivers (J. C. De Schipper et al., 2003).

A growing number of studies suggest that child temperament acts as a moderator between child-care instability and children's well-being. First, children with better emotion regulation competence were less likely to show adjustment problems even in the presence of less predictable and variable daily routine (J. C. De Schipper et al., 2003). Second, children with easy temperament were more likely to show well-being and adapt to the day-care setting when they were in a group with greater availability of trusted caregivers during the day. On the other hand, the availability of trusted caregivers did not prevent children with difficult temperament from having adjustment problems (J. C. De Schipper, Tavecchio, Van IJzendoorn, & Van Zeijl, 2004). Finally, 15-month-old children with difficult temperament were observed to respond less often to peer-initiated contacts when they attended a lower quality child-care setting (Gevers Deynoot-Schaub & Riksen-Walraven, 2006).

Arrangement Multiplicity

A number of studies have shown that many young children (17%–28%) experience multiple concurrent arrangements during the day (Adams et al., 2007; J. C. De Schipper et al., 2003; Early & Burchinal, 2001; Tran & Weinraub, 2006). Children with multiple care arrangements had high levels of internalizing problems and less well-being in the day-care setting per teacher report, even after controlling for gender, temperament, and daily care stability (J. C. De Schipper et al., 2003, 2004).

There is also evidence for moderational relations. First, the availability of parallel arrangements was unrelated to internalizing problems for children with easy temperament. However, for children with difficult temperament, several parallel care arrangements were associated with high levels of internalizing problems (J. C. De Schipper et al., 2004). Second, children with more parallel child-care arrangements had significantly higher rates of insecure attachment when their mothers also showed insensitivity (NICHD ECCRN, 1997). Finally, using data from the NICHD SECCYD, Tran and Weinraub (2006) documented poorest language skills among the 15-month-old infants who were in multiple arrangements (i.e., at least two simultaneous arrangements in addition to mother care) and who also received the lowest quality care for at least 3 to 4 months. On the other hand, infants showed better language performance if they experienced multiple arrangements in high-quality child-care settings.

Long-Term Instability

Center closure and staff turnover are formally not part of the definition of environmental chaos; however, they deserve mention because these factors may contribute to chaos by creating disruptions in children's daily experiences in child care, requiring a series of simultaneous changes such as the separation of children from the child-care staff and peers as well as from familiar physical surroundings (Kershaw, Forer, & Goelman, 2005; Whitebook et al., 1998). Staff turnover and center closure may also result from a low-quality child-care environment combined with low staff wages (Whitebook & Sakai, 2003).

Longitudinal studies have documented high staff turnover rate over time (Whitebook et al., 1998; Whitebook & Sakai, 2003). Research suggests that staff turnover relates to caregivers' less positive attitude toward children in daycare centers and higher social withdrawal and aggression in children while controlling for center quality (J. C. De Schipper et al., 2003; Howes & Hamilton, 1993). On the other hand, negative consequences of staff stability have also been noted, especially in countries where the economic climate is poor and alternative employment opportunities are insufficient such that "high stability of unskilled caregivers within an unfavorable and hectic ecological context may even perpetuate the low-quality consequences" (Koren-Karie, Sagi-Schwartz, & Egoz-Mizrachi, 2005, p. 124).

Ambient Background Noise

Reading and language acquisition has been affected by noise exposure in almost all studies among elementary-school-age children (Evans & Lepore, 1993). However, there is a striking paucity of research on young children attending noisy child-care centers where children acquire preacademic skills. The few available studies suggest a link between noise and attention level among young children. Hambrick-Dixon (2002) found an unexpected pattern of results such that noisy test conditions did not interfere with the discrimination ability of African American preschool-age children who attended child-care centers near subway trains. On the other hand, children who attended child-care centers located away from noise sources performed best when they were tested in quiet conditions. These findings suggest that young children exposed to noise learn to habituate and tune out distracting high noise levels. However, the long-term consequence of this coping skill in the face of chronic exposure to noise was the deterioration of young children's attention skills (Hambrick-Dixon, 2002). It is possible that children not only tune out noise but also the vocal signals of their caregivers and peers, thereby losing their ability to attend efficiently to information from their environment (Evans & Lepore, 1993).

Maxwell and Evans (2000) investigated the relation between exposure to chronic noise and prereading skills among 4-year-old children who were attending a child-care center that was noisy because of poor acoustical design. Evaluations of children's linguistic, cognitive, and motivational competence were conducted before and after the sound attenuation in the classrooms. Children in the quiet condition, compared with the children in the noisy condition, had better cognitive and language discrimination skills as well as less vulnerability to learned helplessness when they were required to work on an unsolvable puzzle. These results imply the necessity of interior acoustic design corrections in child-care centers to reduce noise generated within the building. Unfortunately, evaluation of ambient background noise is often not considered as part of the quality assessment in child-care settings.

Future Research Agenda

Available evidence suggests that even when certain structural quality indices are in place, such as a low child–adult ratio, child-care centers may still suffer from poor process quality, especially those serving poor children (Head Start programs are an exception). Therefore, future researchers must consider a larger model of the child-care environment to address the interplay among multiple chaos and quality indices. For example, do any structural quality indices appear to attenuate the positive effects of favorable child–adult ratios in child-care centers serving low-income families? One potential candidate is teachers' low levels of educational attainment (Abbott-Shim, Lambert, & McCarty, 2000; Early et al., 2006). If teachers have inadequate professional training, they would be less likely to show sensitive and responsive interactions over time, especially if they have children with adjustment problems, who are likely to interrupt the routines, increase classroom commotion, and impede teachers' effective classroom instruction (Abbott-Shim et al., 2000). Therefore, future studies of community "high-risk" child care are needed to provide in-depth assessment of various domains such as teacher and child characteristics (e.g., behavior problems, developmental delays) to identify risk factors that might override the effects of protective factors such as favorable child–adult ratios and the underlying mechanisms for their effect on poor-quality teacher–child interaction.

Another less investigated area is the underlying mechanisms between high levels of crowding and the occurrence of developmentally less facilitative transactions between children and their caregivers. Why high child–adult ratios adversely influence patterns of caregiver–child transactions is a critical issue for further research. The linkages between environmental chaos and patterns of proximal processes may be mediated by crowding leading to a greater sense of fatigue and inefficacy in caregivers, greater sensitivity to daily stress, withdrawal from social interactions, or deterioration in cognitive processes such as difficulty in decision making and attending to social information (Corapci & Wachs, 2002; Evans, Rhee, Forbes, Allen, & Lepore, 2000; Lepore, Evans, & Palsane, 1991; further discussion on moderators and mediators of chaos is also presented in chaps. 3, 4, and 7, this volume).

In addition, the relative strength of multiple environmental quality and chaos indices needs to be addressed. The application of statistical approaches such as cluster analysis may provide more insight into the nature of child-care ecologies (LoCasale-Crouch et al., 2007). This technique allows researchers to identify child-care classroom profiles on the basis of a common set of chaos and quality indices including caregiver-, classroom-, and center-level variables. Such designs would allow a more rigorous test of the child-care environment context and shed light on which particular child-care profiles are more or less powerful in the prediction of caregiver–child social interactions and adjustment outcomes for different age groups. Results from such a study would also have the potential to influence policymakers to embark on a set of policy changes for higher quality child care as a result of increased awareness of the configuration among the different environmental components.

The investigation of relations among child-care crowding, attachment, and cortisol increases might be another interesting area for future research. One line of research has shown increases in cortisol levels in children in child care, especially in crowded child-care classrooms, compared with children at home (Legendre, 2003; Vermeer & Van IJzendoorn, 2006). A second line of research has shown that the association between child-care providers' responsiveness and child attachment security was attenuated by group size and child–adult ratio (Ahnert et al., 2006). It is possible that crowding may undermine the sense of protection in children, who in turn exhibit greater vulnerability to elevated cortisol levels. In the face of high stress, the expected benefit of caregiver sensitivity may not be realized for those children. However, an empirical focus on the mediating role of elevated cortisol level between child-care crowding and attachment quality to child-care providers has not received much attention in current literature.

On the methodological side, to date most research on crowding has been based on correlational design. Because crowding indices such as the child–adult ratio or group size are potentially less sensitive to the influences of teacher–child interactions, correlational results imply that structural features such as crowding in child care contribute to changes in caregivers' sensitivity, stimulation, and positive affect. For obvious ethical reasons, researchers cannot use an experimental condition that involves exposing children to crowded child-care classrooms that are known to have negative influences. However, researchers might be able to compare the quality of caregiver–child interactions before and after a change of child–caregiver ratio as a result of financial circumstances or changes in child-care policy through natural or quasi-experimental studies as exemplified by the study by Maxwell and Evans (2000).

In terms of the measurement issues, chaos has typically been assessed at the entire classroom level per teacher report and by direct observation of classroom crowding. Future research is warranted to assess chaos as experienced by individual children in the same classroom. If repeated behavioral observations of each target child are conducted during specific activities, the number of peers and adults around the child, the child's presence in a noisy area versus in a relaxing corner in the classroom, or the child's routine in terms of arrival and departure from child care can be identified. To predict variability in the exposure to chaos in child care, researchers should also focus on possible moderating factors, such as child's gender, temperamental differences in affect and activity, and the chaos level at home.

Finally, crowding has typically been measured by using child–adult ratios via questionnaires or interviews with directors or staff and verified by observations of randomly selected classrooms on a single day (Fuller et al., 2004; Phillips, Voran, Kisker, Howes, & Whitebook, 1994; Scarr et al., 1994). Recent research has documented that ratio measurements can be improved by measuring child–adult ratios in multiple classrooms on multiple days during longer observation periods (Le, Perlman, Zellman, & Hamilton, 2006). A final recommendation would be to use multilevel modeling strategies with nested data that allow researchers to separate the shared variance in classes that is due to child-care quality from that due to unique characteristics such as a teacher with a particular style (Melhuish, 2001).

Conclusions

This chapter has documented that children's developmental outcomes in child-care settings take place within a person (e.g., child temperament), process (e.g., teacher–child interaction patterns), and context (e.g., child-care quality and child-care chaos) framework (Bronfenbrenner, 1989). We restricted the present review mainly to the physical microenvironmental features of the child care. Other critical exo- and macrosystem level forces, such as demographic transitions over time (e.g., birth rate, maternal employment), provincial public policy, economic conditions of the country, cultural attitudes toward early child care, and rates of urbanization and industrialization, act on child-care quality (Cochran, 1997). Further discussion of chaos beyond the physical microsystem level is presented in this volume with a specific focus on demographic trends (see chap. 2), on economic disruption (see chap. 8), on work conditions (see chap. 12), and on culture (see chap. 13). The small effect sizes between structural and process quality variables in existing studies imply that addressing any single process will likely have a small impact on the improvement of quality. As noted by Cryer et al. (1999), quality improvement requires multifaceted interventions because of the interrelatedness among the structural variables and the unique effects of classroom, center, and wage structures. Therefore, there is a growing recognition that comprehensive approaches should address changes in all spheres of influence simultaneously.

References

Abbott-Shim, M., & Sibley, A. (1987). *Assessment profile for early childhood programs.* Atlanta, GA: Quality Assistance.

Abbott-Shim, M., Lambert, R., & McCarty, F. (2000). Structural model of Head Start classroom quality. *Early Childhood Research Quarterly, 15,* 115–134.

Adams, G., Tout, K., & Zaslow, M. (2007). *Early care and education for children in low-income families' patterns of use, quality, and potential policy implications.* The Urban Institute and Child Trends Roundtable on Children in Low-Income Families. Retrieved July 20, 2007, from http://www.urban.org/url.cfm?ID=411482

Ahnert, L., Pinquart, M., & Lamb, M. E. (2006). Security of children's relationships with nonparental care providers: A meta-analysis. *Child Development, 74,* 664–679.

Belsky, J., Vandell Lowe, D., Burchinal, M., Clarke-Stewart, K. A., McCartney, K., Owen, M., et al. (2007). Are there long-term effects of early child care? *Child Development, 78,* 681–701.

Bradley, R. H., & Vandell, D. L. (2007). Child care and the well-being of children. *Archives of Pediatrics and Adolescent Medicine, 161,* 669–676.

Brauner, J., Gordic, B., & Zigler, E. (2004). Putting the child back into child care: Combining care and education for children ages 3-5. *Social Policy Report, 18*(3), 3–15.

Bronfenbrenner, U. (1989). Ecological systems theory. *Annals of Child Development, 6,* 187–249.

Brooks-Gunn, J., Han, W. J., & Waldfogel, J. (2002). Maternal employment and child cognitive outcomes in the first three years of life: The NICHD Study of Early Child Care. *Child Development, 73,* 1052–1072.

Burchinal, M., Howes, C., & Kontos, S. (2002). Structural predictors of child care quality in child care homes. *Early Childhood Research Quarterly, 17,* 87–105.

Burchinal, M. R., Peisner-Feinberg, E., Bryant, D., & Clifford, R. (2000). Children's social and cognitive development and child-care quality: Testing for differential associations related to poverty, gender, or ethnicity. *Applied Developmental Science, 4,* 149–165.

Burchinal, M. R., Peisner-Feinberg, E., Pianta, R., & Howes, C. (2002). Development of academic skills from preschool through second grade: Family and classroom predictors of developmental trajectories. *Journal of School Psychology, 40,* 415–436.

Clarke-Stewart, K. A., Lee, Y., Allhusen, V. D., Kim, M. S., & McDowell, D. J. (2006). Observed differences between early childhood programs in the U.S. and Korea: Reflections of "developmentally appropriate practices" in two cultural contexts. *Journal of Applied Developmental Psychology, 27,* 427–443.

Clarke-Stewart, K. A., Vandell, D. L., Burchinal, M., O'Brien, M., & McCartney, K. (2002). Do regulable features of child-care homes affect children's development? *Early Childhood Research Quarterly, 17,* 52–86.

Clifford, R. M., Barbarin, O., Chang, F., Early, D., Bryant, D., Howes, C., et al. (2005). What is pre-kindergarten? Characteristics of public pre-kindergarten programs. *Applied Developmental Science, 9,* 126–143.

Cochran, M. (1997). Fitting early childcare services to societal needs and characteristics. In M. E. Young (Ed.), *Early child development: Investing in our children's future* (pp. 159–186). Amsterdam: Elsevier.

Committee on Early Childhood, Adoption, and Dependent Care. (2005). Quality early education and child care from birth to kindergarten. *Pediatrics, 115,* 187–191.

Corapci, F., & Wachs, T. D. (2002). Does parental mood or efficacy mediate the influence of environmental chaos upon parenting behavior? *Merrill-Palmer Quarterly, 48,* 182–201.

Cryer, D., Tietze, W., Burchinal, M., Leal, T., & Palacios, J. (1999). Predicting process quality from structural quality in preschool programs: A cross-country comparison. *Early Childhood Research Quarterly, 14,* 339–361.

De Schipper, E. J., Riksen-Walraven, J. M., & Geurts, S. A. E. (2006). Effects of child-caregiver ratio on the interactions between caregivers and children in child-care centers: An experimental study. *Child Development, 77,* 861–874.

De Schipper, E. J., Riksen-Walraven, J. M., & Geurts, S. A. E. (2007). Multiple determinants of caregiver behavior in child care centers. *Early Childhood Research Quarterly, 22*, 312–326.

De Schipper, J. C., Tavecchio, L. W. C., Van IJzendoorn, M. H., & Linting, M. (2003). The relation of flexible child care to quality of center day care and children's socio-emotional functioning: A survey and observational study. *Infant Behavior and Development, 26*, 300–325.

De Schipper, J. C., Tavecchio, L. W. C., Van IJzendoorn, M. H., & Van Zeijl, J. (2004). Goodness-of-fit in center day care: Relations of temperament, stability, and quality of care with the child's adjustment. *Early Childhood Research Quarterly, 19*, 257–272.

Duncan, G. J., & Brooks-Gunn, J. (1999). *Consequences of growing up poor.* New York: Russell Sage Foundation.

Early, D. M., Bryant, D. M., Pianta, R. C., Clifford, R. M., Burchinal, M. R., Ritchie, S., et al. (2006). Are teachers' education, major, and credentials related to classroom quality and children's academic gains in pre-kindergarten? *Early Childhood Research Quarterly, 21*, 174–195.

Early, D. M., & Burchinal, M. R. (2001). Early childhood care: Relations with family characteristics and preferred care characteristics. *Early Childhood Research Quarterly, 16*, 475–497.

Evans, G., & Lepore, S. (1993). Household crowding and social support: A quasi-experimental analysis. *Journal of Personality and Social Psychology, 65*, 308–316.

Evans, G., Rhee, E., Forbes, C., Allen, K. M., & Lepore, S. J. (2000). The meaning and efficacy of social withdrawal as a strategy for coping with chronic residential crowding. *Journal of Environmental Psychology, 20*, 335–342.

Fuller, B., Kagan, S. L., Loeb, S., & Chang, Y. (2004). Child care quality: Centers and home settings that serve poor families. *Early Childhood Research Quarterly, 19*, 505–527.

Gallagher, J. J., Rooney, R., & Campbell, S. (1999). Child care licensing regulations and child care quality in four states. *Early Childhood Research Quarterly, 14*, 313–333.

Gevers Deynoot-Schaub, M. J. J. M., & Riksen-Walraven, J. M. A. (2005). Child care under pressure: The quality of Dutch centers in 1995 and 2001. *Journal of Genetic Psychology, 166*, 280–296.

Gevers Deynoot-Schaub, M. J., & Riksen-Walraven, J. M. (2006). Peer contacts of 15-month-olds in childcare: Links with child temperament, parent-child interaction and quality of child care. *Social Development, 15*, 709–729.

Hambrick-Dixon, P. J. (2002). The effects of exposure to physical environmental stressors on African American children: A review and research agenda. *Journal of Children and Poverty, 8*, 23–34.

Harms, T., Clifford, R. M., & Cryer, D. (1998). *Early Childhood Environment Rating Scale–revised.* New York: Teachers College Press.

Howes, C. (1997). Children's experiences in center-based child care as a function of teacher background and adult:child ratio. *Merrill-Palmer Quarterly, 43*, 404–425.

Howes, C., & Hamilton, C. E. (1993). The changing experience of child care: Changes in teachers and in teacher-child relationships and children's social competence with peers. *Early Childhood Research Quarterly, 8*, 15–32.

Kershaw, P., Forer, B., & Goelman, H. (2005). Hidden fragility: Closure among licensed child-care services in British Columbia. *Early Childhood Research Quarterly, 20*, 417–432.

Kontos, S., Howes, C., Shinn M., & Galinsky, E. (1995). *Quality in family child care and relative care.* New York: Teachers College Press.

Kontos, S., & Wachs, T. D. (2000). *Life in early childhood programs scale.* Unpublished manuscript, Department of Child Development and Family Studies, Purdue University, West Lafayette, IN.

Koren-Karie, N., Sagi-Schwartz, A., & Egoz-Mizrachi, N. (2005). The emotional quality of childcare centers in Israel: The Haifa study of early childcare. *Infant Mental Health Journal, 26*, 110–126.

Lamb, M. E. (1998). Nonparental child care: Context, quality, correlates, and consequences. In W. Damon (Series Ed.) & I. E. Sigel & K. A. Renninger (Vol. Eds.), *Handbook of child psychology: Vol. 4. Child psychology in practice* (5th ed., pp. 73–133). New York: Wiley.

Le, V., Perlman, M., Zellman, G. L., & Hamilton, L. S. (2006). Measuring child–staff ratios in child care centers: Balancing effort and representativeness. *Early Childhood Research Quarterly, 21*, 267–279.

Legendre, A. (2003). Environmental features influencing toddlers' bioemotional reactions in daycare centers. *Environment and Behavior, 35*, 523–549.

Lepore, S. J., Evans, G., & Palsane, M. N. (1991). Social hassles and psychological health in the context of chronic crowding. *Journal of Health and Social Behavior, 32*, 357–367.

Li-Grining, C. P., & Coley, R. L. (2006). Child care experiences in low-income communities: Developmental quality and maternal views. *Early Childhood Research Quarterly, 21*, 125–141.

LoCasale-Crouch, J., Konold, T., Pianta, R., Howes, C., Burchinal, M., Bryant, D., et al. (2007). Observed classroom quality profiles in state-funded pre-kindergarten programs and associations with teacher, program, and classroom characteristics. *Early Childhood Research Quarterly, 22*, 3–17.

Loeb, S., Fuller, B., Kagan, S. L., & Carrol, B. (2004). Child care in poor communities: Early learning effects of type, quality, and stability. *Child Development, 75*, 47–65.

Love, J. M., Harrison, L., Sagi-Schwartz, A., Van IJzendoorn, M. H., Ross, C., Ungerer, J. A., et al. (2003). Child care quality matters: How conclusions may vary with context. *Child Development, 74*, 1021–1033.

Matheny, A. P., Jr., Wachs, T. D., Ludwig, J. L., & Phillips, K. (1995). Bringing order out of chaos: Psychometric characteristics of the confusion, hubbub, and order scale. *Journal of Applied Developmental Psychology, 16*, 429–444.

Maxwell, L., & Evans, G. W. (2000). The effects of noise on preschool children's prereading skills. *Journal of Environmental Psychology, 20*, 91–97.

Melhuish, E. C. (2001). The quest for quality in early day care and preschool experience continues. *International Journal of Behavioral Development, 25*, 1–6.

Montie, J. E. (2001). Structural characteristics of early childhood settings: A review of the literature. In P. P. Olmsted, J. E. Montie, J. Claxton, & S. Oden (Eds.), *The IEA preprimary project phase 2: Early childhood settings in 15 countries: What are their structural characteristics?* (pp. 13–55). Ypsilanti, MI: High/Scope Press.

Montie, J. E., & Olmsted, P. P. (2001). Closing thoughts. In P. P. Olmsted, J. E. Montie, J. Claxton, & S. Oden (Eds.), *The IEA preprimary project phase 2: Early childhood settings in 15 countries: What are their structural characteristics?* (pp. 293–304). Ypsilanti, MI: High/Scope Press.

Montie, J. E., Xiang, Z., & Schweinhart, L. J. (2006). Preschool experience in 10 countries: Cognitive and language performance at age 7. *Early Childhood Research Quarterly, 21*, 313–331.

NICHD Early Child Care Research Network. (1996). Characteristics of infant child care: Factors contributing to positive caregiving. *Early Childhood Research Quarterly, 11*, 269–306.

NICHD Early Child Care Research Network. (1997). The effects of infant child care on infant–mother attachment security: Results of the NICHD Study of Early Child Care. *Child Development, 68*, 860–879.

NICHD Early Child Care Research Network. (2000). Characteristics and quality of child care for toddlers and preschoolers. *Applied Developmental Science, 4*, 116–135.

NICHD Early Child Care Research Network. (2002). Child-care structure → process → outcome: Direct and indirect effects of child-care quality on young children's development. *Psychological Science, 13*, 199–206.

NICHD Early Child Care Research Network. (2003). Early child care and mother–child interaction from 36 months through first grade. *Infant Behavior and Development, 26*, 345–370.

NICHD Early Child Care Research Network. (2004). Type of child care and children's development. *Early Childhood Research Quarterly, 19*, 203–230.

NICHD Early Child Care Research Network. (2006). Child-care effect sizes for the NICHD Study of Early Child Care and Youth Development. *American Psychologist, 61*, 99–116.

Peisner-Feinberg, E. S., & Burchinal, M. R. (1997). Relations between preschool children's child care experiences and concurrent development: The cost, quality, and outcomes study. *Merrill-Palmer Quarterly, 43*, 451–477.

Phillips, D. A., Mekos, D., Scarr, S., McCartney, K., & Abbott-Shim, M. (2000). Within and beyond the classroom door: Assessing quality in child care centers. *Early Childhood Research Quarterly, 15*, 475–496.

Phillips, D. A., Voran, M. N., Kisker, E., Howes, C., & Whitebook, M. (1994). Child care for children in poverty: Opportunity or inequity? *Child Development, 65*, 472–492.

Phillipsen, L. C., Burchinal, M. R., Howes, C., & Cryer, D. (1997). The prediction of process quality from structural features of child care. *Early Childhood Research Quarterly, 12*, 281–303.

Pianta, R., Howes, C., Burchinal, M., Bryant, D., Clifford, R., Early, D., et al. (2005). Features of pre-kindergarten programs, classrooms, and teachers: Do they predict observed classroom quality and child-teacher interactions? *Applied Developmental Science, 9*, 144–159.

Rao, N., Koong, M., Kwong, M., & Wong, M. (2003). Predictors of preschool process quality in a Chinese context. *Early Childhood Research Quarterly, 18*, 331–350.

Scarr, S., Eisenberg, M., & Deater-Deckard, K. (1994). Measurement of quality in child care centers. *Early Childhood Research Quarterly, 9*, 131–151.

Shonkoff, J., & Phillips, D. (2000). *From neurons to neighborhoods: The science of early childhood development*. Washington, DC: National Academy Press.

Sims, M., Guilfoyle, A., & Parry, T. S. (2006). Children's cortisol levels and quality of child care provision. *Child: Care, Health & Development, 32*, 453–466.

Tran, H., & Weinraub, M. (2006). Child care effects in context: Quality, stability, and multiplicity in nonmaternal child care arrangements during the first 15 months of life. *Developmental Psychology, 42*, 566–582.

Vandell, D. L. (2004). Early child care: The known and the unknown. *Merrill-Palmer Quarterly, 50*, 387–414.

Vermeer, H. J., & Van IJzendoorn, M. H. (2006). Children's elevated cortisol levels at daycare: A review and meta-analysis. *Early Childhood Research Quarterly, 21*, 390–401.

Votruba-Drzal, E., Levine Coley, R., & Chase-Lansdale, P. L. (2004). Child care and low-income children's development: Direct and moderated effects. *Child Development, 75*, 296–312.

Wachs, T. D. (1991). Environmental considerations in studies with nonextreme groups. In T. D. Wachs & R. Plomin (Eds.), *Conceptualization and measurement of organism-environment interaction* (pp. 44–67). Washington, DC: American Psychological Association.

Wachs, T. D., & Corapci, F. (2003). Environmental chaos, development and parenting across cultures. In C. Raeff & J. Benson (Eds.), *Social and cognitive development in the context of individual, social, and cultural processes* (pp. 54–83). New York: Routledge.

Wachs, T. D., Gurkas, P., & Kontos, S. (2004). Predictors of preschool children's compliance behavior in early childhood classroom settings. *Journal of Applied Developmental Psychology, 25*, 439–457.

Whitebook, M., Howes, C., & Phillips, D. (1998). *Worthy work, unlivable wages: The National Child Care Staffing Study, 1988–1997*. Washington, DC: Center for the Child Care Workforce. Retrieved July 20, 2007, from http://www.ccw.org/pubs/worthywork.pdf

Whitebook, M., & Sakai, L. (2003). Turnover begets turnover: An examination of job and occupational instability among child care center staff. *Early Childhood Research Quarterly, 18*, 273–293.

6

Chaos Outside the Home: The School Environment

Lorraine E. Maxwell

This chapter addresses the potential for chaos in one setting of a child's microsystem, the school. I look at the physical environment of the school setting as a context for proximal processes. In particular, I examine the environmental conditions that may constitute chaos in the school setting and how these conditions potentially affect the proximal processes necessary for healthy child development.

A chaotic microsystem is one characterized by high levels of noise, crowding, confusion, instability of school or residence, changes in adult caregivers or friends, and low levels of structure and regularity of routines. When one or more of these conditions are present, children's development may be negatively affected. However, at what point does a setting become chaotic? How is a vibrant school different from a chaotic school? Must all of the conditions for chaos be present to define a school as chaotic? Are there other factors, environmental or social, that might contribute to a chaotic setting? For the purposes of discussion, chaos in the school setting may be thought of as environmental conditions that interfere with the individual's (the child's) sustained and progressively more complex interactions with, and activity in, the learning environment of a school.

In a school setting, children must engage with teachers and other adults, materials in the classroom, other children, and their own work. Therefore, features of the environment that might inhibit engagement in a school setting would also potentially inhibit development. Healthy engagement with the environment involves sustained and progressively more complex interactions. Chaotic circumstances would make such engagement difficult. Therefore, chaos as an attribute of the school environment would most likely inhibit engagement with, and activity in, the school setting.

Chaos may exist when there is too much noise, too many people, too many changes on a daily basis, or too little stability. Chaos may also be an objective or a subjective experience, or both.

Noise

Noise—that is, too much unwanted sound—in a school setting potentially contributes to chaos by creating distractions from, or interfering with, learning activities. The link between noise and classroom chaos may be explained by the relationship of noise to communication in the classroom, annoyance, attention and memory, motivation, and stress. Sources of noise in a school setting can be external (e.g., airplane traffic, road traffic, trains) or internal (e.g., classroom mechanical and ventilation equipment, lighting, use of materials and finishes with poor acoustical properties, classroom conversations, furniture, corridor noise, inappropriate adjacencies).

Good communication in the classroom, both teacher–student and student–student, is essential for learning. Noise can interfere with speech perception and thus make communication difficult or impossible. If students cannot hear or understand the teacher, they will not receive the benefit of the lesson and may eventually lose interest in the lesson. Noisy classrooms' effect on communication may also affect the way teachers teach or the choice of type of learning activities. If noise levels make communication difficult, teachers may choose to limit the amount of time students spend in group activities because such activities generate more noise than do solitary ones. Teachers may suffer from voice strain and fatigue in an attempt to communicate with students in a noisy classroom. Noisy classrooms may also affect the selection of teaching and learning activities. Teachers may resort to more paper-and-pencil instruction (e.g., worksheets) or computer-aided instruction rather than verbal presentations. Though these activities are not inherently inferior teaching strategies, chronically noisy classrooms may limit teachers' choices of learning activities. A critical feature of a microsystem is that it is a place where more complex interactions take place. The ability to be able to communicate with others is essential. Noise can inhibit communication in the school setting.

Noise is also a source of annoyance in classrooms and schools. In this instance, annoyance can be described as the discrepancy between the existing sound conditions and the activity or activities in which the individual wishes to engage. This relationship has most often been examined in adults (Evans & Hygge, 2006; Kryter, 1994; Lercher & Kofler, 1996), but some researchers have investigated it with respect to children. Children have reported being annoyed by community noise generated by chronic airport noise or road traffic (Evans, Bullinger, & Hygge, 1998; Haines, Stansfeld, Job, Berglund, & Head, 2001; Hygge, Evans, & Bullinger, 2002). In addition, chronic aircraft noise and nearby subway train noise have been reported by children as interfering with doing classroom work and with thinking (Bronzaft & McCarthy, 1975; Haines & Stansfeld, 2000).

Each of these studies focused on external noise sources, mainly transportation-related noise. Other research has examined annoyance and internal noise sources, including noise generated by other people, mechanical and ventilation systems, and the design of classroom acoustical features. Adolescents (ages 13–20) reported that other students' chatter, sounds from the corridor, the scraping sound of furniture on the classroom floor, and noise from the ventilation system were the major sources of annoying noise in the classroom (Bonman & Enmarker, 2004; Lundquist, Holmberg, & Landström, 2000). Classroom chatter was annoying regardless of the sound level, meaning that whispering was just as annoying as louder conversations. Bonman and Enmarker noted that annoyance was task dependent, with noise being most disruptive when students were doing a difficult task, specifically math.

Individual differences were found for noise annoyance in the classroom. Students who rated themselves as more sensitive to noise were more likely to be annoyed by the noise sources in the classroom, such as conversations or scraping sounds as furniture was moved. However, this model for the relation of sensitivity to annoyance was mediated by adaptation to noise. Adaptation was a self-rated measure of how quickly a student became accustomed to noise. Those students who adjusted more quickly were less likely to be annoyed by the noise. No gender differences were reported.

One of the ways in which students compensated for the annoyance was to abandon their own tasks and join in the chatter or conversation. This compensation for noise is obviously not optimal for a classroom learning situation. The means used to adjust to noise and the general annoyance itself both contribute to making noise problematic in the microsystem. Students may be distracted from participating in progressively more complex activities in school as a result of exposure to noise.

The work by Bonman and Enmarker (2004) provides a view of the ways in which noise is annoying to students; however, their work was with older youth. Elementary-school-age children may have different views of noise as a source of annoyance, and their adjustment or response to the annoyance may be different. Though research indicates that younger children are annoyed by noise (Evans, Bullinger, & Hygge, 1998; Haines et al., 2001; Hygge, Evans, & Bullinger, 2002), the topic of noise and annoyance in this age group requires further investigation. One small study of elementary-school-age children in New York City suggests that children of this age are indeed disturbed by some of the same sources of classroom noise as are older students. Fourth-grade children reported that they could not concentrate at times on schoolwork because of other children talking (Simon, 2005).

Noise in a school setting also interferes with attention and memory. Many people have experienced situations in which competing auditory (and perhaps visual as well) sources made it difficult if not impossible to pay attention. Competing stimuli often make a setting chaotic. As the youth in Bonman and Enmarker's (2004) study noted, paying attention, especially to a difficult or complex task such as a math problem, was difficult when others nearby were chatting. The subsequent noise was annoying and interfered with attention. Noise exposure makes it difficult to pay attention to or learn complex material (Evans & Lepore, 1993).

Noise's interference with attention is related primarily to the negative effects on auditory discrimination and speech perception. When noise levels are high it becomes difficult for children (and adults) to understand speech, resulting in difficulty to pay attention. Poor speech perception also makes maintaining good communication in the classroom difficult. Though children may develop adaptation strategies for dealing with noise, such as tuning out the noise, this becomes more difficult over time. This method may work on a short-term basis, but if noisy classrooms are a chronic occurrence children may learn to tune out more than noise—namely, speech that is part of the lesson and classroom activity (Evans & Hygge, 2006). Chronic noise exposure also interferes with long-term memory, especially for complex or difficult material (Hygge, 1993).

Learning in noisy classrooms for children who are being instructed in their nonnative language, who have particular difficulty with attention, or who have learning disabilities or autism spectrum conditions may be even more compromised than is learning for children who do not have these conditions (Alcántara, Weisblatt, Moore, & Bolton, 2004). School is where children are supposed to engage in progressively more complex activities, yet noise makes it difficult to pay attention and learn and retain difficult or complex material. It seems apparent that in light of the impact that general noise exposure has on attention and long-term memory, academic achievement will be negatively affected for children with learning disabilities as well as for typically developing children. These effects are most pronounced with chronic noise exposure.

Crowding

Crowding is another potential contributor to chaos in the school setting. A highly stimulating environment such as a classroom requires a high level of psychic energy to maintain the level of concentration necessary for learning. When there are too many students in a classroom, not enough space, and a variety of tasks or activities related to the lesson, trying to pay attention to one's own schoolwork can create a situation of cognitive overload for children (Evans, 2006). Once children become overloaded they lose the ability to concentrate and stay focused on the task at hand. Some students may become disruptive. In fact, disruptive and aggressive behavior, especially in boys, is likely to increase in crowded classrooms (Maxwell, 2003a; Saegert, 1982). Not only does this disrupt the lesson for the children involved in the behavior, but it also makes it difficult for other students to pay attention. Some children in crowded classrooms may withdraw from classroom participation (Loo & Smetana, 1978), which leads to a reduction in important interactions among the students and between students and classroom activities. Whether because of disruptive behavior, lack of engagement with the activities, or attention shutdown resulting from cognitive overload, classroom crowding ultimately affects academic achievement. Lower reading scores are associated with high classroom density (Maxwell, 2003a). Children living in crowded homes also do poorly in school in terms of academic achievement (Evans, Lepore, Shejwal, & Palsane, 1998). Though chaos in the home is not the topic of this chapter, the findings related to crowding in the home and school signal the potential for interaction effects. As such, a model of cumulative risk is suggested. For example, Maxwell (1996) found that crowded day-care classrooms exacerbated negative behavioral outcomes for children from crowded homes.

Crowding puts demands on individuals' ability to cope with stimulation and to regulate interactions with their environment. Chaos is a sense of too much happening at once. In a chaotic situation, how does one make decisions about what demands attention or what to do next? Often the easiest thing to do is nothing. Some children exposed to situations that could be described as chaotic may respond by simply dropping out. Some children drop out temporarily in the classroom by daydreaming; it is conceivable that others might eventually drop out permanently by leaving school.

There is, however, evidence that reducing children's chronic exposure to classroom crowding can be beneficial to their cognitive development and academic achievement. Class size, a measure of social density, was manipulated in a statewide study of elementary schools with random assignment in Tennessee. The findings indicate a positive academic achievement benefit to children who were in a smaller class size for the duration of the study than were their peers (13–17 students per class vs. 22–25 students per class; Ehrenberg, Brewer, Gamoran, & Willms, 2001). In addition, there was a particular benefit for minority students. Reducing classroom social density, therefore, has the potential to enhance academic achievement and perhaps encourage children to stay in school. This study also suggests that there are developmental aspects of chaos. Class-size manipulations were in force for kindergarten through third grade. Positive academic effects resulting from class-size reduction in the lower grades carried over when these children were later placed in equal larger classes.

Individual differences among children with respect to crowding are noted in terms of a response to crowded conditions; that is, some children act out whereas others withdraw. Whether some children are more bothered or annoyed by social or spatial crowding has received less attention by researchers. Likewise, individual differences in response to noise have been documented; for example, some children are more easily distracted than are others (Loo & Smetana, 1978). These individual differences may mediate the response to crowding or noise and subsequent effects on the individual.

Though a learning environment must be stimulating enough to engage children, when the stimulation interferes with the ability to effectively interact with other people and with educational materials and activities, then the conditions are contributing to chaos and thus interfering with progressively more complex interactions with individuals and activities. Not only does crowding make it more difficult for the developing child to engage successfully in activities, but the presence of too many people can also inhibit healthy interactions with other people.

Visual Complexity

Excessive stimulation can contribute to chaos. Stimulation in the classroom can come from other people (i.e., crowding), conversations (i.e., noise), and the need to interact in various activities associated with a school setting. Stimulation can also come from visual sources such as posters and displays on the walls and blackboards of a classroom. Indeed, many elementary-school classrooms may seem visually quite chaotic to the casual observer because of all of the displays. The adolescent participants in Bonman and Enmarker's study (2004) commented that posters should be removed from the classroom to reduce the stimulation. This comment is particularly interesting because the researchers were interested in noise and annoyance, not visual stimulation. The youth recommended artwork instead of posters because artwork would be calmer. Perhaps a sense of visual calm would make the noise less annoying.

Younger children may also appreciate less busy classroom walls. In response to a question about the ideal learning place in open-ended interviews about home, school, and neighborhood, an elementary-school-age child stated that only items related to the current lesson should be displayed (Maxwell, 2003b). This child's teacher kept all displays visible for the entire school year. Too much stimulation, either from physical attributes of the surrounding environment or from other people, can inhibit or limit meaningful involvement with desired classroom activities.

Little research, however, has been done on the role of classroom visual complexity and chaos, especially from the viewpoint of the students in the classroom. Perhaps visual complexity is a form of noise. When there are many concurrent conversations or sounds, one is not sure which one is the most important. Similarly, when there are many visual displays in a classroom, students may have difficulty discerning the important information that is being conveyed. In addition, a highly visually complex classroom may be a source of distraction to students and contribute to chaos. High visual complexity may also add to the complexity created by noise or crowding, thereby potentially increasing a sense of chaos. Complex school settings may not support the necessary interactions or activities that microsystems are supposed to encourage. There must be a balance in the classroom between the physical environment and the activities and interactions taking place there. Without this balance, chaos or boredom, or both, ensue.

Social Components of Chaos

Noise, crowding, and visual complexity are features of the physical environment that may contribute to a sense of chaos. Each of these features may be present in the classroom individually or in some combination. Other contributing features to chaos may be less dependent on the physical environment but potentially as damaging. For example, instability of school attendance and low levels of structure and regularity of routines are also characteristics of a chaotic setting. A lack of regular routines, school attendance being one such routine, can contribute to a sense of chaos in children's lives. Excessive absenteeism is also a major contributor to poor academic achievement and eventually dropping out of school (Lamdin, 1996).

Chronic student absenteeism can be affected by school size; large secondary schools tend to have more truancy than do smaller schools (Finn & Voelkl, 1993). Students may feel more anonymous in a large school, which makes it easier to slip through the cracks. In smaller high schools, for example, more students are involved in school extracurricular activities and engage in more prosocial behaviors (Moore & Lackney, 1993). A belief that teachers do not care about them and the existence of a chaotic classroom environment have been cited as conditions that encourage high school students to cut classes or skip school (Duckworth & DeJung, 1989; Roderick et al., 1997). Though absenteeism is not often understood by school administrators as a facility-related issue, chaos may be a contributing factor. Too many students in the school or classroom can contribute to chaos, pushing students away from the classroom and eventually from school. Furthermore, a physically deteriorating school may mediate the relationship between student absenteeism and academic achievement in elementary school children (Durán-Narucki, 2008). Students simply may not enjoy going to a school that is physically deteriorating and just stop attending. In summary, physical conditions in the school may contribute to a sense of chaos which in turn eventually pushes children out of school. Irregular school attendance, however, may contribute to a general level of chaos in children's daily lives.

Other factors not specifically related to chaos in the physical environment may also relate to student achievement, learning, and cognition. Restructuring of social and bureaucratic organization of schools can positively affect academic achievement (Lee & Smith, 1993, 1995). In addition, students in appropriately restructured schools are more socially engaged. Lee and Smith (1993, 1995) do not describe how restructuring related to departmental organization, teacher assignment, and student groupings may also have been coupled with, or affected by, changes in the physical environment. It would certainly be important to look at the relative effects of how school restructuring relates to actual or perceived chaos.

Irregular school attendance may be the result of homelessness or other disruptions in family continuity. For example, children in foster care for an extended period not only have instability in caregivers and the physical home but also most likely attend different schools. This means that these children must adjust to different school routines, different teacher expectations, and different classmates and friends.

In addition to placement in foster care, children can suffer instability in their living arrangements as a result of their own family's homelessness. Homeless families often move between more than one shelter and other makeshift living arrangements. Homelessness can contribute to a host of negative outcomes for children (and their families) apart from the lack of adequate shelter. These outcomes include poor health and nutrition, psychological problems (depression, anxiety, and behavioral problems), and poor academic achievement measured by standardized tests (Rafferty & Shinn, 1991). Though some communities may make an effort to keep homeless children in a stable school situation, these efforts do not always succeed. Parents or other caregivers may not have the financial, physical, psychological, or emotional resources to shield children from the chaos they are experiencing in their lives. The result for children in homeless families or those in foster care is, among other things, a lack of regular and consistent school attendance. It is interesting that although certain attributes of a school setting may contribute to chaotic conditions in the school, not attending school on a regular basis also contributes to chaos in children's daily lives. A question that might be asked is, What happens to the child who lives in a variety of housing situations with changing caregivers but has a stable school setting that is chaotic because of noise, crowding, or other factors?

Social factors combined with physical environmental attributes of the school setting (and possibly home as well) may coexist for many children. Children experiencing discontinuity in family life as a result of foster care placement or homelessness are likely to experience residential crowding. Schools serving low-income communities are also more likely to be crowded. All of these circumstances create a situation of cumulative risk.

Not attending school on a regular basis can contribute to a lack of stable routines in children's lives and chaos. Children's failure to regularly attend school can also result in poor academic achievement and eventually dropping out of school. What is the effect, however, of teacher absenteeism on the school setting and on children's academic achievement? How does teacher absenteeism contribute to chaos in the school setting? Little research has investigated these questions. Some evidence suggests that teacher absenteeism does not have a direct negative effect on academic achievement as measured by passing grades on standardized tests (Ehrenberg, Ehrenberg, Rees, & Ehrenberg, 1989). Teacher absenteeism, however, may have an effect on student absenteeism, which does relate to poor academic achievement. If teacher absenteeism has a high positive correlation to student absenteeism, teacher absenteeism might be related to school chaos. It could be intuitively argued that instability in classroom teachers would have a similar relation to chaos in children's lives as would instability of guardians and caregivers in the home. A succession of substitute teachers in the classroom means not only that children must adjust to different teaching styles and expectations but also that children do not have the opportunity to develop continuity in a teacher–student relationship. The microsystem is supposed to be a place that supports sustained, meaningful, and complex interactions with the environment as well as with people. A constant string of substitute teachers will make it difficult if not impossible for this to happen. Children will not know what, or who, to expect and this can feel chaotic. Therefore, researchers should investigate the role of instability in classroom teachers and the potential role it might have in school chaos.

Chaos and Socioemotional Development

Thus far the discussion around chaos in the school setting has focused on classroom conditions and their relation primarily to cognitive development and academic achievement. Of course, the microsystem is where all development takes place so schools are places that must support socioemotional development (and physical development) as well. Instability in classroom teachers may not be as detrimental to socioemotional development as, say, instability in primary caregivers (i.e., parents or guardians), yet the inability for a child to have a stable relationship with a trusted adult in school may exacerbate family instability.

As stated earlier, crowding in the school can contribute to a sense of chaos. School crowding can have detrimental effects on academic achievement. School crowding may also be related to aspects of socioemotional development. Crowding is not only too many people or not enough space but also the absence of control over social interactions and the inability to achieve privacy when desired. Thus, crowding in a school setting can contribute to children's sense of loss of control and perhaps a sense of helplessness. Helplessness diminishes self-esteem and self-efficacy.

As stated previously, crowding can lead to cognitive overload or overstimulation; time and a place away from other people can help to restore energy and attention, allowing the individual to again engage in essential tasks or activities (Kaplan & Kaplan, 1989). Restorative places also provide opportunities for reflection and the development of self-regulation, a critical part of socioemotional development. Crowded classrooms may interfere with learning self-regulation skills. In an attempt to remove themselves from crowded school situations, children may psychologically withdraw if no other means of escape is available. Not only can withdrawal have academic consequences, it can also damage the development of social interaction skills. Because some children's response to crowding is to act out or become aggressive, whereas others withdraw, chaos may negatively affect socioemotional development.

Likewise, classroom chaos attributed to noise may have consequences for socioemotional development. Noise makes communication difficult and therefore has an impact on social interaction. Noise can also contribute to feelings of irritability, which may also impact the desire for social interaction. Learning acceptable social interaction skills is part of the classroom experience. A noisy classroom can make this difficult. In addition, chronic exposure to noise may contribute to a sense of helplessness (Evans & Stecker, 2004). Helplessness, as stated previously, has implications not only for academic achievement but also for the development of self-esteem and self-efficacy.

The classroom is a critical place for school-related activities, and it is the place where children spend the majority of their school time. But other spaces in a school also have particular importance for their relation to socioemotional development. One such space is the school cafeteria. It is the primary place for children's social interaction outside of the classroom. Mealtime should be pleasant and relaxing; instead, many school cafeterias are noisy and crowded and offer a chaotic atmosphere. It is not surprising that the cafeteria is sometimes cited by children as one of their least favorite places in the school (Simon, Evans, & Maxwell, 2006). This may be because of the noise and crowding that occurs in school cafeterias.

It is possible that corridors, libraries, and auditoriums may contribute to chaotic school environments. Although children may not spend a lot of time in these places (compared with time spent in the classroom), these spaces contribute to the overall school experience. Libraries and auditoriums are school places sometimes used as study halls, yet they can be noisy. It is possible that these areas might serve as places of retreat or restoration away from a crowded or noisy classroom. However, if they are also chaotic then children have no respite. In assessing the role of chaos in the school environment researchers should not neglect the study of spaces outside of the formal classroom.

Individual Differences

Crowding (Evans, 2006; Loo & Smetana, 1978; Maxwell, 2003a), noise (Evans & Lepore, 1993; Haines et al., 2001), and high visual complexity (Bonman & Enmarker, 2004; Maxwell, 2003b) in a school setting; irregular student attendance (Lamdin, 1996); and possibly high teacher turnover and absenteeism (Ehrenberg et al., 1989) can all have negative outcomes for children. Is every crowded or noisy classroom chaotic? Chaos may diminish, interrupt, or change the predictability and timing of proximal processes. A child exposed to a chronically noisy or crowded school may very well exhibit an interruption or change in his or her critical proximal developmental processes. Is the school environment therefore defined as a chaotic environment? What is the threshold for chaos? It can be assumed that chaos has certain characteristics and certain potential outcomes, particularly for children. But is there a value in identifying a school as chaotic rather than merely describing it as noisy or crowded? Defining or describing a school or classroom as chaotic implies that not only must noise or classroom density be reduced but also order must somehow be restored.

Just as there is some evidence of individual differences in the perception of, sensitivity to, and response to noise (Lercher & Kofler, 1996) and crowding (Loo & Smetana, 1978), there may also be individual differences related to the perception of chaos. Calling something "chaotic" implies that there is little or no order and that no one is in control. A crowded or noisy school, however, does not necessarily imply that there is no order or that no one is in control. Children may experience a school setting as chaotic if they have little ability to control their surroundings, gain any privacy, or perceive any order in the school. Adults in the same setting, however, may not perceive the school as chaotic, and some children may not experience a lack of order or a lack of personal control in a crowded or noisy school either. Therefore, chaos may not be defined purely by crowding or noise, or visual complexity, or lack of routines. It is possible that a setting could be perceived as chaotic by one group of participants but not by others.

The investigation of individual differences related to chaos is critical to an understanding of the threshold conditions for chaos. Chaos may exist on a continuum, and this continuum might look different for different types of people. Development is, in part, a function of characteristics of the individual (Bronfenbrenner & Morris, 1998). Individual characteristics play a role in how people respond to situations and environmental experiences. For example, an abstract painting may appear to be a chaotic mix of colors to the casual observer; however, to the artist or another trained eye, there is a discernible pattern and order of color and form. The important variables in this example are the background (including training) of the person viewing the artwork and relationship of the viewer to the artwork. The perception of chaos in a school or other physical environment may depend on one's sensitivity to noise or crowding, tolerance for high complexity or disorder, role in the setting, or required interactions and activities in the setting.

If there are individual differences in perception of chaos given a set of environmental or social conditions, researchers should examine whether these perceptual differences result in different responses and consequences for the individual. For example, a child may state that he or she is not bothered by classroom crowding, but the effects of being in the crowded classroom may still be evident in his or her academic achievement record. Children may not perceive a school setting as chaotic but may nevertheless be affected in negative ways by the chaos. Perceptual differences regarding chaos may also serve as a buffer, dampening the negative effects of crowding.

Reversing Chaos

Another important consideration in the examination of chaos in the school is the adult response to a chaotic environment. If chaos implies a lack of order and control, one way to reduce chaos is to impose control and order in the situation. In a school setting, adults have all of the control. The imposition of order could be useful in achieving a good educational environment, but it could also be stifling. Teachers may choose to alter classroom lessons to combat noise or crowding by reducing conversations and student movement throughout the classroom, but this can potentially do as much harm to the learning environment as the chaos it was meant to correct.

In other words, order and control exist on a continuum as well, and the use of order or control in the classroom or school environment could also have negative outcomes for children. The imposition of a no-talking rule, for example, might restore a sense of order and calm to a noisy cafeteria; however, children's opportunity for practicing and learning socioemotional skills would be diminished. Schools as institutions already have the potential to discourage creativity and "out of the box" responses to problem solving. Children are expected to conform to specific ways of learning and behaving. There is a need to achieve an environment that has structure so that children know what to expect but also to leave room for the exercise and development of individual autonomy. Children have little or no control in a chaotic classroom; however, children will also have little or no control in an overly structured classroom. The classroom must be vibrant but at the same time orderly.

Therefore, for children, the consequences of chaos in the school setting are twofold. On the one hand, a lack of order or control attributed to crowding, high visual complexity, noise, or high teacher absenteeism may have serious consequences for many children in terms of the proximal processes that govern development. On the other hand, attempts to decrease chaos and impose order in the school setting may also have consequences for child development. If chaos is a set of environmental conditions that interfere with the child's sustained and progressively more complex interactions with, and activity in, the environment, then efforts to decrease chaos should not interfere with the child's ability to have sustained interactions with the environment. A research agenda on chaos should consider both of these scenarios.

Conclusions

The relative role of the school as part of the child's microsystem is a critical consideration for inclusion in a research agenda on chaos. Children's lives are not compartmentalized into home and school. The activities and the nature of social interactions may differ in home and in school but circumstances in one setting may, and usually do, have consequences for the other setting. Therefore, understanding the unique role that schools play in providing a setting for proximal processes to take place must be part of the research agenda on chaos.

Schools are institutional settings, and as such they represent, in part, society's values. Schools are places for learning, but they are also places for youth socialization. If schools are chaotic, the long-term consequences for society may be more than low scores on standardized achievement tests. Because chaos implies a lack of order and reliable expectations, chaotic schools may contribute to a sense of helplessness. This could have serious implications for society as segments of the population reach adulthood with little expectation for reliable outcomes based on their own efforts.

Earlier in this chapter, I posed several questions that essentially were asking for a more precise definition of chaos. I propose that chaos exists when individuals perceive a lack of control over the activities or outcomes of the setting because of certain environmental or social factors. Many factors may contribute to this lack of control: too much noise, too many people, too much stimulation from a variety of sources, a lack of continuity of relationships with other people, and finally a lack of reliable routines. Any one of these may have negative effects on the individual but the lack of the individual's ability to control these factors is most crucial in establishing a sense of chaos. Therefore, in part, individual differences will help to define chaos. Development is a function of individual characteristics, the environment, and the place in time. Ultimately, the effects of chaos on child development must be viewed as a function of both physical and social environmental characteristics and individual differences.

References

Alcántara, J. L., Weisblatt, E. J., Moore, B. C., & Bolton, P. F. (2004). Speech-in-noise perception in high functioning individuals with autism or Asperger's syndrome. *Journal of Child Psychology and Psychiatry, 45*, 1107–1114.

Bonman, E., & Enmarker, I. (2004). Factors affecting pupils' noise annoyance in school: The building and testing of models. *Environment and Behavior, 36*, 187–206.

Bronfenbrenner, U., & Morris, P. (1998). The ecology of developmental processes. In W. Damon & R. Lerner (Eds.), *Handbook of child psychology* (5th ed., Vol. 1, pp. 993–1028). New York: Wiley.

Bronzaft, A., & McCarthy, D. (1975). The effect of elevated train noise on reading ability. *Environment and Behavior, 7*, 517–527.

Duckworth, K., & DeJung, J. (1989). Inhibiting class cutting among high school students. *The High School Journal, 72*, 188–195.

Durán-Narucki, V. (2008). School building condition, school attendance and academic achievement in New York City public schools: A mediation model. *Journal of Environmental Psychology, 28*, 278–286.

Ehrenberg, R. G., Brewer, D. J., Gamoran, A., & Willms, J. D. (2001). Class size and student achievement. *Psychological Science in the Public Interest, 2*(1), 1–30.

Ehrenberg, R. G., Ehrenberg, R. A., Rees, D. I., & Ehrenberg, E. L. (1989). *School district leave policies, teacher absenteeism, and student achievement* (Working Paper No. 2874). Cambridge, MA: National Bureau of Economic Research.

Evans, G. W. (2006). Child development and the physical environment. *Annual Review of Psychology, 57*, 423–451.

Evans, G. W., Bullinger, M., & Hygge, S. (1998). Chronic noise exposure and physiological response: A prospective study of children living under environmental stress. *Psychological Science, 9*, 75–77.

Evans, G. W., & Hygge, S. (2006). Noise and performance in children and adults. In L. Luxon & D. Prasher (Eds.), *Noise and its effects* (pp. 549–566). New York: Wiley.

Evans, G. W., & Lepore, S. J. (1993). Nonauditory effects of noise on children: A critical review. *Children's Environments, 10*, 31–51.

Evans, G. W., Lepore, S. J., Shejwal, B. R., & Palsane, M. N. (1998). Chronic residential crowding and children's well-being. An ecological perspective. *Child Development, 69*, 1514–1523.

Evans, G. W., & Stecker, R. (2004). Motivational consequences of environmental stress. *Journal of Environmental Psychology, 24*, 143–165.

Finn, J. D., & Voelkl, K. E. (1993). School characteristics related to student engagement. *Journal of Negro Education, 62*, 249–268.

Haines, M. M., & Stansfeld, S. A. (2000). Measuring annoyance and health in child social surveys. In D. Cassereau (Ed.). *Proceedings of Inter-Noise 2000* (Vol. 3, pp. 1609–1614). Indianapolis, IN: Institute of Noise Control Engineering of the USA.

Haines, M. M., Stansfeld, S. A., Job, R. F. S., Berglund, B., & Head, J. (2001). Chronic aircraft noise exposure, stress responses, mental health, and cognitive performance in school children. *Psychological Medicine, 31*, 265–277.

Hygge, S. (1993). Classroom experiments on the effects of aircraft, traffic, train, and verbal noise on long-term recall and recognition in children aged 12-14 years. In M. Vallet (Ed.), *Proceedings of the Sixth International Conference on Noise as a Public Health Problem*(Vol. 2, pp. 531–538). Nice, France: Institut National de Recherche sur les Transport et leur Sécurité, Bron.

Hygge, S., Evans, G. W., & Bullinger, M. (2002). A prospective study of some effects of aircraft noise on cognitive performance in school children. *Psychological Science, 13*, 469–474.

Kaplan, R., & Kaplan, S. (1989). *The experience of nature: A psychological perspective.* New York: Cambridge University Press.

Kryter, K. (1994). *The handbook of hearing and the effects of noise.* New York: Academic Press.

Lamdin, D. J. (1996). Evidence of student attendance as an independent variable in education production functions. *The Journal of Educational Research, 89*, 155–162.

Lee, V. E., & Smith, J. B. (1993). School restructuring of middle-grade students. *Sociology of Education, 66*, 164–187.

Lee, V. E., & Smith, J. B. (1995). Effects of high school restructuring and size on early gains in achievement and engagement. *Sociology of Education, 68*, 241–270.

Lercher, P., & Kofler, W. W. (1996). Behavioral and health responses associated with road traffic noise exposure along alpine through-traffic routes. *The Science of the Total Environment, 189/190*, 85–89.

Loo, C. M., & Smetana, J. (1978). The effects of crowding on the behavior and perception of 10-year-old boys. *Environmental Psychology and Nonverbal Behavior, 2*, 226–249.

Lundquist, P., Holmberg, K., & Landström, U. (2000). Annoyance and effects on work from environmental noise at school. *Noise & Health, 2*(8), 39–46.

Maxwell, L. E. (1996). Multiple effects of home and daycare crowding. *Environment and Behavior, 28*, 494–511.

Maxwell, L. E. (2003a). Home and school density effects on elementary school children: The role of spatial density. *Environment and Behavior, 35*, 566–578.

Maxwell, L. E. (2003b, July). *The role of home and community in shaping children's self concept.* Paper presented at the meeting of The Community Development Society, Cornell University, Ithaca, NY.

Moore, G. T., & Lackney, J. A. (1993). School design: Crisis, educational performance and design applications. *Children's Environments, 10*, 99–112.

Rafferty, Y., & Shinn, M. (1991). The impact of homelessness on children. *American Psychologist, 46*, 1170–1179.

Roderick, M., Arney, M., Axelman, M., DaCosta, K., Steiger, C., & Stone, S., et al. (1997). *Habits hard to break: A new look at truancy in Chicago's public high schools* (Research brief from the Student Life in High Schools Project). Chicago: School of Social Service Administration, University of Chicago.

Saegert, S. (1982). Environment and children's mental health: Residential density and low income children. In A. Baum & J. E. Singer (Eds.), *Handbook of psychology and health: Vol. 2. Issues in child health and adolescent health* (pp. 247–271). Hillsdale, NJ: Erlbaum.

Simon, N. S. (2005). *Building quality, academic achievement, and self-competency in New York City Public Schools.* Unpublished master's thesis, Cornell University, Ithaca, NY.

Simon, N. S., Evans, G. W., & Maxwell, L. E. (2006, March). *Building quality, academic achievement and the achievement gap.* Twelfth Architecture & Behavior Colloquium, Monte Verita, Switzerland.

7

Viewing Microsystem Chaos Through a Bronfenbrenner Bioecological Lens

Theodore D. Wachs

Chaos at the microsystem level is typically defined on the basis of high levels of noise and crowding, high-context traffic patterns (many people coming and going in a given context), and a lack of physical and temporal structure (nothing has its place, there are few regularly scheduled routines; Wachs, 1989). The increasing interest in environmental chaos has been fueled by previous evidence on the adverse impact of chaotic microsystem environments on children's cognitive–linguistic and socioemotional development and on parenting behaviors (Evans, 2006; Wachs & Corapci, 2003; see also chaps. 3, 4, 5, and 6, this volume). In this chapter I go beyond main-effect studies of chaos and development to integrate the study of environmental chaos at the microsystem level with proposition 2 of Bronfenbrenner's (1999) bioecological model. Proposition 2 states that relations between development and microsystem proximal characteristics, such as environmental chaos, must be interpreted within the more general framework of person, process, context, and time (PPCT).

> The form, power, content and direction of the proximal processes affecting development vary systematically as a joint function of the characteristics of the *developing person*, the *environment*—both immediate and more remote—in which the processes are taking place, the nature of the *developmental outcomes* under consideration, and the social continuities and changes occurring over time during the historical period through which the person has lived. (Bronfenbrenner, 1999, p. 5)

A detailed discussion of proximal processes and the PPCT framework is found in chapter 1 of this volume. Because other chapters in this volume are devoted to higher level contextual chaos (see chaps. 11–15) and temporal chaos (see chaps. 2 and 8), my primary focus is on the microsystem, with particular reference to person and process. However, toward the end of this chapter, I briefly discuss the nature and influence of microsystem chaos as a function of context and time.

Chaos and Person

In methodological terms, the *person* dimension in Bronfenbrenner's PPCT framework refers to the concept of moderation. Moderation occurs when the influence of microsystem characteristics such as environmental chaos systematically varies according to individual characteristics. A central issue in the integration of person and chaos is identification of what individual characteristics are most likely to be involved in moderating the influence of chaos. Though evidence is limited, some candidate characteristics have been identified as actual or potential moderators.

Gender

The overwhelming majority of research on individual difference moderators of environmental chaos involves the question of whether males or females react differently to microsystem chaos. A review of this literature yields a pattern of findings that, at best, is inconsistent. Some studies report that males are more adversely influenced by microsystem chaos than are females (Aiello, Nicosia, & Thompson, 1979; Matheny, 1991; Matheny & Phillips, 2001; Wachs, 1979, 1987; Wachs, Gurkas, & Kontos, 2004), other studies report that females are more sensitive to chaos than are males (Christie & Glickman, 1980; Hambrick-Dixon, 1998), and still others report nonsignificant Gender × Chaos interactions (Evans, 2003; Evans, Lepore, & Allen, 2000; Haines et al., 2001; Pike, Iervolino, Eley, Price, & Plomin, 2006). It could be argued that those studies reporting no moderating influence of gender have insufficient statistical power. This explanation is unlikely given that some studies that do not find Gender × Chaos interactions have very large sample sizes (e.g., the sample in the Pike et al., 2006, study exceeded 5,500). The possibility that the inconsistent pattern of findings may be due to males and females being sensitive to different dimensions of chaos also seems unlikely, given that both significant and nonsignificant findings encompass measures of two major indices of microsystem chaos, namely, home noise and crowding.

Consistent with the terms of proposition 2, one possibility for explaining the inconsistent pattern of findings for Chaos × Gender interactions is the operation of higher order interactions involving more than one person variable. Chronological age would be one additional person variable. In this regard, it is of interest to note that five of the six studies reporting greater male sensitivity to microsystem chaos involved infants and preschool children (Matheny, 1991; Matheny & Phillips, 2001; Wachs, 1979, 1987; Wachs et al., 2004), whereas three of the four studies reporting nonsignificant Gender × Chaos interactions involved preschool- or elementary-school-age children or adults (Evans, 2003; Haines et al., 2001; Pike et al., 2006; the two studies showing more female sensitivity involved kindergarten or school-age children). Though this pattern of findings is not conclusive, it does justify further research to test the hypothesis that greater male sensitivity to environmental chaos is most likely to occur in the early years of life, with declining male sensitivity thereafter (a Gender × Chaos × Age higher order interaction).

Individual Vulnerability

Moderation involving vulnerability can occur either when individuals with higher levels of biological risk, biosocial risk (e.g., difficult temperament), or psychosocial risk are more vulnerable to environmental chaos, or when chaos accentuates the developmental impact of biological risk. Though there is a large literature on Vulnerability × Environment interactions (e.g., diathesis stress), relatively few studies have looked at environmental chaos as part of the moderation equation (Wachs, 2000). Nonetheless, there is some consistency in the findings.

BIOLOGICAL RISK. Results from two studies indicate that the impact of biological risk conditions is accentuated by exposure to environmental chaos. In specific terms, the association between inadequate maternal dietary intake and reduced infant alertness was stronger for infants living in more crowded homes (Rahmanifar et al, 1993), whereas increased noise exposure was linked to attentional deficits for preterm but not for full-term infants (Barreto, Morris, Philbin, Gray, & Lasky, 2006).

BIOSOCIAL RISK (TEMPERAMENT). Adults' reactivity to noise is moderated by individual characteristics such as stimulus sensitivity (Job, 1999; Ramirez, Alvarado, & Santisteban, 2004), neuroticism (Stansfeld, 1992; Turrero, Zuluaga, & Santisteban, 2001), and extraversion (Dornic & Ekehammar, 1990). Studies have also reported that infants with difficult temperaments are more sensitive than are infants with easy temperaments to higher levels of home traffic pattern (Wachs, 1987) and home noise (Wachs & Gandour, 1983). Further, highly active infants living in poorly organized homes show lower object manipulation skills and less understanding of object properties than do highly active infants living in more organized homes (Peters-Martin & Wachs, 1984). In addition, although a test for statistical interaction was nonsignificant, preschool children's classroom compliance was predicted by additive coaction (combined multiple main effects; Rutter, 1983) between child temperament and classroom chaos (Wachs et al., 2004).

PSYCHOSOCIAL RISK. Two twin studies have reported moderation of psychosocial risk influences by home chaos. Results indicated that the impact of discordant treatment by parents on offspring behavioral problems at 4 years of age (Asbury, Dunn, Pike, & Plomin, 2003) and offspring conduct problems and academic achievement at 7 years of age (Asbury, Dunn, & Plomin, 2006) was accentuated in more chaotic homes. In addition, results from a study involving middle-school children from low-income families showed that the impact of exposure to cumulative risk factors, including aspects of home chaos, on child allostatic load (physiological wear and tear on the body) was significantly greater when mothers were low in responsiveness to their children (Evans, Kim, Ting, Tesher, & Shannis, 2007).

Moderation of Genetic Influences by Chaos

The link between nature and nurture has been a prominent feature of Bronfenbrenner's bioecological model (Bronfenbrenner & Ceci, 1994). Results from a large-scale twin study indicated that the heritability of verbal abilities varied as a function of home chaos, with higher heritability found for children living in more chaotic homes (Asbury, Wachs & Plomin, 2005). In a follow-up study with this sample, though there was not a statistically significant interaction between chaos and a composite measure of five DNA markers, there was again evidence for additive coaction with individual differences in intelligence varying as a function of predicted main effects for both DNA markers and chaos (Harlaar et al., 2005).

Future Directions in Process Research on Microsystem Chaos

The evidence base for Chaos × Individual characteristics interactions (moderation), though still relatively small, is intriguing, with some studies reporting significant interactions, others reporting additive coaction, and still others reporting inconsistent or nonsignificant findings. Nonsignificant findings may reflect the fact that the operation of moderating processes typically is identified by significant statistical interaction terms (Evans & Lepore, 1997). It is well known that statistical power demands are greater in attempts to detect interactions than in attempts to detect main effects (Cronbach, 1991; McClelland & Judd, 1993). However, a variety of strategies can be used to increase power or the likelihood of detecting Organism × Environment interactions. The most obvious is increasing sample size. However, the gain in power associated with using a larger sample may be attenuated if increased sample size requires the use of less time-consuming but less sensitive measures of environmental chaos. Other alternatives would be to (a) use a less stringent p value when testing for interactions (e.g., $p < .10$; Rosnow & Rosenthal, 1989), (b) increase range by oversampling at the extremes of individual characteristics or chaotic environments (McClelland & Judd, 1994), (c) use factor-derived aggregate scores of individual characteristics to integrate the contributions of multiple correlated predictors into a single score (Burchinal et al., 2000), or (d) use theory-driven planned comparisons where testing for interactions occurs only within a range of individual characteristics or for subgroups in which moderation is most likely to occur (McCartney Burchinal & Bub, 2006; Rutter & Pickles, 1991).

In addition to the use of more sensitive analytical procedures, I also propose that the search for individual moderators of chaos is more likely to be successful when research designs are closely linked to the terms of Bronfenbrenner's second bioecological proposition. The terms of proposition 2 involving the joint contributions of person, context, and time indicate that looking for first-order Person × Chaos interactions may be insufficient and that higher order interactions involving other moderators may be involved. As discussed previously, one such higher order interaction would involve Chaos × Gender × Age. Of course, in searching for higher order interactions the problem of insufficient power to detect existing interactions becomes even greater than when dealing with simpler first-order interactions. Thus, more powerful and sensitive research designs need to be used, as discussed previously.

Proposition 2 of the bioecological model also stipulates that relations between proximal processes and development may vary according to the outcome being assessed. One implication of this aspect of proposition 2 is the need for selection of outcome variables that may be particularly sensitive to Organism ×

Chaos interactions. For example, results of the Evans, Lepore, Shejwal, and Palsane (1998) study indicate that males were more adversely influenced by crowding when blood pressure was the outcome, whereas females were more adversely influenced by crowding when measures of learned helplessness were the outcome. Thus, it may be useful to include outcomes where there are known gender differences, particularly in regard to stress, when looking for Gender × Chaos interactions. The same point may also apply when choosing measures of chaos. As McCall (1991) discussed, existing gene–environment interactions (moderation) may be masked when gene–environment covariation is also occurring. It is interesting that in the Haarlar et al. (2005) study, where results indicated additive coaction but not interactions between DNA markers and environmental chaos, results also indicated a negative correlation between the composite DNA markers and the parent report measure of environmental chaos used in this study. Although virtually no evidence is available, I would speculate that parent report measures of perceived chaos are more likely to covary with specific genetic markers than would objective observational measures such as rooms-to-people ratio or ambient noise level. If so, the latter markers of chaos may be more appropriate in future studies on Gene × Chaos interactions.

Chaos and Process

Process refers to mechanisms through which environmental chaos acts to influence developmental outcomes. Such mechanisms are usually identified through testing for mediation (McCartney et al., 2006). Though most evidence on mediation involves the microsystem, in light of proposition 2 it is important to recognize that chaos can also mediate or be mediated by higher order exosystem and macrosystem levels of the environment (see chaps. 12, 13, 14, and 15, this volume). In addition, though most research has focused on contextual mediators of chaos, biological mediators may also be operating. For example, the relation of chaos to development may be mediated by chaos-driven alterations in brain development (Chang & Merzenich, 2003), alterations in neurotransmitter metabolism (Repetti, Taylor, & Saxbe, 2007), alterations in the functioning of the limbic–hypothalamic–pituitary–adrenal axis (Sanchez, Ladd, & Plotsky, 2001), or alterations in child cardiovascular reactivity (Evans et al., 2007; see also chaps. 8 and 14, this volume).

Microsystem Characteristics

With regard to microsystem mediators, the most consistent evidence involves social support. Social support has been shown to mediate between housing density and indices of adult socioemotional functioning in four studies (Evans, Rhee, et al., 2000; Evans & Lepore, 1993; Evans, Palsane, Lepore, & Martin, 1989; Lepore, Evans, & Schneider, 1991), with density eroding social support and reduced social support resulting in an increased level of psychological problems.

A second potential mediator is parenting behaviors. Results on the question of whether quality of parenting mediates the influence of environmental chaos are less consistent than are results for social support. Findings from some studies indicate that parental responsivity to infant vocalizations (Evans, Maxwell, & Hart, 1999), parent–child conflict (Evans et al., 1998), and level of positive parental emotional reactivity (Valiente, Lemery-Chalfant, & Reiser, 2007) function as mediators of home chaos. However, in other studies home chaos has been found to predict child developmental outcomes even after controlling for the influence of parenting behaviors (Coldwell, Pike, & Dunn, 2006; Wachs, 1990). Results also indicate that relations between level of preschool classroom chaos and level of child compliance were not mediated by teachers' use of control strategies (Wachs et al., 2004). Because of differences in methodology across studies, including age of child and type of parenting assessment, direct comparison across studies is difficult. Potential hypotheses as to why some parenting studies found evidence for mediation while others did not are considered in the next section.

Future Directions in Chaos Process Research

Within the framework of Bronfenbrenner's PPCT model, mediation of the influence of chaos on developmental or behavioral outcomes may well be multidimensional, incorporating the impact of moderating factors, multistep mediating processes, or mediating patterns varying as a function of outcomes (specificity). For example, Evans and Lepore (1993) have identified a two-stage mediating process, such that increased density leads to greater social withdrawal, which in turn leads to lower giving of and response to social support. Similarly, Valiente et al. (2007) have shown that relations between chaos and child behavioral problems were mediated by both parental emotionality and child temperament. Both the terms of Bronfenbrenner's PPCT model and these types of findings emphasize the need to identify and test specific mechanisms through which chaos alters the quality of parenting.

A number of potential and testable candidate process mechanisms have been suggested. On the basis of research with both child (Evans et al., 1999) and adult populations (Evans & Lepore, 1993), Evans has proposed two potential mechanisms. The first, which I label the *time allotment hypothesis*, proposes that in crowded homes parents have less time available to interact with their children. At present, little direct evidence is available on the validity of this hypothesis, though qualitative findings suggest that relations between density and parental time allotment can vary as a function of cultural characteristics and norms (Fuller, Edwards, Vorakitphokatorn, & Sermsri, 1993; Wachs & Corapci, 2003). For example, the reduced levels of maternal object or language stimulation or maternal responsivity seen in more crowded homes in North American culture (e.g., Evans et al., 1999; Wachs, 1993) may be attenuated in cultures where primary caregivers also include older siblings. In these cultures more siblings may mean higher household density but also more caregivers available to interact with young children.

The second hypothesis suggested by Evans, which I label the *social withdrawal hypothesis*, postulates that high density increases parental social withdrawal, which in turn results in alterations in parenting behavior. This hypothesis can be extended to include measures of parental responsivity, given that low parental responsivity may be a marker for high parental withdrawal.

Wachs and Corapci (2003) proposed four additional underlying process mechanisms. The *habituation hypothesis* assumes that parents ultimately habituate to continued exposure to environmental chaos and environmental habituation adversely impacts parenting behavior. This hypothesis can be viewed as the adult version of the tuning out hypothesis, which was used as an initial explanation of why exposure to noise adversely impacted young children's speech perception (Evans & Hygge, 2007). Though there is evidence that adults do not completely habituate to ambient noise (Griefahn, 1991), at present there is little direct evidence for the validity of this hypothesis for parenting. The *fatigue hypothesis* postulates that continued exposure to chaos increases parental physical fatigue, which reduces the adequacy of parenting behaviors. The *self-efficacy hypothesis* is based on the assumption that continued exposure to uncontrollable noise and crowding reduces parental self-efficacy beliefs, which in turn alter parenting behavior. This hypothesis can be extended to include measures of parental learned helplessness, given that learned helplessness may be a marker for reduced parental self-efficacy beliefs. Finally, the *negative emotionality hypothesis* postulates that continued exposure to environmental chaos increases the level of negative parent emotional reactivity (including depression), which in turn adversely impacts parenting. Although treated separately, the terms of these six hypotheses may be linked (e.g., parental habituation or fatigue may lead to social withdrawal; social withdrawal may reduce parental self-efficacy beliefs).

Table 7.1 summarizes the evidence for the four hypotheses. With regard to the social withdrawal hypothesis, though evidence supports the validity of this hypothesis with regard to adult behavior, results for parenting are not supportive of this hypothesis. In one study cited in Table 7.1, higher density was associated with more support to offspring by parents. The authors of this study suggested that higher support may reflect a compensatory reaction by parents (Fuller et al., 1993). Results of a second study cited in Table 7.1 did not support a mediating influence of social support on links between chaos, parenting, and offspring development. Similarly, despite consistent evidence that in chaotic home environments parents are lower in responsivity to their children (Wachs & Corapci, 2003), the available evidence does not support the hypothesis that reduced parental responsivity actually mediates observed relations between chaos and development (e.g., Evans et al., 2007).

With regard to the fatigue hypothesis, some studies shown in Table 7.1 suggest that noise exposure can increase parental fatigue through increasing the risk of parental sleep problems. However, consistent with findings previously cited on moderation of chaos, other studies suggest that this effect occurs only for individuals who are high in noise sensitivity or only under conditions of intermittent noise. In light of recent findings by Brown and Low (2008), it is possible that the adverse developmental consequences associated with living in a chaotic home may be mediated by chaos acting to increase the risk of children's sleep problems, which could directly influence children's functioning or increase the risk of parental sleep problems and fatigue. With regard to another dimension of the microsystem, namely, schools, qualitative data indicate that teach-

Table 7.1. Hypothesized Process Mechanisms Underlying Chaos-Driven Alterations in Parenting

Hypothesized mechanism	Available evidence
High levels of density result in increased social withdrawal, which can reduce the quality of parenting and the level of social support available to parents.	High density predicts greater adult social withdrawal (Evans, Palsane, Lepore, & Martin, 1989; Evans & Lepore, 1993; Lepore, Evans, & Palsane, 1991). Greater home density is associated with more parental support to offspring (Fuller, Edwards, Vorakitphokatorn, & Sermsri, 1993). Perceived social support does not mediate the chaos→ parenting→ child outcome pathway (Evans, Lepore, Shejwal, & Palsane, 1998; Wachs & Camli, 1991). Though maternal responsivity moderates chaos, it does not meet the criteria for mediation (Evans, Kim, Ting, Tesher, & Shannis, 2007).
Continued exposure to noise and crowding increases physical fatigue, which reduces parenting competence.	Exposure to ambient noise sources leads to increased adult sleep problems (Kageyama et al., 1997; Langdon & Buller, 1977), but this effect occurs primarily among women with higher levels of sensory sensitivity (Nivison & Endresen, 1993), or when noise is intermittent (Griefahn, 1991).
In chaotic environments parents believe that they have little control over their environment, which can result in a lower level of parental self-efficacy, or parental learned helplessness, which in turn reduces the quality of parenting.	Reduced parental efficacy beliefs were related to measures of noise chaos, but criteria for mediation were not met. However, the relation of home noise chaos to parental efficacy beliefs and the relation of efficacy beliefs to parenting varied as a function of infant temperament. (Corapci & Wachs, 2002). On tasks assessing learned helplessness adults perform more poorly as a function of exposure to noise or crowding (Evans & Stecker, 2004). However, it is not known whether helplessness serves to mediate chaos→parenting.
Continued exposure to environmental chaos increases parental negative emotionality (e.g., increased irritability, anxiety, depression), which disrupts the quality of parenting.	Objective and perceived measures of home crowding predict greater parental psychological distress, and perceived crowding predicts more parental discipline and greater parental support of children. Parental psychological distress fully mediates the crowding–discipline link and partially mediates the crowding–support link (Fuller et al., 1993). Relations between crowding and parent object stimulation remain significant after partialling out an index of parental mental health (Wachs & Camli, 1991). Parental self-reported mood was unrelated to indices of environmental chaos (Corapci & Wachs, 2002). Parent report of home chaos was related to higher parental self-reported depression; higher parental depression was related to greater use of harsh discipline. Harsh discipline predicts more offspring behavior problems and lower offspring cognitive performance (Pike, Iervolino, Eley, Price, & Plomin, 2006). Home chaos is unrelated to parental negative emotional reactivity but is related to lower positive parental emotional reac-

tivity. The path from reduced parental positive emotional reactivity is mediated through children's effortful control (Valiente, Lemery-Chalfant, & Reiser, 2007).

---

ers in noisier schools report feeling more fatigued (Evans & Hygge, 2007), though evidence is lacking on whether teacher fatigue actually mediates relations between school noise and pupil academic performance.

Available evidence on the self-efficacy hypothesis indicates that though ambient noise reduces parental efficacy beliefs and both noise and crowding increase adult feelings of learned helplessness, either the overall criteria for mediation were not met or mediation was not directly tested. However, relations between noise and parental efficacy beliefs, and between efficacy beliefs and parenting, may be moderated by infant temperament. The importance of bringing in individual characteristics such as temperament when attempting to understand mediating mechanisms linking parenting to environmental chaos is underscored by evidence on the negative emotionality hypothesis. In the Valiente et al. (2007) study reported in Table 7.1, temperament enters as an intervening step, with the link between home chaos, parent emotionality, and offspring outcomes being further mediated by individual differences in child effortful control. In addition, congruent with the terms of Proposition 2, the evidence from several studies shown in Table 7.1 also indicates that relations between chaos and parent negative emotionality may vary across different domains of chaos.

One conclusion that can be drawn from the overall pattern of findings shown in Table 7.1 is the need for studies that test whether the process mechanisms described earlier actually function as mediators between chaos and parenting. A second conclusion is that such future studies should be designed to test for moderated mediation of chaos (Evans & Lepore, 1997), using individual parent and child characteristics such as parental sensory sensitivity and child temperament as moderators. The findings shown in Table 7.1 also emphasize the need to consider that multilevel mediators are likely to be involved in the pathways between chaos and parenting. Integrating person characteristics and multilevel mediators into the paths from chaos to parenting to development is consistent with the terms of Bronfenbrenner's bioecological model.

Chaos and Context

Context usually refers to higher levels of the environment (the macrosystem), such as culture or poverty, which can act to moderate either the nature or the influence of proximal processes. The relation of chaos to both culture and poverty is systematically covered in chapters 13 and 14, respectively. To reduce overlap, I limit my discussion of chaos and context within a PPCT framework to a major disjunction between predictions derived from Proposition 2 of Bronfenbrenner's theory and what the evidence reveals.

Bronfenbrenner's Proposition 2 states that the nature and influence of proximal processes such as microsystem chaos should systematically vary across dimensions of the macrosystem, such as culture. The construct validity of proposition 2 is supported by evidence documenting cultural differences in the perception of when an environment is overcrowded (Evans, Lepore, & Allen, 2000). However, a number of studies from non-Western developing countries have reported patterns of findings on chaos and development or chaos and parenting that are similar to what has been reported in Western developed countries (for a review of this research, see Wachs & Corapci, 2003). The discrepancy between what is predicted by Bronfenbrenner's theory and the database on context and chaos is problematic. Delineating specific cultural–contextual characteristics that define what is and what is not a chaotic environment, or that can attenuate the link between culture and microsystem chaos, is a critical research issue. Space restrictions preclude further discussion of this issue in this chapter, but a detailed analysis of how cultural characteristics can influence the nature and impact of environmental chaos is found in chapter 13 of this volume.

Chaos Across Time

Within the PPCT framework, *time* can refer to historical time, individual life course time, or cumulative effects of proximal processes across time. A discussion linking chaos to historical time is presented in chapter 2, and evidence on the influence of temporal chaos at higher levels of the environment is found in chapter 8. At the microsystem level a fundamental implication of Proposition 2 is the need to study chaos as a cumulative phenomenon. At least during the infancy and preschool period, microsystem chaos is clearly a relatively stable phenomenon. When measured by parental perceptions, the 12-month stability of parental scores on CHAOS (Confusion, Hubbub, and Order Scale) is a healthy $r = .74$ in infancy (Matheny & Phillips, 2001) and $r = .69$ in the preschool years (Petrill, Pike, Price, & Plomin, 2004). In research from my laboratory using repeated home observations during the 2nd year of life, the mean 12-month stability of dimensions of chaos involving crowding, noise, and home traffic patterns is a respectable $r = .66$. The relative stability of microsystem chaos may be one reason chaos can act to mediate the level of developmental changes. Evidence for this is shown in an article by Petrill et al. (2004) indicating that the stability of verbal and nonverbal abilities between 3 and 4 years of age is partially mediated by level of environmental chaos in the home, even after controlling for family social class.

A number of articles by Evans and colleagues (Evans, 2003; Evans et al., 2007; Evans & Marcynszyn, 2004) have linked environmental chaos to the concept of allostatic load. *Allostatic load* refers to the strain on physiological function resulting from the need to adapt to chronic stress. The more chronic or cumulative the stress, the greater the allostatic load over time (Repetti et al., 2007). To the extent that exposure to environmental chaos produces stress and a need to adapt to chaos-driven stress, there should be an increasing cumulative level of allostatic load with longer exposure to chaos. The empirical research on cumulative concurrent exposure to multiple biosocial risk factors, including chaos (Corapci, 2008; Evans & Marcynszyn, 2004; Stansfeld et al., 2005), is consistent with the larger body of research reporting that as the number of risk factors encountered at a given time increases, developmental competence decreases (Sameroff, Gutman, & Peck, 2003).

Though there are both empirical and conceptual reasons for predicting that the effects of continued exposure to microsystem chaos will lead to cumulative performance deficits, there has been surprisingly little actual longitudinal research testing whether the adverse developmental consequences of continued exposure to chaos increase as exposure time increases. Available studies have indicated that young children do not adapt to continued ambient noise exposure (Cohen, Evans, Krantz, Stokols, & Kelly, 1982), that there is a cumulative decline in reading performance as exposure to noise continues (Cohen, Glass, & Singer, 1973; Hygge, Evans, & Bullinger, 2002), and that the impact of household density on children's cognitive performance emerges only over time (Richter, 1989). Studies with young adults also have reported a cumulative increase in social withdrawal (Evans & Lepore, 1993) and psychological distress (Lepore, Evans, & Schneider, 1991) as length of residence in more crowded environments increases.

This pattern of findings emphasizes the need for more longitudinal studies on the cumulative effects of exposure to microsystem chaos. One important contribution of longitudinal studies would be to establish the directionality of relations between chaos and parenting. For example, does perceived crowding cause increased parental conflict, or does conflict increase parental perceptions of their environment as crowded; do high home noise levels compromise parental feelings of self-efficacy, or are less efficacious parents less able to control household noise? Longitudinal studies would also be valuable for exploring another time dimension in the PPCT framework noted earlier, namely, whether the impact of environmental chaos on developmental competence varies as a function of the child's chronological age. It has been suggested that the impact of environmental chaos on development may be greater when specific abilities are in the process of being developed, as opposed to after abilities have come on line (Manlove, Frank, & Vernon-Feagans, 2001). If this theory is correct, the impact of chaos on development also would depend on the timing of emergence of specific cognitive and socioemotional competencies. In light of existing knowledge on the developmental trajectories of different competencies, it may be possible to predict when a given cognitive (e.g., language) or socioemotional domain (e.g., voluntary self-regulation) would be most likely to be influenced by exposure to environmental chaos.

Conclusions

In this chapter I have attempted to place microsystem chaos within the overall framework of Bronfenbrenner's bioecological model, with specific reference to the person, process, context, and time dimensions of this model. My conceptualization of how microsystem chaos fits within the PPCT framework is shown in Figure 7.1. A fundamental aspect of Figure 7.1 is that understanding the nature of relations between microsystem chaos and development will likely require more than just a simple, main effect chaos → development model (see also chap. 3). Rather, the influence of microsystem chaos will depend, in part,

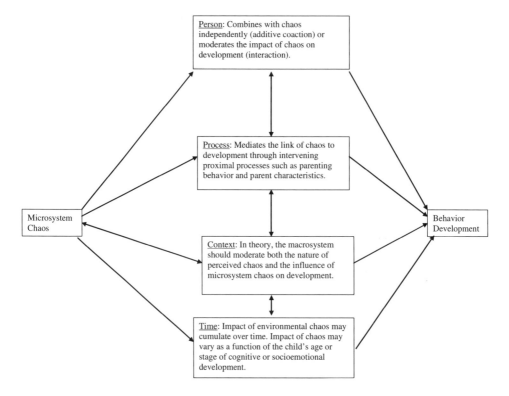

Figure 7.1. Microsystem chaos within Bronfenbrenner's person, process, context, and time framework.

on person characteristics such as temperament, age, and prior history of exposure to chaos or nonchaotic psychosocial risks and supports, as well as the larger macrosystem context within which the individual develops. Further, the influence of microsystem chaos on development is likely to be indirect, involving the impact of chaos on intervening processes that in turn result in developmental variability. Finally, chaos forms part of a larger environmental system that can function to either increase or decrease the level of microsystem chaos encountered by a child (e.g., poverty, refugee status). Given the complexity of the constructs involved, I consider Figure 7.1 as nothing more than an oversimplified sketch summarizing the various ways in which microsystem chaos translates into developmental variability. However, what this sketch illustrates is how placing microsystem chaos within the PPCT framework offers both a compatible conceptual home and a springboard for future research on chaos and development.

References

Aiello, J., Nicosia, G., & Thompson, D. (1979). Physiological, social, and behavioral consequences of crowding on children and adolescents. *Child Development, 50*, 195–202.

Asbury, K., Dunn, J., Pike, A., & Plomin, R. (2003). Nonshared environmental influences on individual differences in early behavioral development: A monozygotic twin differences study. *Child Development, 74*, 933–943.

Asbury, K., Dunn, J., & Plomin, R. (2006). Birthweight-discordance and differences in early parenting relate to monozygotic twin differences in behaviour problems and academic achievement at age 7. *Developmental Science, 9*, 22–31.

Asbury, K., Wachs, T. D., & Plomin, R. (2005). Environmental moderators of genetic influence on verbal and nonverbal abilities in early childhood. *Intelligence, 33*, 643–661.

Barreto, E., Morris, B., Philbin, M., Gray, L., & Lasky, R. (2006). Do former preterm infants remember and respond to neonatal intensive care unit noise? *Early Human Development, 82*, 703–707.

Bronfenbrenner, U. (1999). Environments in developmental perspective. In S. Friedman & T. D. Wachs (Eds.), *Measuring the environment across the life span* (pp. 3–30). Washington, DC: American Psychological Association.

Bronfenbrenner, U., & Ceci, S. (1994). Nature-nurture reconceptualized in developmental perspective: A bioecological model. *Psychological Review, 101*, 568–586.

Brown, E., & Low, C. (2008). *Chaotic living conditions and sleep problems predict children's responses to academic challenge.* Paper submitted for publication.

Chang, E., & Merzenich, M. (2003, April 18). Environmental noise retards auditory cortical development. *Science, 300*, 498–502.

Christie, D., & Glickman, C. (1980). The effects of classroom noise on children: Evidence for sex differences. *Psychology in the Schools, 17*, 405–408.

Cohen, S., Evans, G., Krantz, D., Stokols, D., & Kelly, S. (1982). Aircraft noise and children: Longitudinal and cross-sectional evidence on adaptation to noise and the effectiveness of noise abatement. *Journal of Personality and Social Psychology, 40*, 331–345.

Cohen, S., Glass, D., & Singer, J. (1973). Apartment noise, auditory discrimination and reading ability in children. *Journal of Experimental Social Psychology, 9*, 407–422.

Coldwell, J., Pike, A., & Dunn, J. (2006). Household chaos—links with parenting and child behaviour. *Journal of Child Psychology and Psychiatry, 47*, 1116–1122.

Corapci, F. (2008). The role of child temperament on Head Start preschoolers' social competence in the context of cumulative risk. *Journal of Applied Developmental Psychology, 29*, 1–16.

Corapci, F., & Wachs, T. D. (2002). Does parental mood or efficacy mediate the influence of environmental chaos upon parenting behavior? *Merrill-Palmer Quarterly, 48*, 182–201.

Cronbach, L. (1991). Emerging views on methodology. In T. D. Wachs & R. Plomin (Eds.), *Conceptualization and measurement of organism-environment interaction* (pp. 87–104). Washington, DC: American Psychological Association.

Dornic, S., & Ekehammar, B. (1990). Extraversion, neuroticism, and noise sensitivity. *Personality and Individual Differences, 11*, 989–992.

Evans, G. (2003). A multimethodological analysis of cumulative risk and allostatic load among rural children. *Developmental Psychology, 39*, 924–933.

Evans, G. (2006). Child development and the physical environment. *Annual Review of Psychology, 57*, 423–451.

Evans, G., & Hygge, S. (2007). Noise and performance in children and adults. In L. Luxon & D. Prasher (Eds.), *Noise and its effects* (pp. 549–566). London: Wiley.

Evans, G., Kim, P., Ting, A., Tesher, H., & Shannis, D. (2007). Cumulative risk, maternal responsiveness and allostatic load among young adolescents. *Developmental Psychology, 43*, 341–351.

Evans, G., & Lepore, S. (1993). Household crowding and social support: A quasi-experimental analysis. *Journal of Personality and Social Psychology, 65*, 308–316.

Evans, G., & Lepore, S. (1997). Moderating and mediating processes in environment behavior research. In G. Moore & R. Marans (Eds.), *Advances in environment, behavior and design* (Vol. 4, pp. 255–285). New York: Plenum Press.

Evans, G., Lepore, S., & Allen, K. (2000). Cross-cultural differences in tolerance for crowding: Fact or fiction? *Journal of Personality and Social Psychology, 79*, 204–210.

Evans, G., Lepore, S., Shejwal, B., & Palsane, M. (1998). Chronic residential crowding and children's well being. *Child Development, 69*, 1514–1523.

Evans, G., & Marcynszyn, L. (2004). Environmental justice, cumulative environmental risk, and health among low- and middle-income children in Upstate New York. *American Journal of Public Health, 94*, 1942–1944.

Evans, G., Maxwell, L., & Hart, B. (1999). Parental language and verbal responsiveness to children in crowded homes. *Developmental Psychology, 35*, 1020–1024.

Evans, G., Palsane, M., Lepore, S., & Martin, J. (1989). Residential density and psychological health. *Journal of Personality and Social Psychology, 79*, 994–999.

Evans, G., Rhee, E., Forbes, C., Allen, K., & Lepore, S. (2000). The meaning and efficacy of social withdrawal as a strategy for coping with chronic crowding. *Journal of Environmental Psychology, 20*, 204–210.

Evans, G., & Stecker, R. (2004). Motivational consequences of environmental stress. *Journal of Environmental Psychology, 24*, 143–165.

Fuller, T., Edwards, J., Vorakitphokatorn, S., & Sermsri, S. (1993). Household crowding and family relations in Bangkok. *Social Problems, 40*, 410–430.

Griefahn, B. (1991). Environmental noise and sleep. Review—Need for further research. *Applied Acoustics, 32*, 255–268.

Haines, M., Stansfeld, S., Brentnall, S., Head, J., Berry, B., Jiggins, M., & Hygge, S. (2001). The West London Schools Study: The effects of chronic aircraft noise exposure on child health. *Psychological Medicine, 31*, 1385–1396.

Hambrick-Dixon, P. (1998). The effect of elevated subway train noise over time on Black children's visual vigilance performance. *Journal of Environmental Psychology, 8*, 299–314.

Harlaar, N., Butcher, L., Meaburn, E., Sham, P., Craig, I., & Plomin, R. (2005). A behavioural genomic analysis of DNA markers associated with general cognitive ability in 7-year-olds. *Journal of Child Psychology and Psychiatry, 46*, 1097–1107.

Hygge, S., Evans, G., & Bullinger, M. (2002). A prospective study of some effects of aircraft noise on cognitive performance in school children. *Psychological Science, 13*, 469–474.

Job, R. (1999). Noise sensitivity as a factor influencing human reaction to noise. *Noise and Health, 3*, 57–68.

Kageyama, T., Kabuto, M., Nitta, H., Kurokawa, Y., Taira, K., Suzuki, S., & Takemoto, T. (1997). A population study on risk factors for insomnia among adult Japanese women: A possible effect of road traffic volume. *Sleep, 20*, 963–971.

Langdon, F., & Buller, I. (1977). Road traffic noise and disturbance to sleep. *Journal of Sound and Vibration, 50*(1), 13–28.

Lepore, S. J., Evans, G., & Palsane, M. (1991). Social hassles and psychological health in the context of chronic crowding. *Journal of Health and Social Behavior, 32*, 357–367.

Lepore, S., Evans, G., & Schneider, M. (1991). Dynamic role of social support in the link between chronic stress and psychological distress. *Journal of Personality and Social Psychology, 61*, 899–909.

Manlove, E., Frank, T., & Vernon-Feagans, L. (2001). Why should we care about noise in classrooms and child care settings? *Child & Youth Care Forum, 30*, 55–64.

Matheny, A. (1991). Children's unintentional injuries and gender: Differentiation by environmental and psychosocial aspects. *Children's Environments Quarterly, 8*(3/4), 51–61.

Matheny, A., & Phillips, K. (2001). Temperament and context: Correlates of home environment with temperament continuity and change, newborn to 30 months. In T. D. Wachs & G. Kohnstamm (Eds.), *Temperament in context* (pp. 81–102). Hillsdale, NJ: Erlbaum.

McCall, R. (1991). So many interactions, so little evidence. Why? In T. D. Wachs & R. Plomin (Eds.), *Conceptualization and measurement of organism-environment interaction* (pp. 142–161). Washington, DC: American Psychological Association.

McCartney, K., Burchinal, M., & Bub, K. (2006). Best practices in quantitative methods for developmentalists. *Monographs of the Society for Research in Child Development, 71*(3, Serial No. 285).

McClelland, G., & Judd, C. (1993). Statistical difficulties of detecting interactions and moderator effects. *Psychological Bulletin, 114*, 376–390.

Nivison, M., & Endresen, I. (1993). An analysis of relationships among environmental noise, annoyance and sensitivity to noise, and the consequences for health and sleep. *Journal of Behavioral Medicine, 16*, 257–276.

Peters-Martin, P., & Wachs, T. D. (1984). A longitudinal study of temperament and its correlates in the first 12 months. *Infant Behavior & Development, 7*, 285–298.

Petrill, S., Pike, A., Price, T., & Plomin, R. (2004). Chaos in the home and socioeconomic status are associated with cognitive development in early childhood: Environmental mediators identified in a genetic design. *Intelligence, 32*, 445–460.

Pike, A., Iervolino, A., Eley, T., Price, T., & Plomin, R. (2006). Environmental risk and young children's cognitive and behavioral development. *International Journal of Behavioral Development, 30*, 55–66.

Rahmanifar, A., Kirksey, A., Wachs, T. D., McCabe, G., Sobhy, A., Galal, O., et al. (1993). Diet during lactation associated with infant behavior and caregiver/infant interaction in a semirural Egyptian population. *Journal of Nutrition, 123*, 164–175.

Ramirez, J., Alvarado, J., & Santisteban, C. (2004). Individual differences in anger reaction to noise. *Individual Differences Research, 2*, 125–136.

Repetti, R., Taylor, S., & Saxbe, D. (2007). The influence of early socialization experiences on the development of biological systems. In J. Grusec & P. Hastings (Eds.), *Handbook of socialization* (pp. 124–152). New York: Guilford Press.

Richter, L. (1989). Household density, family size and the growth and development of Black children: A cross-sectional study from infancy to middle childhood. *South African Journal of Psychology, 19*, 191–198.

Rosnow, R., & Rosenthal, R. (1989). Statistical procedures and the justification of knowledge in psychological science. *American Psychologist, 44*, 1276–1284.

Rutter, M. (1983). Statistical and personal interactions. In D. Magnusson & V. Allen (Eds.), *Human development* (pp. 295–320). New York: Academic Press.

Rutter, M., & Pickles, A. (1991). Person–environment interaction. In T. D. Wachs & R. Plomin (Eds.), *Conceptualization and measurement of organism-environment interaction* (pp. 105–141). Washington, DC: American Psychological Association.

Sameroff, A., Gutman, L., & Peck, S. (2003). Adaptation among youth facing multiple risks: Prospective research findings. In S. Luthar (Ed.), *Resilience and vulnerability: Adaptation in the context of childhood adversities* (pp. 364–391). New York: Cambridge University Press.

Sanchez, M., Ladd, C., & Plotsky, P. (2001). Early adverse experience as a developmental risk factor for later psychopathology: Evidence from rodent and primate models. *Development and Psychopathology, 13*, 419–449.

Stansfeld, S. (1992). Noise, noise sensitivity and psychiatric disorder. *Psychological Medicine: Monograph Supplement, 22*, 1–44.

Stansfeld, S., Berglund, B., Clark, C., Lopez-Barrio, I., Fischer, P., Ohrstrom, E., et al. (2005, June 4). Aircraft and road traffic noise and children's cognition and health: A cross-national study. *The Lancet, 365*, 1942–1949.

Turrero, A., Zuluaga, P., & Santisteban, C. (2001). Joint effect of noise, personality and environmental factors on the intelligibility of speech. *Methods of Psychological Research Online, 6*, 175–197.

Valiente, C., Lemery-Chalfant, K., & Reiser, M. (2007). Pathways to problem behaviors: Chaotic homes, parent and child effortful control, and parenting. *Social Development, 16*, 249–267.

Wachs, T. D. (1979). Proximal experience and early cognitive-intellectual development: The physical environment. *Merrill-Palmer Quarterly, 25*, 3–41.

Wachs, T. D. (1987). Specificity of environmental action as manifest in environmental correlates of infant's mastery motivation. *Developmental Psychology, 23*, 782–790.

Wachs, T. D. (1989). The nature of the physical micro-environment: An expanded classification system. *Merrill-Palmer Quarterly, 35*, 399–420.

Wachs, T. D. (1990). Must the physical environment be influenced by the social environment in order to influence development: A further test. *Journal of Applied Developmental Psychology, 11*, 163–178.

Wachs, T. D. (1993). Nature of relations between the physical and social microenvironment of the two year old child. *Early Development and Parenting, 2*, 81–87.

Wachs, T. D. (2000). *Necessary but not sufficient: The respective roles of single and multiple influences on individual development.* Washington, DC: American Psychological Association.

Wachs, T. D., & Camli, O. (1991). Do ecological or individual characteristics mediate the influence of the physical environment upon maternal behavior? *Journal of Environmental Psychology, 11*, 249–264.

Wachs, T. D., & Corapci, F. (2003). Environmental chaos, development and parenting across cultures. In C. Raeff & J. Benson (Eds.), *Social and cognitive development in the context of individual, social and cultural processes* (pp. 54–83). New York: Routledge.

Wachs, T. D., & Gandour, M. (1983). Temperament, environment and 6-month cognitive-intellectual development: A test of the organismic specificity hypothesis. *International Journal of Behavioral Development, 6*, 135–152.

Wachs, T. D., Gurkas, P., & Kontos, S. (2004). Predictors of preschool children's compliance behavior in early childhood classroom settings. *Journal of Applied Developmental Psychology, 25*, 439–457.

8

The Role of Temporal and Spatial Instability in Child Development

Clyde Hertzman

As was discussed in chapter 1 of this volume, temporal and spatial instability are fundamental dimensions of environmental chaos. They are also issues in child development, in two distinct but interlocking ways. The first issue is whether, how, and to what extent temporal or spatial instability in the individual child's life affects his or her development across three key enabling domains: language and cognitive, social and emotional, and physical. These domains are key because they are strongly influenced by early childhood environments and, in turn, influence health, well-being, learning, and behavior throughout the life course (Keating & Hertzman, 1999). The second issue is whether instability in the communities where children grow up, live, and learn has systematic impacts on children's development (Xue, Leventhal, Brooks-Gunn, & Earls, 2005). As shown later in this chapter, the underlying mechanisms of influence overlap and thus deserve coverage in any treatment of temporal–spatial instability.

In the first section of this chapter, I present some lines of evidence and supposition that address the question of whether temporal–spatial instability is a significant determinant of child development, and whether it has been adequately studied to date. In the second section, I develop several theoretical bases for believing that temporal–spatial instability *should* influence child development, and how this should occur. These theoretical perspectives raise questions that go well beyond the existing empirical research base in this field and help inform the final section, where I propose future directions for research on instability and child development.

Is Temporal–Spatial Instability an Important Influence on Child Development?

The sections that follow establish the importance of temporal–spatial instability in terms of the experience of the child and, also, from the perspective of community characteristics. It makes the point that *both* matter when considering the role of temporal–spatial instability in early childhood.

Temporal–Spatial Instability in a Child's Life

Thomas (1999) pointed out that by 1990 American parents were, on average, available to their children 10 hours per week less than they had been in 1980 and 40% less than they had been in 1965. Furthermore, a characteristic shared among the vast majority of violent juveniles is family instability, indicating that family stability can trump negative contextual influences that might otherwise lead to a child's violent behavior. The association between temporal–spatial instability brought about by family disruption and behavioral dysfunction is said to be the most consistent finding in the literature.

The early waves of the Canadian National Longitudinal Survey of Children and Youth (NLSCY) in the late 1990s demonstrated that behavioral and receptive language delays in development by age 4 or 5 rose monotonically with the number of residential moves experienced by the child. Forty-one percent of preschool children had changed residence at least once before starting kindergarten in their 5th year of life (Kohen, Hertzman, & Wiens, 1998). However, in that study, as in others, it is difficult to distinguish residential moves for reasons of economic opportunity from those for reasons of economic necessity, weak ties to the community, or family breakup. Moves for the latter three reasons presumably would be associated with greater developmental risk than would moves for the former reason.

In chapter 2 of this volume, Lichter and Wethington made the counterintuitive point that family instability for reasons of economic necessity did not increase during the 20th century. Rather, they contended that the primary source of family instability has migrated from the macrosystem to the microsystem, because in recent years parent-initiated family breakups have replaced parental death, unemployment, and other uncontrollable factors as the principal causes of family instability. Put in context of the conventional wisdom, as stated in the previous paragraph, that internal family disruption is the principal risk factor for the development of violent behavior, violent behavior should be on the rise in North America. Yet it is not. In fact, decline in violent behavior since the early 1990s has been much commented on and is the subject of an intriguing review (Reyes, 2007) suggesting that reductions in child lead exposure and legalization of abortions during the 1970s are the main explanatory factors. It may be that internal family disruption predicts children's later violent behavior but does so at a lower prevalence rate in a "low lead, high abortion, high parental distraction" world than in the world that preceded it. In other words, the macrosystem sets the basic prevalence rate over historical time, whereas the microsystem sets the relative risks in current time. I believe that this phenomenon is worthy of an historical inquiry of its own.

With regard to the microsystem, Forman and Davies (2003) examined relationships among family instability and adolescents' psychological functioning, using family models of children's emotional security, in a sample of 220 young adolescents and their primary caregivers. Primary caregiver reports of family instability were positively associated with multiple informant measures of adolescent internalizing and externalizing symptoms. Fomby and Cherlin (2007) claimed that each breakup, divorce, remarriage, or new cohabitation carries with it increased risk for the developing child. In a nationally representative sample of American children, they found that those who go through frequent transitions are more likely to have behavioral problems than are children raised in stable families.

Cavanagh and Huston (2006) investigated the association between family instability, its timing, and children's peer relationships and social competencies during elementary school in a sample of more than 1,000 children drawn from the National Institute of Child Health and Human Development Study of Early Child Care and Youth Development. Nearly 40% of the sample experienced at least one family transition between birth and the end of fourth grade: 15% experienced instability in early childhood only, another 11% experienced instability during the elementary school years only, and about 12% experienced change throughout both developmental stages. Overall, children who experienced instability had lower teacher reports of peer competency and popularity, higher reports of externalizing behaviors in the classroom, and more self-reported loneliness. Most important, family instability experienced in early childhood conferred more risk than that experienced during school years only. This latter observation is consistent with recently published work suggesting that residential instability in a narrow age window, during years 2 and 3 of life, is associated with alterations in the expression of DNA sequences governing hypothalamic–pituitary–adrenal (HPA) axis function, which plays a crucial role in stress biology. These effects were found to operate through epigenetic mechanisms and were not mediated through economic circumstances, life stress, or health practices. The epigenetic effect suggests that residential instability may have the capacity to "get under the skin," presumably through psychosocial transmission mechanisms, to influence brain and biological development (Miller & Chen, 2007).

Other studies using census and other administrative data have also shown that residential instability is associated with lower school readiness (Chase-Lansdale & Gordon, 1996; Hertzman, MacLean, Kohen, Dunn, & Evans, 2002; Kershaw, Irwin, Trafford, & Hertzman, 2005), early behavioral and emotional problems (Brooks-Gunn, Duncan, Klebanov, & Sealand, 1993; Chase-Lansdale & Gordon, 1996; Duncan, Brooks-Gunn, & Klebanov, 1994), and adolescent sexuality and childbearing (Ku, Sonenstein, & Pleck, 1993).

In the Economy, Security, Community Study of social capital in Canada, the British Columbia–based research collaborators oversampled adults in seven sawmill communities in British Columbia and interviewed them twice, 2 years apart. Those who moved out of the community between the first and second interview were, according to a variety of measures, less involved in their community at the time of first interview than were those who stayed. This finding validated the supposition that transient families tend to have been less integrated into the community they have left than have residentially stable families and, in all likelihood, will be less integrated into the community into which they move. Moreover, this tendency may have been exacerbated by a selection factor: namely, that less involved families could have been more transient to begin with and thus less involved. In the former case, transience leads to lower involvement; in the latter case, lower involvement promotes transience. Together, they may form a vicious cycle.

Also in British Columbia, focus groups were conducted by community development workers whose job it was to strengthen local child development services, following the initiation of federal–provincial transfer payments for early child development in 2001. The findings revealed a common refrain that many families could benefit from local programs and services, but they could not be found by service providers, or do not, on their own, find the programs and services because necessary two-way links are hard to establish with highly transient families. This situation is often exacerbated by barriers of access resulting from distrust and social distance between service providers and families as well as language, money, transportation, and time-conflict barriers.

In a seemingly extreme case, a study of several hundred homeless Victoria-area families with young children documented a lifestyle of "couch surfing" during the winter and living in parks during the summer; the children were weakly attached to school, if at all (Kasting & Artz, 2005). The phenomenon of young children being weakly attached to school may not be confined to homeless families. A data linkage study in Winnipeg, Manitoba, was able to merge records of children enrolled in the province's universal medical insurance program (in Canada, referred to as "Medicare") with school records, organized according to the socioeconomic status (SES) of each neighborhood. This exercise demonstrated dramatic differences in school success rates by Grade 3. On the basis of classroom denominators, it appeared as though approximately 83% of the children from low SES neighborhoods had passed the language arts test. However, when all the children enrolled in the Medicare system were included, the pass rate dropped to 50% (Brownell et al., 2006). This result is significant because the Medicare enrollment file is a quasi-census of the population and provides the true denominator for a geographic area. Figure 8.1 illustrates this, showing how, as one goes from the wealthiest to the poorest neighborhoods, the proportion of children who failed to show up at school and take the test rises from 12% to more than 40%. One of the key contributors to this difference was transiency and subsequent weak attachment to school—a 15% de facto dropout rate by Grade 3 among children who, according to the up-to-date Medicare records, lived in an inner-city area but according to school records were not currently active in any school (Brownell et al., 2006).

Notwithstanding observations of the type already discussed, temporal–spatial instability is rarely considered in studies of neighborhood characteristics or of child development. A recently developed Standardized Neighborhood Deprivation Index in the United States did not end up including residential instability because it did not align with the principal component in the analysis (Messer et al., 2006). This situation points to three problems characteristic of temporal–spatial instability that lead to its being underrepresented in child development research. First, it is technically difficult to find highly transient families and, without direct contact, difficult to reconstruct residential histories of families with children. Second, the potential influence of temporal–spatial instability on child development may be closely associated with other well-studied factors, such as low income. Finally, its effects may be ambiguous, such as the difference between residential moves for the purposes of economic opportunity versus those for reasons of social isolation, family breakdown, or economic deprivation.

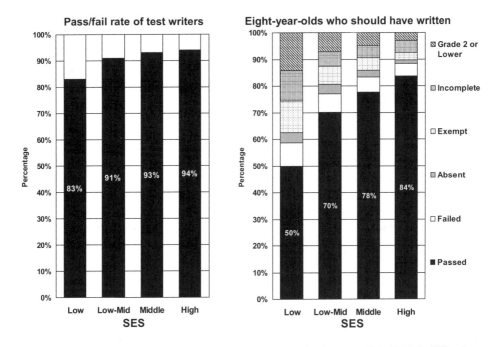

Figure 8.1. Performance on Grade 3 language arts standards test 1998–1999, by Winnipeg socioeconomic status group. From "Is the Class Half Empty? A Population-Based Perspective on Socio-Economic Status and Educational Outcomes," by M. D. Brownell, N. P. Roos, R. Fransoo, L. L. Roos, A. Guèvremont, L. MacWilliam, et al., 2006, *IRPP Choices*, *12*(5), p. 14. Copyright 2006 by the Institute for Research on Public Policy. Reprinted with permission.

The list of studies that should have taken temporal–spatial instability into account, but did not, is long and varied. Here are four examples. Recent evidence of the role of childhood SES on physical functioning half a century later has come from the British 1946 birth cohort, but temporal–spatial instability was not considered in the analysis (Guralnick, Butterworth, Wadsworth, & Kuh, 2006). Despite its focus on neighborhood and family influences, the 18-year follow-up of the Ontario Child Health Study did not consider residential transiency, school transiency, or neighborhood instability in explaining educational attainment (Boyle, Georgiades, Mustard, & Racine, 2007). The first evaluation of the Sure Start program in the United Kingdom makes no mention of neighborhood instability or the influence of residential transiency on child development, despite the fact that Sure Start is meant to be a community-strengthening program (Belsky, Melhuish, Barnes, Leyland, & Romaniuk, 2006). Similarly, the recent evaluation of New Hope, an American program to increase parent employment and reduce poverty, did consider the effects of the gain and then loss of child-care vouchers on stability of child care, and the importance of the program for increasing the sustainability of children's daily routines, but did not consider other aspects of temporal–spatial instability when judging its impact on children's school achievement, motivation, and social behavior (Huston et al., 2005; Lowe, Weisner, Geis, & Huston, 2005).

Finally, two recent reviews—the first was of 15 randomized controlled trials (RCTs) of prevention of childhood mental health problems in the United States, Canada, Australia, and the United Kingdom (Waddell, Peters, Hua, McEwan, & Garland, 2007), and the second was of 17 RCTs of early child development programs with relevance for children's mental health in Canada (Waddell, Hua, McEwan, Garland, & Peters, 2007)—demonstrated the degree to which child development, mental health, and community resources have come to be seen as related constructs in the clinical community. Eleven of 15 RCTs in the first group and all 17 in the second group either relied on community-based contacts to find target families and children for intervention or used a community development model to shift norms of children's development as a strategy for primary prevention of mental health problems. None, however, dealt directly with issues of temporal–spatial instability or systems of follow-up for unstable families.

Temporal–Spatial Instability in the Community

There is also the prospect that residential instability, at the level of the whole community, may adversely affect a community's capacity to provide resources or a nurturant environment for child development. British Columbia serves as an interesting laboratory to examine this issue because it has a large variety of geographically isolated resource communities whose economic fate varies dramatically with global commodity prices. This reality creates a variety of natural experiments in community instability. To date, the research team at the Human Early Learning Partnership has identified two contrasting patterns of community response to economic shock. The first is the partial-layoff scenario. In this case, layoff usually occurs by seniority and skill level. Those with the least seniority and lowest skill levels tend to be the first laid off and, disproportionately, leave the community, ironically leaving behind a residue of more privileged, highly skilled workers and their families. (In terms of selection effects, it cannot be ruled out that more transient workers are less likely to have seniority or higher skills, so in part their leaving the community may be a function of their being a more transient population per se.) The second pattern is the complete shutdown scenario. In this case, the most economically active members of the community are the first to leave, and the community tends to become a reservoir for those who are less economically active. As time goes on, the community tends to attract other low SES families as rents drop to a level that allows social assistance payments and fixed incomes to be stretched the furthest. These observations suggest that community stability, as a determinant of child development, may be complex and contingent on the specific details of what has made a community transient or stable. Instability may be associated with both the voluntary and involuntary aspects of the economic base of a community: its housing and physical desirability, its role as a base for new immigrants, or other factors that select for more or less transient individuals and families.

If, as was suggested earlier, the temporal–spatial characteristics of communities' capacity to nurture children depend on their individual socioeconomic trajectories, then the best analytical approach would be one that treats communities as individuals, characterizes them according to whether or not they nurture well, and works backward to understand which community factors make a difference to their capacity to nurture. This is the approach the research team at the Human Early Learning Partnership is taking in British Columbia. In 2004, under the auspices of the Human Early Learning Partnership (HELP), we became the first jurisdiction in the world to complete a total population-based assessment of early child development (ECD). Assessments were done during the kindergarten year, and ECD was measured using the Early Development Instrument (EDI), in which kindergarten teachers fill out a detailed checklist about each child in their class based on five scale measures of development: physical well-being, social competence, emotional maturity, language and cognitive development, and communication and general knowledge. Data included approximately 44,000 kindergarten children from all walks of life across the province (Kershaw, Irwin, Trafford, & Hertzman, 2005); between 90% and 100% of kindergarten-age children from at least one school entry cohort were included from every geographic school district.

The literature about the effects of neighborhoods on child development most frequently relies on data that are reported using census boundaries or other administrative units of analysis (Burton & Jarrett, 2000). The convenience of census or other survey boundaries comes with costs, however, including the fact that census boundaries often do not match local perceptions of neighborhood divisions. In response, HELP worked closely with communities to benefit from local knowledge in determining neighborhood boundaries that more accurately reflect the daily experience of very young children. Local ECD coalition representatives were invited to draw lines on maps of their area to signal the presence of perceived divides in their community. Whereas some local coalitions opted to maintain the census or another existing boundary system, others opted for dramatically different breakdowns than those used for survey data collection. This exercise allowed us to make observations about rates of vulnerability for the kindergarten population for every residential neighborhood in the province. Starting with roughly 100 census variables (organized according to our neighborhood boundaries by Statistics Canada), we produced best fit regression models for the five EDI scales, as well as a summary variable of vulnerability on one or more of the five EDI scales. These models depict the relationship between SES and vulnerability rates across all 480 British Columbian neighborhoods. In each case, seven or fewer census variables entered the model. Table 8.1 shows the proportion of neighborhood variation in EDI vulnerability accounted for by these models.

This explanatory power is considerably higher than that attributed to the combined influence of neighborhood and family SES effects in traditional analytic studies because the focus here is on ecological correlations for which neighborhood rates of vulnerability—not individual children—are the unit of analysis. The six best fit models collectively identify 19 neighborhood SES indicators to be significant predictors of the share of children at risk of developmental delays in British Columbia as measured by EDI scores. These cover a range of constructs, such as neighborhood income characteristics, immigration and ethnic mix, level of child care provided by men, occupational characteristics of men

Table 8.1. Variation in EDI Vulnerability by British Columbia Neighborhood "Explained" by Neighborhood Census Characteristics

EDI scale	Variance explained (%)
Physical health and well-being	33.8
Social competence	20.9
Emotional maturity	23.4
Language and cognitive	27.2
Communication skills and general knowledge	46.9
One or more EDI vulnerabilities	42.7

Note. EDI = Early Development Instrument. Data from Kershaw, Irwin, Trafford, and Hertzman (2005).

and women, residential transiency, and proportion of the population that is aboriginal (native Canadian). Neighborhoods with high levels of residential transiency were associated with high levels of vulnerability on "one or more scales" of the EDI. Of interest here, the "percent of nonimmigrant movers" (based on census reporting of the proportion of those who moved into the neighborhood between 1996 and 2001) was an independent predictor of increased vulnerability on one or more scales of the EDI (Kershaw et al., 2005). Virtually all of the high-transiency neighborhoods in the province are high vulnerability on the EDI. This observation raises analytical issues that are touched on in the discussion of future directions of research in the final section of this chapter.

Theoretical Bases for Believing That Temporal–Spatial Instability Should Influence Child Development

The state of a child's development is, to a large extent, an emergent property of the quotidian aspects of life—that is, the qualities of stimulation, support, and nurturance that the child experiences moment to moment and day to day—and the way the child alters the intimate environment through his or her participation in it. This transactional perspective (Irwin, Johnson, Dahinten, Henderson, & Hertzman, 2007) has started to replace the traditional maturational perspective, which emphasized the child reaching developmental milestones simply by virtue of his or her age, averaging out the role of the environment in development. In contrast, the transactional perspective recognizes that all the random, commonplace interactions of daily life have the capacity to influence the development of the child. It is in this sense that environmental chaos and development most closely overlap. At the same time, the reality of sensitive periods in brain development is a reminder that not all aspects of daily life have an equal capacity to influence development at any given age. Children's brains and biologies are predisposed to develop specific competencies during temporally circumscribed windows of opportunity, according to biological hierarchies from the foundational (e.g., face recognition) to the sophisticated (e.g., decoding emotions from facial expressions; Keating & Hertzman, 1999).

If temporal–spatial instability is to play a role in child development, it must do so by influencing the transactional environment of the child at the given age when a specific window of developmental opportunity opens. Is it reasonable to suppose that there are windows of opportunity in young children's lives that are amenable to influence from temporal–spatial instability? The answer to this question is yes. Several lines of evidence supporting this view were mentioned earlier, but there are more. One is the prospect that children's mental construction of physical or social space will be disrupted and that this, in turn, will have adverse developmental consequences. As young as they can be reliably interviewed on the subject (in our hands, 5 to 6 years old), children can articulate a view of their social space. They can describe where they do and do not belong and how extensive this zone of mastery and belonging may be, in terms of space and people (Irwin, Siddiqi, & Hertzman, 2007). Their narratives reflect the range of physical and social spaces that they are able to occupy and, thus, in which they can access opportunities for physical and social development. Moreover, these space definitions are influenced by parental consciousness, especially concerning danger and safety. Similarly, we have shown that, in the Canadian NLSCY, whereas neighborhood factors did not seem to be associated with language and cognitive or behavioral development at 0 to 3 years, by age 4 or 5 the neighborhood context was demonstrably associated with development (Kohen, Brooks-Gunn, Leventhal, & Hertzman, 2002).

Those who provide care to young children have long assumed that instability in the care contexts of young children—characterized by multiple different care contexts in an average week—has a negative impact on their development (for further discussion on instability in child-care settings, see chap. 5, this volume). Evidence from studies of preschool children shows that adapting to novel social contexts is stressful for young children (Davis, Donzella, Krueger, & Gunnar, 1999) as measured by the function of children's HPA axis at the beginning of a new year when they meet a new combination of children and adults for the first time. Although it has been little investigated, at least one study has shown that conscious management of the transition into kindergarten—through information sent home to parents, parent orientation sessions, teacher home visits, children visiting the classroom prior to enrollment, and shortened school days at the beginning of the year—can improve early academic achievement, especially for low SES children (Schulting, Malone, & Dodge, 2005).

What might be the biological character of the windows of opportunity at issue here? It has been speculated that certain *candidate systems* serve as transducers between society and human biology; that is, among all the ways that experience can theoretically get under the skin, certain biological pathways better conform to what is understood about humans as a social species than do others. Candidate systems must meet four basic criteria: the short- and long-term function of the system is influenced by quotidian experiences (often early in life); the system continues to respond to quotidian experiences throughout the life course; the system, if dysfunctional, has the biological capacity to influence health, well-being, learning, and behavior; and differential functioning of the system across the life course, to the extent that outcomes are affected, derives from the quotidian. Four candidate systems meet these criteria and are worthy of mention here: the HPA axis and its role in cortisol secretion; the autonomic nervous system in association with epinephrine and norepinephrine; the development of memory, attention, and other executive functions in the prefrontal cortex; and the systems of social affiliation involving the primitive amygdala and locus coeruleus with accompanying higher order cerebral connections, mediated by serotonin and other hormones.

As a notable example, D'Anguilli et al. (2008) investigated the relationship among SES, performance, and the neural correlates of auditory selective attention by comparing event-related potentials (ERPs) in lower and higher SES preadolescent children during a task in which they had to attend to two types of pure tones but ignore two other types. The hypothesis was that at comparable performance levels, higher SES children would easily ignore distracters (the unattended, irrelevant tones) whereas lower SES children would attend equally to distracters and target tones. Indeed, they found that ERP waveform differences between attended and unattended tones were significantly different in the higher SES but not in the lower SES group. Although the groups did not differ in reaction times or accuracy, electroencephalographic power analysis revealed that the high and low SES children recruited different neural processes to achieve the task. Lower SES children seemed to deploy supplementary resources to attend to irrelevant information, consistent with what D'Anguilli et al. called the *ear to the ground hypothesis*; that is, lower SES children may, on average, experience more chaos, disorganization, and threat in their environment than do higher SES children. This may have direct effects on learning. By devoting more processing capacity to tune out irrelevant information, these children may have less available capacity to process higher order information. On the other hand, by attending more to irrelevant information, these children may end up attending less to school-related relevant information. If so, biological embedding of a more distractible executive function system would be adaptive for the lower SES children in their daily lives but would likely make school success more difficult to achieve over the long run.

Just as the case for temporal–spatial instability as an important factor in child development cannot be separated from windows of opportunity, it also cannot be separated from the case that context matters. From the perspective of the child, instability could theoretically influence development as a result of the change in context for daily living. In the most direct way, temporal–spatial instability means a change in immediate physical surroundings and disruption of daily routines. In more indirect ways, it means a change in context that changes the character of the patterns of influence on the child.

The contextual effects on children's development have been characterized according to modes of transmission: neighborhood resources, collective socialization, contagion, competition, relative deprivation, and stress due to toxins or safety concerns (Connor & Brink, 1999; Jencks & Mayer, 1990; Shonkoff & Phillips, 2000). From the perspective of the physical environment, Evans (2006) identified behavioral toxicology, noise, crowding, housing and neighborhood quality, natural settings, schools, and day-care settings. By focusing on the question of what constitutes a supportive community, other investigators have produced slightly different lists: quality of social services, socialization by adults, peer influences, social networks, exposure to crime and violence, physical distance and isolation (Ellen & Turner, 1997) or safety and cohesion, increased participation in community activities, and high levels of collective efficacy (Connor & Brink, 1999).

Although each of these approaches produces different lists and, therefore, different measurement issues, each must influence development over the long term in one or more of three possible life course effects: latency, cumulative, or pathway. *Latency* refers to relationships between an exposure at one point in the life course and the probability of expressions years or decades later, irrespective of intervening experience. *Cumulative* refers to multiple exposures over the life course whose effects on health combine. These may be either multiple exposures to a single recurrent factor (e.g., chronic poverty or persistent smoking) or a series of exposures to different factors. Finally, *pathways* represent dependent sequences of exposures in which exposure at one stage of the life course influences the probability of other exposures later in the life course, as well as associated expressions.

The episodic quality of temporal–spatial instability relates to developmental windows of opportunity, hierarchical unfolding of developmental competencies, candidate systems, and life course effects in a unique way that can best be understood through the experiment of marbles in a bowl (Pierson, 2003). The experiment begins with two marbles in a bowl, one white and one black. Wearing a blindfold, the participant removes one marble from the bowl. Next, the experimenter replaces it with two marbles of the same color. Now, with three marbles in the bowl, the process is repeated. However, at this point, two of three marbles are the color of the first one removed, so the probability of removing a marble of that color goes up with the second trial. With repeated trials, it is most likely that the bowl will become disproportionately filled with marbles of the color of the first one removed. In other words, the initial random selection is the most important one and ends up having a profound and enduring impact on the long-term trajectory. By the end of the experiment three critical influences have emerged: an effect of the initial selection, an effect inherent in the principles of ongoing selection, and an effect based on the relative prevalence of black versus white marbles in the bowl at any given time. Together, these three influences convert a small episode at the outset into a trajectory of change over the course of the experiment, resulting in a unique and increasingly irreversible outcome state. This understanding is very important when juxtaposed against the notions of sensitive periods in brain development and hierarchies of developmental competence. If temporal–spatial instability, because of its episodic nature, occurs during a specific window of developmental opportunity, it may influence specific aspects of competence, analogous to the selection of the first marble, thus predisposing this aspect of competence to follow a particular path of development. In more general terms, the relevant theoretical questions here parallel the three critical influences identified earlier. Does temporal–spatial instability have the capacity to cause an initial perturbation to a child's development analogous to the first marble selection (Gibbons et al., 2007)? Can the aspects of temporal–spatial stability and instability in a young child's life have a large enough influence to represent a principle of selection? Finally, do children subject to temporal–spatial instability live in worlds of systematically different experience and opportunity, over the long term, than do other children?

Future Directions

In this section I describe three future directions for research on child development in relation to temporal–spatial instability: the developmental biology of instability, appropriate data systems for studying unable families, and approaches to studying the role of unstable communities in nurturing children.

Developmental Biology of Instability

It can be seen from the earlier discussion that there is a need to create a developmental biology of temporal–spatial instability wherein studies of how experience gets under the skin focus on children living in unstable circumstances. Here, the objective would be to understand how, and to what extent, frequent changes in the environments of daily living influence the development of the candidate systems mentioned earlier. From what is already understood about developmental biology and the nature of temporal–spatial instability, it is possible to frame a coherent research agenda, including both animal and human studies, much in the way that the primate research of investigators such as Suomi (2006) and Sapolsky (2005) has closely paralleled and informed existing research on population health and human development.

It is not the purpose of this chapter to propose a comprehensive research agenda but rather to outline its direction. Thus, the basic frame of reference for both animal and human studies would be a two-by-two contingency based on stability or instability of intimate caregiver(s) or the physical–social context. It is anticipated that differences would quickly emerge between the stable–stable and the unstable–unstable groups, in one of two ways. Either the unstable–unstable condition will produce, early in life, a higher proportion of those with a hyperactive HPA axis and diminished capacity for focused attention in the prefrontal cortex, or the two groups will have the same proportions of these "phenotypes," but in the unstable–unstable condition their developmental trajectories will tend to be less successful than in the stable–stable condition. The next stage of exploration would be the stable–unstable and the unstable–stable conditions. Here, it is anticipated that stable intimate caregivers will compensate for a certain amount of instability in the physical–social context. But the question is, how much? Animal studies clearly need to be done first, because it will be necessary to vary the frequency, duration, and intensity of context changes, as well as the level of contact with the primary caregiver, to establish plausible dose–effect relationships and tipping points and identify plausible analogues to the human condition.

Data Systems for Studying Temporal–Spatial Instability

The primary barriers to successfully studying the effects of temporal–spatial instability on development and health were stated earlier. In brief, they are the difficulty of finding highly transient families and reconstructing residential histories of families with children without direct contact; the prospect that temporal–spatial instability may be closely associated with, and subsidiary to, other well-studied factors such as low income; and the fact that its effects may be ambiguous, such as the difference between residential moves for the purposes of economic opportunity versus those for reasons of social isolation, family breakdown, weak community ties, or economic deprivation Thus, there is a need to recruit large enough samples such that transiency from the full range of causes can be studied on its own terms. I argue that each of the barriers identified here supports future directions in research that build best on population-based, person-specific, longitudinal data systems.

POPULATION-BASED DATA. Doing valid research on temporal–spatial instability and child development is inseparable from creating data systems that identify highly transient families and children and are able to follow them despite frequent moves. Current problems include the prospect that highly transient families will be less likely to volunteer for study and to be found on follow-up. The example of the Winnipeg study presented earlier gives the basic counterfactual. Without a quasi-census of the population in question it would have been impossible to identify the children who were not attached to the school system. Also, when prevalence samples are selected in a given community, the prevalence–incidence bias will dictate that highly transient children will be underrepresented, even in the initial sample, unless complex sampling frames are used to counteract the influence of residence time on the probability of sampling. Consider the following simple scenario: A local community has 50% owner-occupied family housing where all the children are stable occupants over a 2-year interval. The other 50% of family housing is rental, and the average length of stay of the families with children is 6 months. A sample of children is drawn for study at one point in time, yielding a ratio of 50% stable and 50% unstable children. Yet, over a 2-year period, there are three complete turnovers of unstable children, making the actual period prevalence ratio 4:1, rather than 1:1 as identified in the sampling frame. The only efficient means to overcome these problems is with fully population-based data systems.

LONGITUDINAL, PERSON-SPECIFIC DATA. Child longitudinal studies in general, and birth cohort studies in particular, are the primary source of valid information on developmental trajectories and the only basis for studying the determinants of "successful" and "unsuccessful" developmental trajectories. However, just as these studies need to be complemented by studies of developmental biology (to understand the mechanisms of human development) and intervention (to understand how to improve developmental trajectories), so too is there a need to complement them with population-based data flows for the reasons presented earlier. Over the long term, finding ways to merge population-based and life course methodologies should provide the most valid and reliable data system for studying the influence of temporal–spatial instability on child development.

For a successful example of this approach, with implications for child development and mental health, consider the child follow-up of the British Columbia sawmill workers cohort. We originally created a cohort of 28,794 British Columbia male sawmill workers to study the carcinogenic effects of exposures to chlorophenol antisapstains. Personnel records were accessed from 14 sawmills for workers who had worked, for at least 1 year, in one of these mills anytime between 1950 and 1998 (Ostry et al., 2006). The number of episodes of unemployment, job mobility (classified as upward, downward, or stable), and occupation (manager, skilled trades, machine operator, or unskilled) were obtained. Historical estimates of job control, demands, noise, and social support were obtained from experienced job evaluators (two union and two management) who worked across the industry and also from panels of senior workers at each mill. The cohort was then linked to the file of all births in British Columbia to identify all of the children of these workers born in British Columbia between 1952 and 2000. There were 37,827 children in the cohort, of whom 19,833 satisfied the eligibility criterion that their father had at least 1 year of work in a study sawmill while the child or children were aged 0 to 16. With probabilistic linkage techniques the children's cohort was then linked to the British Columbia Linked Health Database (BCLHDB), consisting of person-specific, longitudinal records for all British Columbians. Of the members of the children's cohort, 88% were found in the BCLHDB, allowing access to data on deaths, hospital discharges, and other medical encounters for the years 1985 through 2001. Using ICD9 codes available in the hospital discharge database and the death file, it was possible to identify all suicide cases (completed, or attempted in a serious enough way to generate an admission to hospital) that occurred between January 1985 and March 31, 2001. Children's exposure to adverse family circumstances was assessed by applying father's exposure (in terms of job mobility, unemployment experience, and exposure to control, psychological demand, physical demand, social support, and noise) during each year of the child's life from age zero to the end of the 16th year. Also available were father's marital status and ethnicity. Finally, from the BCLHDB it was determined whether a father had a completed or attempted suicide, a mental health diagnosis, or an alcohol-related diagnosis prior to the child's attempted or completed suicide.

Of the 19,833 children in the cohort, 252 attempted or committed suicide between 1985 and 2001. After controlling for father's sociodemographic characteristics, occupation at the time prior to the child's suicide (attempted or completed), mental health diagnosis, and alcohol-related mental health diagnosis prior to the child's suicide attempts or completion, we found the following: (a) male children of fathers with low duration of employment at a study sawmill while their child or children were less than age 16 had greater odds of attempting suicide than did children of fathers with high duration of employment, (b) female children of fathers who experienced low job control at their sawmill jobs during the first 16 years of their child's life had significantly greater odds of attempting suicide, and (c) male children of fathers employed in sawmill jobs with low psychological demand showed significantly greater odds of completing suicide (Ostry et al., 2006).

Unfortunately, we did not attempt to collect residential information on this cohort. Nevertheless, when loss to follow-up due to transiency through linkage to population-based records was minimized, the factor most associated with family instability (duration of employment) did not disappear but came to the top of the list of risk factors for suicide in the offspring. At the same time, it should be noted that when we attempted a live follow-up of 2,000 sawmill workers, those with short work histories were, by far, the most difficult to find and were underrepresented in the follow-up. These results are important because they make the case that temporal–spatial instability could well be a much more important determinant of child development that we currently believe, because most current study methodologies bias against sampling unstable families at baseline and finding them during follow-up.

As of this writing, we are attempting to create a database of developmental trajectories for all British Columbian children. The EDI records for British Columbia are to be linked forward in time to individual school records and backward in time to birth, medical, hospital, and pharmaceutical records to create skeleton developmental trajectories for the whole population. Individual trajectories are then being reaggregated by neighborhood or school to properly nest individuals within contexts of interest. As of this writing, three significant advances have already been made. First, we have matched a unique British Columbia Personal Education Number to 94% of the EDI assessments carried out between 2000 and 2005, making it possible to relate EDI to school progress on an individual basis. Second, we have matched 97% of the Personal Education Numbers to a Personal Health Number, allowing, in principle, a merge between school, birth, medical, and EDI records. Finally, we have established the physical infrastructure and ethical framework that will allow us to link records on a person-specific, population-based but anonymous basis. Together, these advances mean that we will be able to create a developmental trajectory for each British Columbian child for whom an EDI has been completed, starting with birth experience (e.g., birth weight, small for gestational age), proceeding to early health status (inferred through medical and hospital services utilization), to state of development at school entry (using the EDI), to progress through school. From these data we will be able to identify basic success and failure trajectories for British Columbian children.

A first analysis has already come from this data system, underlying the importance of school and residential transiency for children. The Foundation Skills Assessment (FSA) is a standardized test designed to measure the reading comprehension and numeracy skills of fourth- and seventh-grade students in British Columbia. It is administered in all public and independent schools across the province each year. Through the data linkage system described earlier, we have been able to merge EDI and FSA scores on an individual basis for approximately 7,000 students in four geographic school districts. From the student-level data file we were able to trace residential and school changes from kindergarten to Grade 4. In our analyses, higher rates of vulnerability on all scales of the EDI (physical, socioemotional, and language, cognitive, communication), higher rates of "failure to meet expectations" on the FSA (i.e., showing up, writing, and failing the FSAs) and, also, higher rates of "not succeeding" on the FSA (either failing to meet expectations or not showing up to write the test for any reason) were all found to be associated with multiple school and residential moves in the kindergarten to Grade 4 period. For example, 34% of children who changed school between kindergarten and Grade 4 were vulnerable on one or more scales of the EDI in kindergarten, compared with 28% of those who did not change schools. Nineteen percent of children who changed schools failed the numeracy FSA and 27% failed the reading, compared with 15% and 22%, respectively, for those who did not change schools. In general, the relationship between EDI and FSA was similar for school movers and those who did not change schools. That is, the increased risk of failing either the numeracy or reading test was similar when vulnerable and nonvulnerable school movers were compared with one another, as it was for vulnerable and nonvulnerable nonmovers. On the other hand, the pattern for residential moves was quite different from school moves. Thirty-four percent of those who did not change residential address between kindergarten and Grade 4 were vulnerable on the EDI at kindergarten, whereas, of those who changed residences two or more times, 37% were vulnerable. In other words, the difference was small at baseline. By Grade 4, however, considerable differences had emerged. Twenty-five percent of children who moved had failed the numeracy FSA, compared with 18% of the nonmovers. For reading, the difference was 30% versus 25%. The risk of failure was increased among both the EDI vulnerable and nonvulnerable children. Eighteen percent of nonvulnerable movers failed the numeracy test, compared with 12% of the nonvulnerable nonmovers, whereas 37% of the vulnerable movers and 29% of the vulnerable nonmovers failed. For reading, the pattern was similar.

Thus, one or more school moves and two or more residential moves increased the risk of numeracy and reading failure by Grade 4, but through different mechanisms. School changes operated at baseline, on the proportion of children who came to school vulnerable. This would appear to be a selection effect. On the other hand, residential moves operated prospectively, between kindergarten and Grade 4, increasing the risk that both EDI vulnerable and nonvulnerable children would fail.

Role of Unstable Communities in Nurturing Children

Earlier in this chapter, I presented the observation that the percentage of nonimmigrant movers was an independent predictor of increased vulnerability on one or more scales of the EDI, but not on any of the individual EDI scales. In other words, it was a weak predictor of community vulnerability despite the fact that virtually all of the high-transiency neighborhoods in the province register as highly vulnerable on the EDI. This situation highlights a problem: that community instability is closely correlated with low community SES. The key future research direction is how to disentangle these factors. As described earlier, we have already characterized all British Columbian neighborhoods according to SES on the 2001 census, and we recently supplemented this with data from the 1998 and 2004 tax-filer database, to examine socioeconomic dynamics of neighborhoods. Doing this will in turn allow us to compare SES and EDI outcomes by neighborhood to identify where children are performing much better or worse than predicted, given neighborhood SES characteristics. Relative to the line of best fit, such neighborhoods are deemed *off-diagonal* because actual rates of vulnerability deviate widely from predictions based on local SES.

Our plan is to study pairs of adjacent (off- and on-diagonal) neighborhoods because these have the potential to provide insight into community-level characteristics and processes that either buffer or exacerbate the influence of SES in the early years. The first task is to determine whether routinely collected data on residential mobility help to explain (in a statistical sense) some of the off-diagonal neighborhoods. That is, within a given socioeconomic strata of neighborhood, are children in the highly mobile neighborhoods more likely to be vulnerable on the EDI than those in low-mobility neighborhoods? Next, by conducting projects in the same geographic location, and whenever possible with the same children, we will strengthen our capacity to describe the different familial and social settings in which individual children play, participate, and learn and illuminate how children's psychobiological characteristics mediate or moderate their personal experiences of these settings.

By bringing together these community studies with the population-based, person-specific, longitudinal data resources described earlier, we hope to be able to better account for the influence of temporal–spatial instability on child development than has existed until now.

Summary and Conclusions

Temporal and spatial instability, in both the family and the community, influences child development. Yet, to date, these issues have been underrepresented in the child development research literature and largely ignored in policy discussions. This, despite the fact that those working with young children and their families at the local level often speak of family and community instability as key challenges in their work. Innovations in research design described in this chapter, as well as the new insights into the role of social factors on the developmental biology of childhood, should help raise consciousness of the role temporal and spatial instability plays in the early years. Whether or not these insights will lead to helpful social change is, at this point in time, a matter for speculation.

References

Belsky, J., Melhuish, E., Barnes, J., Leyland, A. H., Romaniuk, H., & the National Evaluation of Sure Start Research Team. (2006, June 16). Effects of Sure Start local programmes on children and families: Early findings from a quasi-experimental, cross-sectional study. *British Medical Journal, 332*, 1476–1481.

Boyle, M. H., Georgiades, K., Mustard, C., & Racine, Y. (2007). Neighborhood and family influences on educational attainment: Results from the Ontario Child Health Study follow-up 2001. *Child Development, 78*, 168–189.

Brooks-Gunn, J., Duncan, G. J., Klebanov, P. K., & Sealand, N. (1993). Do neighbourhoods influence child and adolescent behavior? *American Journal of Sociology, 99*, 353–395.

Brownell, M. D., Roos, N. P., Fransoo, R., Roos, L. L., Guèvremont, A., MacWilliam, L., et al. (2006). Is the class half empty? A population-based perspective on socio-economic status and educational outcomes. *IRPP Choices, 12*(5), 1–30.

Burton, L. M., & Jarrett, R. L. (2000). In the mix, yet on the margins: The place of families in urban neighborhood and child development research. *Journal of Marriage and the Family, 62*, 1114–1135.

Cavanagh, S. E., & Huston. A. (2006, August). *Family instability and children's friendships and social competencies in elementary school.* Abstract presented at the American Sociological Association Annual Meeting, Montreal, Quebec, Canada.

Chase-Lansdale, L. P., & Gordon, R. A. (1996). Economic hardship and the development of five- and six-year-olds: Neighbourhood and regional perspectives. *Child Development, 67,* 3338–3367.

Connor, S., & Brink, S. (1999). *Understanding the early years—Community impacts on child development* (Working Paper No. W-099-6E). Ottawa, ON: Applied Research Branch, Strategic Policy, Human Resources Development Canada.

D'Anguilli, A., Herdman, A., Stappels, D., Weinberg, J., Grunau, R., & Hertzman, C. (2008). Children's event-related potentials of auditory selective attention vary with their socioeconomic status. *Neuropsychology, 22,* 293–300.

Davis, E. P., Donzella, B., Krueger, W. K., & Gunnar, M. R. (1999). The start of a new school year: Individual differences in salivary cortisol response in relation to child temperament. *Developmental Psychobiology, 35,* 188–196.

Duncan, G. J., Brooks-Gunn, J. P., & Klebanov, P. K. (1994). Economic deprivation and early childhood development. *Child Development, 65,* 296–318.

Ellen I., & Turner, M. (1997). Does neighbourhood matter? Assessing recent evidence. *Housing Policy Debate, 8,* 833–866.

Evans, G. W. (2006). Child development and the physical environment. *Annual Review of Psychology, 57,* 423–451.

Fomby, P., & Cherlin, A. (2007). Family instability and child well-being. *American Sociological Review, 72,* 181–204.

Forman, E. M., & Davies, P. T. (2003). Family instability and young adolescent maladjustment: The mediating effects of parenting quality and adolescent appraisals of family security. *Psychology, 32,* 94–105.

Gibbons, F. X., Yeh, H. C., Gerrard, M., Cleveland, M. J., Cutrona, C., Simons, R. L., et al. (2007). Early experience with racial discrimination and conduct disorder as predictors of subsequent drug use: A critical period hypothesis. *Drug and Alcohol Dependence, 88*(Suppl.), S27–S37.

Guralnick, J. M., Butterworth, S., Wadsworth, M. E. J., & Kuh, D. (2006). Childhood socioeconomic status predicts physical functioning a half century later. *The Journals of Gerontology: Series A. Biological Sciences and Medical Sciences, 61,* 694–701.

Hertzman, C., MacLean, S., Kohen, D., Dunn, J. R., & Evans, T. (2002). *Early development in Vancouver: Report of the Community Asset Mapping Project (CAMP).* Vancouver, BC: Human Early Learning Partnership.

Huston, A. C., Duncan, G. J., McLoyd, V. C., Crosby, D. A., Ripke, M. N., Weisner, T. S., & Eldred, C. A. (2005). Impacts on children of a policy to promote employment and reduce poverty for low-income parents: New Hope after 5 years. *Developmental Psychology, 41,* 902–918.

Irwin, L. G., Johnson, J., Dahinten, S., Henderson, A., & Hertzman, C. (2007). Examining how contexts shape children's perspectives of health. *Child: Care, Health & Development, 33,* 353–359.

Irwin, L. G., Siddiqi, A., & Hertzman, C. (2007). *Early child development: A powerful equalizer.* Final report for the World Health Organization Commission on the Social Determinants of Health. Vancouver, British Columbia, Canada: Human Early Learning Partnership.

Jencks, C., & Mayer, S. (1990). The social consequences of growing up in a poor neighbourhood. In L. E. Lynn & M. G. McGeary (Eds.), *Inner city poverty in the United States* (pp. 111–186). Washington, DC: National Academy Press.

Kasting, C., & Artz, S. (2005). Homeless outreach projects for single-parent families. What happens to the children? *Irish Journal of Applied Social Studies, 6,* 27–50.

Keating, D. P., & Hertzman, C. (Eds.). (1999). *Developmental health and the wealth of nations.* New York: Guilford Press.

Kershaw, P., Irwin, L. G., Trafford, K., & Hertzman, C. (2005). *The British Columbia atlas of child development.* Victoria, BC: Western Geographical Press.

Kohen, D. E., Brooks-Gunn, J., Leventhal, T., & Hertzman, C. (2002). Neighbourhood income and physical and social disorder in Canada: Associations with young children's competencies. *Child Development, 73,* 1844–1860.

Kohen, D. E., Hertzman, C., & Wiens, M. (1998). *Environmental changes and children's competencies* (Working Paper W-98-25E). Ottawa, ON: Human Resources Development Canada.

Ku, L., Sonenstein, F. L., & Pleck, J. H. (1993). Neighborhood, family and work: Influences on the premarital behaviors of sexual males. *Social Forces, 72*, 479–503.

Lowe, E., Weisner, T. S., Geis, S., & Huston, A. C. (2005). Child-care instability and the effort to sustain a working daily routine: Evidence from the New Hope ethnographic study of low-income families. In C. R. Cooper, C. T. Garcia Cool, W. T. Bartko, H. Davis, & C. Chatman (Eds.), *Developmental pathways through middle childhood: Rethinking contexts and diversity as resources* (pp. 121–144). Mahwah, NJ: Erlbaum.

Messer, L. C., Laraia, B. A., Kaufman, J. S., Eyster, J., Holzman, C., & Culhane, J., et al. (2006). The development of a standardized neighborhood deprivation index. *Journal of Urban Health, 83*, 1041–1062.

Miller, G., & Chen, E. (2007). Unfavorable socioeconomic conditions in early life presage expression of proinflammatory phenotype in adolescence. *Psychosomatic Medicine, 69*, 402–409.

Ostry, A., Tansey, J., Maggi, S., Dunn, J., Hershler, R., Chen, L., & Hertzman, C. (2006). The impact of fathers' physical and psychosocial work conditions on attempted and completed suicide among their children. *BMC Public Health, 6*, 77–86.

Pierson, P. (2003). Big, slow-moving, and invisible: Macrosocial processes in the study of comparative politics. In J. Mahoney & D. Reuschemeyer (Eds.), *Comparative historical analysis in the social sciences* (pp. 177–207). Cambridge, MA: Cambridge University Press.

Reyes, J. W. (2007). *Environmental policy as social policy? The impact of childhood lead exposure on crime* (Working Paper No. 13097). Washington, DC: National Bureau of Economic Research.

Sapolsky, R. M. (2005). The influence of social hierarchy on primate health. *Science, 308*, 648–652.

Schulting, A. B., Malone, P. S., & Dodge, K. A. (2005). The effect of school-based kindergarten transition policies and practices on child academic outcomes. *Developmental Psychology, 41*, 860–871.

Shonkoff, J., & Phillips, D. (2000). *From neurons to neighborhoods: The science of early childhood development*. National Research Council, Institute of Medicine. Washington, DC: National Academy Press.

Suomi, S. J. (2006). Risk, resilience, and gene x environment interactions in rhesus monkeys. *Annals of the New York Academy of Sciences, 1094*, 52–62.

Thomas, J. C. (1999). *How juvenile violence begins: The root causes of juvenile violence*. Colorado Springs, CO: Focus on the Family.

Waddell, C., Peters, R. V., Hua, J. M., McEwan, K., & Garland, O. M. (2007). Preventing mental disorders in children: A systematic review to inform policy-making. *Canadian Journal of Public Health, 98*, 166–173.

Waddell, C., Hua, J. M., McEwan, K., Garland, O. M., & Peters, R. V. (2007). Preventing mental disorders in children: A public health priority. *Canadian Journal of Public Health, 98*, 174–178.

Xue, Y., Leventhal, T., Brooks-Gunn, J., & Earls, F. J. (2005). Neighbourhood residence and mental health problems of 5- to 11-year-olds. *Archives of General Psychiatry, 62*, 554–563.

Part III

Chaos at the Mesosystem Level

9

From Home to Day Care: Chaos in the Family/Child-Care Mesosystem

Robert H. Bradley

Bronfenbrenner's ecological model (Bronfenbrenner & Morris, 1998) offers a rich and powerful means of understanding the complex set of forces implicated in human development. Even so, it remains difficult to fully characterize how mesosystems function to impact individual lives. Part of the difficulty derives from the fact that most of the key contexts in which we live are dynamic systems, changing in real and historic time. Consequently, in this chapter I am circumspect in my remarks about how chaos at home or in child care might affect what happens in the other setting (i.e., mesosystem relations) and how those relations might affect child well-being. Specifically, I try to place child care in historic context and describe the role it tends to play at the outset of the 21st century, with specific attention to the factors that inform parental choices about child care. I then discuss communication processes between parents and caregivers in child-care settings and how the partnership between parents and caregivers may affect the quality of care children receive. I then discuss parental involvement in child care and the continuity of children's experience at home and child-care settings. That discussion is followed by a consideration of how time spent in child care affects child well-being and how chaos in home or child care may affect the quality of a child's experience in the other setting. In the final section, I focus on the future of home/child-care relations and what is needed by way of new research to more fully explicate these relations.

Child Care and Family Life at the Start of the 21st Century

Child care (also known as *nonparental care*) has become normative for American children (Overturf, 2005). Most children spend a significant amount of time being cared for by adults other than their parents. Many families rely on a patchwork of child-care arrangements to assist them with child rearing (Bradley & Vandell, 2007); thus, child care is neither a consistent nor a stable experience for many children (National Institute of Child Health and Human Development Early Child Care Research Network [NICHD ECCRN], 2005a). Instability of child-care arrangements, both in and through time, may represent a type of chaos at the mesosystem level. It is certainly perceived that way by many parents and scholars (Shpancer, 2002). What is not clear is how much such shifting of caregivers matters in terms of what happens at home. The number of different child-care arrangements children experienced during the first 3 years of life was not associated with attachment security in the NICHD Study of Early Child Care (NICHD ECCRN, 1997a). Neither was child care instability associated with caregiver-reported or -observed measures of self-regulatory problems in children; however, it was associated with increased behavior problems reported by mothers (NICHD ECCRN, 1998a). That said, this latter finding has to be taken with a grain of salt because the mothers themselves were likely more stressed given the difficulties in finding stable care.

Child care can function as a support to families in two major ways that might affect the impact of home chaos on the well-being of children. First, access to stable, affordable care can enable families to maintain stable employment and thus provide for the basic needs of children. That may reduce the likelihood children would have to live in crowded home environments, and it may increase the likelihood that potentially protective routines at home could be maintained; however, no data are specifically available regarding either possibility. Second, child care can be a source of information about child rearing, such as might occur during drop-off and pickup or through formal information provided by day-care providers (Phillips, McCartney, & Sussman, 2006). Parents might be able to obtain useful information from care providers regarding how to deal with the aftermath of chaos at home. Programs such as Early Head Start emphasize this latter role in their service to low-income families. There is some evidence that parents do view child care as a source of informational and emotional support (Britner & Phillips, 1995), but at any given time a substantial minority indicate they would like to find another caregiving arrangement (Hofferth, Brayfield, Deich, & Holcomb, 1991).

Parent and family characteristics play a role in the choices families make about child care (Holloway & Reichart-Erickson, 1989; NICHD ECCRN, 1997b; Pugello & Kurtz-Costes, 1999). Family wealth, parental employment, and geography, perhaps more than most factors, help determine whether and what kind of child-care arrangements families make. However, changes in household composition and employment circumstances (i.e., family instability) may also play a significant role. As the number of women employed in the workforce has increased, so too has the number of children placed in formal caregiving arrangements (McCartney, 2007). Families with more economic resources tend to select center care more often (Blau, 2001; Hofferth & Wissoker, 1992). Mothers who earn greater incomes are more likely to place their children in care earlier and have their children in care for longer hours (Lebowitz, Klerman, & Waite, 1992). That said, there is an uneven relation between family income and child-care quality, partly owing to the fact that some low-income families have access to good-quality subsidized child care (Phillips, Voran, Kisker, Howes, & Whitebook, 1994; Waite, Leibowitz, & Witsberger, 1991) and partly because there is limited high-quality care in some communities (Cost, Quality, and Outcomes Team, 1995; NICHD ECCRN, 2000a). A variety of parent and family characteristics help determine the kind of relationships parents will have with caregivers (Berghout et al., 1996), including stress reactions that may result from chaos at home. As discussed in chapter 12 of this volume, adults who experience traumatic events or chronic adversity tend to feel less agentic, tend to be less effective in social situations, tend to make poorer decisions, and tend to do less well in managing the moment-to-moment circumstances of their lives. Their encounters with child-care providers are thus likely to be less fruitful in general than are those of parents not faced with high levels of home chaos. On the other hand, they may be more likely to reach out for support—sometimes with a favorable outcome for both themselves and their children.

Decision making about child care varies as a function of family demographics (e.g., marital status, education) and parental awareness (Fulmer, 1997; Leslie, Ettenson, & Cumsille, 2000). Han (2004) found that family choice of type of care to use was strongly determined by the mother's work situation. Families frequently changed child-care arrangements when the mother changed jobs or work hours; that is, instability in work circumstances is often connected to instability in child-care arrangements and sometimes inconsistent attendance at child care (see chap. 12, this volume). Such inconsistent attendance at child care could reduce the potentially positive (buffering) benefits of the supportive routines a child might experience in a well-organized and well-managed child-care setting. As it happens, mothers with nonstandard work hours often turn to fathers or other family members for care. However, if they have standard work hours, they more often select child-care centers.

Relatively little is known about the calculus families use to determine the particular nonparental care their children receive. Studies show that parents look for nurturant caregivers and for caregivers who are like themselves (Clarke-Stewart, Gruber, & Fitzgerald, 1994; Hofferth & Phillips, 1991). Whether chaos at home reduces the likelihood parents will select highly nurturant care is unknown, though such a finding would be consistent with what is generally known about the impacts of chronic adversity on human decision making (Conger & Donnellan, 2007). The same pattern pertains to parenting self-efficacy: The more self-efficacious one feels as a parent, the more one relies on formal information sources, the more one reads to a child, and the less one is involved in nonparental caregiving (Coleman & Karraker, 1997). To the extent that home chaos degrades feelings of self-efficacy, it becomes less likely parents in such circumstances will consistently involve themselves effectively with the child's circumstances in child care (Corapci & Wachs, 2002).

For years experts have argued that parents are not particularly skilled consumers of child care (Gormley, 1995). Parents in both the United States and Germany indicate that they value the same qualities in child care as do professionals, but they tend to rate caregivers and caregiving situations more positively on these dimensions than do trained observers (Cryer, Tietze, & Wessels, 2002; van IJzendoorn, Tavecchio, Stams, Verhoerven, & Reling, 1998). Very often parents wind up relying on informal networks in making choices about where to place their children, or they make assumptions that a brand-name enterprise or a center affiliated with a social organization such as a church is likely to provide good care (Gable & Cole, 2000). Chaos at home could in some cases fracture connections with social networks, but it could also increase the use of supports from them. In either event, it could change how parents handle their children's experience regarding child care.

How Communication Between Parents and Caregivers Affects Children and Family Life

Parents and day-care providers speak relatively little to each other on an ongoing basis (Hand & Wise, 2006; Horowitz, 1984; Leavitt, 1995; Powell, 1978; Winkelstein, 1987), citing limited time as a major reason for the lack of extended dialogue (Hayden, De Gioia & Hadley, 2003). Parents and caregivers often view the frequency and quality of communication differently (Knopf & Swick, 2007; Powell, 1978). Parents tend to focus more narrowly on issues related to the care of their child, and they are sensitive to whether caregivers listen to them and respect their concerns (Swick, 2004). The frequency and quality of communications between parents and child-care providers could well be a barometer of the extent to which chaos in either setting is affecting the adults in those settings. Stress can lead to hyper- or hypovigilance, but it often means inconsistency of effort directed to particular activities and a degradation in quality of attention to what may be viewed as nonessential activities. Careful listening and precise articulation of the information one wants to convey (both hallmarks of good communication) are likely victims of stress connected to high levels of chaos.

Even when parents and caregivers communicate frequently, it does not mean they agree on all matters pertaining to child care. Research demonstrates that differences arise in many areas, including discipline practices, the value of certain kinds of play, sleep routines, and self-care issues such as toileting (Feagans & Manlove, 1994; Hughes & MacNaughton, 2002; Nelson & Garduque, 1991). Van IJzendoorn et al. (1998) found only modest attunement between parents and caregivers in the area of childrearing attitudes and beliefs. In effect, there was discontinuity between the two microsystems. Unknown at this point is whether chaos in either system may increase discontinuity between them. However, there is reason to suspect that instabilities at home or in child care could undermine relationships between child-care providers and parents, leading to negative consequences for children. For example, some evidence indicates that when caregivers were more authoritarian and less supportive of mothers, children were less at ease in the child-care setting (van IJzendoorn et al., 1998).

Agreements between parents and caregivers are often brokered by children. If a child seems happy and functions well within the child-care environment, the relationship between caregiver and parent is likely to be a supportive one for the family. If not, the parents are more likely to change care arrangements, with stresses coming to the child in the aftermath. In effect, if chaos at home produces maladaptive behavior in the child while at day care, relationships between parent and child-care provider could be degraded. In some cases, suspecting that their child may be responding to the stresses of home chaos, parents may press caregivers to treat their child differently so as to ameliorate the effects of those stresses, perhaps irritating care providers who already feel overwhelmed with child-care responsibilities. Likewise, if chaos at child care produces maladaptive behavior in the child while at home, the parent–caregiver relationship could be degraded if parents try to press for changes in the child-care setting; of course, a positive response is also possible. At present, far too little is known about the nature or effects of parent–caregiver partnerships and what role children play in them.

The Partnership Between Home and Child Care

Although parents and child-care providers do not always form optimal partnerships, child-care settings have long been viewed as places that can support family life. Some child-care providers offer materials for families to use in helping their children, and guidelines are available for setting up such resources as toy and book lending libraries (Rettig, 1998). Such efforts might be especially helpful for families beset by frequent moves (i.e., residential instability) or poverty that leads to high levels of crowding. That said, the stresses connected to instability and crowding may make it difficult for parents to fully use any assistance offered from child-care providers (Bronfenbrenner, 1974). To date, there is not much evidence that parents use caregivers as resources for child-rearing advice (Kontos & Dunn, 1989) or as models of how to care for children (Clarke-Stewart, 1991). Some agencies provide informational handouts, group meetings, and workshops. There is at least some evidence that these efforts have helped reduce health risks, though there is less evidence that it has actually changed parenting practice (Administration for Children and Families, 2002; Benasich, Brooks-Gunn, & Clewell, 1992; Bowman, 1997).

Many providers of educational child care have well-developed curricula for improving literacy and teaching concepts considered important in school. In homes where few routines are established for the management of daily life (one aspect of chaos), the likelihood of reading regularly to children or engaging in other learning activities regularly with children is diminished. Thus, there is little likelihood of positive feedback to child-care providers, either via child behavior while at child care or from parental report, on this potentially positive mesosystem function. Relatively few studies have examined parental reactions to the implementation of new approaches to learning, behavior management, or daily routines (McClow & Gillespie, 1998). We recently initiated a new intervention model aimed at educating and supporting parents as part of educational child care services. Called Brief Parenting Interventions, it is based on research showing that short interventions often work as well or better than interventions of longer duration (Bakermans-Kranenburg, van IJzendoorn, & Bradley, 2005). It is also based on the premise that parents will engage more fully in interventions that address self-identified needs and that do not require lengthy commitments, given the stresses and limited resources they have—an intervention philosophy that would seem especially valuable in working with families living in chaotic conditions. Finally, it derives from the conviction that child-care providers will feel more comfortable and effective working with parents on issues the parents have chosen rather than working on preselected topics. We are in our 2nd year of program development and implementation and, thus far, we can offer only anecdotes regarding program success.

One of the difficulties in making generalizations about the family/child-care mesosystem is that relations can be very different depending on the circumstances of the child or the family. Parent–caregiver relationships can be quite different for households with members who have disabilities or serious medical problems, for families that are homeless, or in other conditions where chaos connected with family life tends to be quite high (Chan & Sigafoos, 2001; Swick & Bailey, 2004). Such situations can lead to inconsistent attendance at child care (in effect, creating a kind of chaos as regards child care itself) with the potential to disrupt whatever positive benefits a child might accrue from regularly attending high-quality child care; obviously, such potential losses are diminished if the child-care environment itself is not of high quality. In cases in which family members require greater support or where family life is disrupted, caregivers are often called on to give additional attention to a child or family situation. In these cases, the provision of high-quality care could substantially reduce stress and improve the parent's own capacities to provide high-quality care for a child. Again, however, during times of stress it may be difficult for families to fully uptake the support offered. Moreover, there are times when home and child care can be at loggerheads, such as when either suspects the other of child abuse or neglect. Research on these situations is very limited.

Parental Involvement in Child Care

It is assumed that parental involvement in child care should increase the quality of care children receive and thereby foster children's development. There is evidence that increased parental involvement is linked to higher quality care (Endsley, Minish, & Zhou, 1993; Ghazvini & Readdick, 1994), although issues remain regarding the exact nature of this relation (Shpancer, 2002). There is also evidence that parental involvement in preschool classrooms (educational day care) is beneficial for parents and children, including those attending Head Start (Lamb-Parker et al., 2001; Marcon, 1999; Miedel & Reynolds, 1999). However, much of the research on the topic is weak in that it depends on retrospective reporting by either parents or teachers. Castro, Bryant, Peisner-Feinberg, and Skinner (2004) conducted a more detailed study of 1,491 parents and 59 teachers from four Head Start programs that involved the keeping of daily logs by classroom teachers. A fairly high percentage of parents (76%) volunteered at least once in the classroom, but the vast majority of those (59%) volunteered only once or twice. About one third of parents attended parent meetings and about one fourth attended classroom meetings. Employed mothers were less involved as were mothers with more barriers to their involvement. Parents who reported doing more things with their children at home were recorded as being more involved in Head Start. Classrooms rated as having the highest quality had the highest number of volunteers, though there were no differences in the total hours of volunteer time. Despite very little research on relations between household chaos and parental involvement in child care per se, the broader literature on the topic of family adversity strongly suggests lower levels of involvement (Evans, 2004).

There has long been interest in facilitating parental involvement in child care and early education. Most of the efforts have been directed to increasing participation of mothers (a maternal template). Head Start and other governmental agencies have more recently made a push to involve fathers as well.

Relatively little is known about the success of these efforts, for either mothers or fathers (Stile & Ortiz, 1999). The focus on involving fathers is part of the changing landscape of family life and consequent adjustments in social institutions. Because of the increasing importance of maternal employment to family life, societal pressures are moving men to play a greater role in child care (McBride & Rane, 1997). Thus, child-care providers are having to adjust to interacting with fathers more. Child-care staff are to some degree ambivalent regarding this new situation and uncertain how to meet the challenge. Dads are not moms, even if they too provide attentive care to their children and view child care as a supportive service to them and their children. In light of research suggesting that paternal parenting is more affected by context than is maternal parenting, chaos at home could affect the type and quality of paternal involvement in child care (Lewis & Lamb, 2003). Changes in patterns of employment for mothers or fathers could change who drops children off or picks children up from child care, with associated impacts on parent and child-care provider relationships, opportunities to share information, and what child-care providers make of the child's situation.

Continuity From Home to Child Care

Organizations such as the National Association for the Education of Young Children typically tout the value of continuity between home and child care (Powell, 1990) on the assumption that children are likely to make better adjustments to child care and feel less stress generally if the circumstances at home and child care are similar. However, the issue of continuity is a complex one (Shpancer, 2002). What does continuity mean with respect to nonparental care? It could mean something as literal as caregivers behaving toward a child in essentially the same way as parents behave at home (i.e., persistence of particular behaviors), but more typically it means maintaining the essence of treatment although the precise behavior might be different (Radke-Yarrow, 1989). In child-care arrangements that resemble home settings (e.g., nanny care, care by relatives, and child-care homes), the affordances tend to be similar to those present in parental care. However, in child-care centers the affordances tend to be different. In fact, the physical arrangements, the material resources, the orchestration of child activity, the adult-to-child interactions, and the child-to-child interactions in center care tend to be quite different than those observed in family households and family child care (Melhuish, Mooney, Martin, & Lloyd, 1990). Instructional materials and materials used to make things are more commonly found in child-care centers, for example (Clarke-Stewart, 1991).

The fact is, not much research has been done on the degree of continuity between home and child care. We recently compared the experiences children received in their own homes with treatment received in home-like settings by caregivers using the Child Care HOME (Home Observation for Measurement of the Environment) Inventory when children were 6, 15, and 36 months old. We looked at both care received by relatives (most often grandparents) and care received by professional caregivers in family child-care homes. Two general findings emerged. First, the scores were slightly lower for nonparental care—not in all ways but in several (responsiveness, learning stimulation, toys and materials). Second, there were only modest correlations between HOME scores obtained in the child's own home and Child Care HOME scores obtained in nonparental care. Correlations between parental responsiveness and responsiveness by other providers ranged from .05 to .32. Correlations between parental harshness and harshness from alternate caregivers ranged from .01 to .16. The correlations between learning stimulation provided by parents and learning stimulation provided by others ranged from .14 to .32. The only correlations that tended to be of greater magnitude were those between toys and materials available at home and toys and materials available in the alternate care setting (.28–.47). Bornstein and his colleagues (M. Bornstein, personal communication, March 18, 2008) observed similar findings using measures of adult–child interaction based on extensive observation in the two venues: Parents provided somewhat higher quality care, but there was low correlation between parental care and caregiving provided by others.

Bits and pieces of research speak to the notion of carryover from home to child care, with findings offering no consistent picture of what does carry over or what it means for children. For example, evidence indicates that if parents maltreat their children, the children are more likely to be viewed less favorably by child-care personnel and rejected by peers (Haskett & Kistner, 1991; Howes & Eldridge, 1985). Hence, if chaos at home leads to more negative parenting, children's experiences in child care could also be of lower quality. Somewhat by contrast, being from a middle-class home increases the likelihood of secure caregiver–child attachment, but the quality of infant–caregiver attachment was independent of the quality of infant–mother and infant–father attachments (Goosens & van IJzendoorn, 1990). There is interesting anecdotal information on cross-context transfer in such areas as self-care (bowel control) and learning-related activity (reading together), but details on how behaviors and activities in one domain carry over and influence the other are largely lacking. At this point, the jury is still out on the value of continuity for children, largely owing to the small number of actual studies.

It is interesting that a couple of studies suggest that discontinuity between what children experience at home and what they experience at day care can sometimes be of value. Provost, Garron, and LaBarre (1991) found that children's social play was better when the number and types of play objects diverged at home and in child care. Likewise, children had fewer behavior problems when there was greater discrepancy between mother and caregiver control strategies (Erwin, Sanson, Amos, & Bradley, 1993). One could imagine the potential value of discontinuity in experience at home and in child care in situations in which chaos in the form of lack of routines in one is countered by positive routines in the other, clutter and noise in one is countered by tidy arrangements and relative quiet in the other, and lack of supportive and stimulating care in one is countered by supportive and stimulating care in the other. Maxwell (1996) found that children fared worse when they experienced crowding both at home and in day care. There is also some evidence that "stimulus shelters" (i.e., places where children can be removed from dangerous or disquieting circumstances) can be effective in reducing the negative impacts of adverse circumstances (Bradley, 2007; Wachs, 2000).

In general, discontinuity in experiences at home and child care is probably helpful in promoting well-being when the chaos in one system is countered by stability and organization in the other. However, in situations in which factors connected to chaos (e.g., noise, disorganization) are at extremely low levels in one setting, there might actually be some advantage to having somewhat higher (however, no more than moderate) levels of chaos associated with the same factors in the other setting. Take, for example, a situation in which a child is in a highly structured child-care environment where all activities and actions are rigidly prescribed. It may be helpful to have a somewhat looser structure at home so as to allow the child a greater degree of relaxation. Similarly, suppose a child's home environment is essentially noise-free as well as free from high levels of any sensory input, except as freely chosen by members of the household. Perhaps, then, the child would benefit by being in a child-care setting that has at least a moderate level of noise and other sources of sensory distraction. Such a setting may prompt a child to develop skills to cope with these more ordinary circumstances of sensory input. In such cases, moderate levels of chaos may provide the kind of challenge needed to foster coping skills and self-regulatory competence (a kind of "positive chaos"; see chap. 11, this volume). More research is clearly needed to determine what impact such discontinuities have not only on children's well-being but also on what meaning children make of such discontinuities.

Child Care and the Well-Being of Children

Some emerging evidence indicates that being in high-quality child-care (or early education) environments may help offset the negative impact of living in adverse environments in terms of children's language and cognitive development (Caughy, DePietro, & Strobino, 1994; McCartney, Dearing, & Taylor, 2003; Peisner-Feinberg et al., 2001; Schliecker, White, & Jacobs, 1991). However, the findings pertaining to the effects of quality are not fully consistent (Belsky et al., 2007; Burchinal, Peisner-Feinberg, Bryant, & Clifford, 2000; NICHD ECCRN, 2000b, 2001b), likely because of differences in samples examined and statistical procedures used. In one of the more salient examinations of the issue from the standpoint of chaos, data from the NICHD Study of Early Child Care (2002) were used to look at the interaction of family risk and both quality and amount of child care as they pertain to children's behavior problems, prosocial behavior, and language. The findings indicated that dual risk (high family risk and low-quality child care) may undermine the development of prosocial behavior; in effect, chaos in both places may well increase the likelihood of negative impacts on children. A second study showed that children from low-income home environments (presumably more chaotic) benefit from high-quality child care if their homes are at least moderately stimulating; in effect, high-quality child care may be of relatively little benefit to children's cognitive development and achievement when children have almost no stimulation at home (Votruba-Drzal, Coley, & Chase-Lansdale, 2004). On a related matter, being in long hours of care may be troublesome for children who come from families where mothers provide insensitive care (an outcome often connected to homes higher in chaos); that is, there is evidence that long hours in care, coupled with insensitive parenting, can increase the likelihood of attachment insecurity (NICHD ECCRN, 1997a). These effects tended to dissipate as the children entered school (NICHD ECCRN, 2005b). That said, there is little evidence that time spent in child care tends to dilute the impact of parental care on children's behavior and development more generally (NICHD ECCRN, 1998b); however, the literature on this latter issue is quite limited at present. What is far clearer is that time spent in child care, especially large- group care, carries an increased risk of common communicable illnesses and otitis media, though the differences between children reared at home and those with substantial child-care experience tend to be minimal after about age 3 (Ball, Holberg, Aldoiun, Martinez, & Wright, 2002; Lu et al., 2004; Nafstad, Hagen, Oie, Magnus, & Jaakkola, 1999; NICHD ECCRN, 2001a, 2003). To the extent that chaos increases stress and stress compromises immune function, then children are more likely to succumb to exposures to common pathogens (Segerstrom & Miller, 2004; see also chap. 8, this volume). In a study of 9- to 12-year-old children, Johnston-Brooks, Lewis, Evans, and Whalen (1998) found that crowding was associated with increased cardiovascular reactivity and school absence connected to illness. All in all, findings point more to main effects of both home and child care, with a small number of findings suggestive of multiplicative effects when both settings are extreme. Further examination of this latter issue is critical given that families with high levels of economic and social risk are more likely to select child care of low quality except in cases in which subsidized care is provided.

The Impact of Chaos on Family/Child-Care Mesosystem Relations

As may be obvious from my earlier comments about weak information on mesosystem relations generally, it is difficult to know how chaos impacts family/child-care relations. That said, let me divide my efforts to address this issue into two sections. In this section, I offer some observations about likely relations between environmental chaos as described by Evans, Eckenrode, and Marcynyszyn (see chap. 14) and Wachs (see chap. 7), and in the following section I offer some observations on chaos more broadly conceived (for a more detailed review of child-care chaos, see chap. 5).

Environmental chaos, as conceived by Matheny, Wachs, Ludwig, and Phillips (1995), includes such features as high levels of crowding, traffic, and background noise (extreme levels of other sensory input may be included as well in some cases), lack of temporal and structural regularity, instability of household composition or child-care providers, and residential moves or movements from room to room in child care. The idea that extreme levels of social or physical stimuli or high irregularity in one microsystem can affect what happens in the second seems reasonable as a general proposition. However, the mechanisms through which this happens are not well characterized, nor is it clear how consistent the impacts are likely to be for different components of environmental chaos. Most studies of child care have not focused on chaos within the child-care experience per se. In addition, almost no studies have specifically examined how chaos in child care (of whatever sort) penetrates back into the child's home life and what impact that has on child well-being. Such studies need to be given high priority given the prevalence of child care in modern family life and governmental concerns about ensuring sufficient quality in child care. Likewise, despite many studies on various factors associated with chaos in the family, too few have focused on how different forms of chaos affect child-care use and involvement of parents in child care, or how chaos in family life affects the quality or consistency of the child-care experience.

To the extent that environmental chaos produces significant stress reactions in children, adults are more likely to take notice. However, the likelihood that adults would exhibit any particular response connected to their awareness of the child's stress reactions is much more difficult to predict—and it is conditioned on what is feasible. Children's stress reactions can elicit protective responses, including efforts by a parent or child-care provider to inform the other about the situation in hopes that the other will take positive, appropriate actions on behalf of the child. Children's stress reactions can also elicit nurturing responses, but these are less likely to affect what goes on in the other environment because they are unlikely to be known by the other except indirectly via reduced stress reactions on the part of the child. On the other hand, children's stress reactions sometimes elicit controlling responses on the part of the adult. Controlling responses too are unlikely to be known to caregivers in the other setting except as they are manifest in children's reactions to the adult's controlling behavior. Finally, children's stress reactions sometimes evoke no response from adults and, again, would affect adult behaviors in the other setting only through their impact on the child. Of particular concern is whether some children may be more adversely impacted by chaos than others (e.g., children with difficult temperaments or children with health problems; see chap. 7). Children high in stress reactivity for whatever reason or low in competence for whatever reason are more at risk as a consequence of exposure to chaos in either system and are more likely to be affected by how the second system functions. Thus, bad relationships between parents and child-care providers or inadequacies in their communications could have greater consequences for vulnerable children.

As stated earlier, communications between home and child care tend to be limited and not of high quality. Research suggests that communications between home and caregivers in family day care may be a little better than those between home and caregivers in center care, so one might assume a somewhat greater likelihood of impacts flowing from one microsystem to the second for children in family day care—but that is simply a conjecture. That said, it is difficult to imagine that some elements of environmental chaos identified by Matheny et al. (1995) would prompt parents or caregivers to communicate (e.g., crowding, noise levels, general routines). So, to the extent that communication is the mechanism through which chaos in one system prompts adjustments in the other system, strong relations seem unlikely. Changes in group membership (e.g., divorce or introduction of a stepparent in the case of families, changes in providers in the case of child care) or changes in location seem more likely to lead to communication. Disruptions or modifications in routines that flow from employment-related changes in the family would also be more likely to evoke communication (see chap. 12, this volume). Conditions in the home such as crowding, instability, and poor family routines are likely to penetrate into the child-care setting to the extent that they provoke visible disturbances in child well-being, particularly in the form of negative mood, inattention, and behavioral maladjustment. As children get older, they are more likely to communicate with child-care providers concerning these conditions.

Chaos in child care is more likely to affect what happens at home than vice versa for two reasons: First, parents are more likely to be aware of chaos in child care than child-care providers are to be aware of chaos in individual families. Parents bring their children to child care and are more likely to ask questions both about the normal patterns of care their children receive and about the changes in care they see than child-care providers are to inquire about home life. Second, parents typically have a vested interest in the well-being of their child and thus would be prone to making adjustments to chaos in child care to the extent they become aware of it. Child-care providers typically are more focused at the level of the group, facility, or agency, and thus are not as likely to make significant changes in response to their awareness of chaos in the home life of a particular child. For these proffered "best guesses," three caveats are important. First, research on these issues is minimal. Second, certain child-care providers have a mission to serve not just the child but the family as well (e.g., Early Head Start, providers who serve many Temporary Assistance for Needy Families recipients). In such agencies, awareness of chaos in the family is more likely to evoke change in what happens in connection with child care. Third, parents who live in homes that are highly chaotic may be less sensitive to or aware of chaos in child care because of depression, distraction, or habituation to chaotic circumstances.

Researchers who have extensively studied families, child care, and early education point to the difficulties of deriving accurate estimates of "environmental effects" from such complex dynamic systems. The difficulties inherent in studying any one of these microsystems are magnified when studying the interplay between them. Consequently, any effort to derive very precise estimates of how chaos in the family or chaos in child care affects what happens in the other would be time-consuming and labor intensive, particularly in view of the fact that life in both systems plays out differently in each individual case through time. Lots of cases and a significant amount of time would be needed to derive more than gross estimates of population parameters. That said, there are now good measures of both child care and family chaos that can serve as useful starting points.

Will the Chaos Be Unbroken (or the Transformation Be Completed)?

Child care has become part of the fabric (some might call it the patchwork quilt) of daily life for many families in the United States. As such, one is hard pressed to determine how much and what aspects of a children's development are attributable to home or day care or their interface. Child care has become a pervasive experience, one that is connected to an array of other circumstances currently pervading life in North America, Europe, Japan, Australia, and elsewhere.

The child-care experiences of today's children and their families insinuate themselves into the lives of the children and families who have not had direct experience with child care. Just ask the kindergarten teacher who has been teaching for 30 years, and she will tell you that what goes on in the classroom now is different, partly because of the child-care and preschool experiences that so many children are bringing to school. It has become part of the culture. In light of a statement offered earlier in this chapter (i.e., Bronfenbrenner's notion of the chronosystem; Bronfenbrenner & Morris, 1998), it is perhaps more accurate to say that life with child care is *becoming* part of the culture, for the transformation is still taking place. Child care, although now normative in a statistical sense, is not yet quite normative in a sociopolitical sense. That is, society has not yet come full circle in understanding how to manage what it means for personal life, family life, work life, and life in society more generally. Many parents, especially mothers, are still struggling to manage multiple roles and competing demands. Many children are still struggling to evaluate their circumstances relative to the circumstances of others who have not been in child care, after-school care, and the like. Many employers have not developed policies and procedures for dealing with the realities of modern family life. Many social institutions (e.g., churches) are still struggling to redefine what it means to live a proper life and what role their institution should play in supporting family life. Through time societies tend to develop new ways of comfortably and productively managing the realities of life, ways that include useful socialization, informational, and support processes (see chap. 13). Unfortunately, those new ways tend to lag the new realities, leaving members of a society to struggle in the meanwhile. In that regard, it is perhaps worth noting that throughout human history there has been a common response to the kind of stresses families are now facing. For instance, many societies have shown a tendency to delay childbirth and reduce the number of children per family during times of severe adversity, especially when children are not critical to family survival. These trends have emerged again over the past two generations in most industrialized societies. Despite signs of movement on most fronts, changes are not yet complete and research on how to most expeditiously move forward is minimal.

For some families, child care functions as a bulwark against the chaos that threatens them. For others, it is part of the chaos with which they must contend. There is a particular need to develop information on how to assist parents with the stresses connected to employment, school, and child care. As well, there is a need to develop information on how to best reformulate policies and procedures in the workplace so that they promote healthy families and optimal development in children while maintaining job productivity. Bronfenbrenner's model sets a frame for this endeavor. Now comes the hard job of gathering pertinent descriptive and qualitative data, developing theoretical propositions, and testing them.

In light of the prevalence and deemed importance of child care, it is striking how little research has been devoted to characterizing the family/child-care mesosystem. Part of the neglect likely derives from the complexity and dynamism of the system. Extensive data would need to be collected both in and through time to document how the system functions or how it is implicated in children's development. Research done to date suggests that, on average, parental characteristics and experiences at home tend to matter more in children's lives than what children experience in child care—useful information, as far as it goes, but not the full story. More needs to be known about the interface between home and child care; what happens on the journey to and fro; how ideas, motivations, and competencies are transformed by joint experience; and how negotiations between the systems at every level reverberate through the life course in large and small ways in different lives. Or, as Paul Harvey used to say, we need to know "the rest of the story." In the short haul, I would focus on children and families who appear most vulnerable: the temperamentally difficult child, the child with autism, the family with a recently incarcerated member, the family that is bouncing between jobs that provide limited and unstable income. That is where the interface is likely to matter most.

Summary

As greater and greater numbers of women entered the workforce in the late 20th century, nonparental child care became a fixture in family life in most industrialized societies. Although nonparental care has been commonplace throughout human history, the forms it has taken most recently (e.g., care in centers, family child-care homes, nannies) has little precedent. This, together with other rapid transformations in family life, has made it difficult for parents to comfortably and confidently engage the child-care system. Parents often struggle to negotiate their work and child-rearing responsibilities. Likewise, the formal system of child care has not yet fully determined how best to serve families. In effect, there has not yet been enough time for modern societies to evolve the most productive practices and structures to deal with the new realities of family life as they pertain to children's care. Family and child-care systems and the relations between them often do not function maximally to support family life or child well-being. Neither do they function maximally to support those who provide nonparental care. As a consequence, chaos within or between systems is often not well accommodated. Research, both basic and policy-relevant, is needed to help parents, practitioners, and policymakers formulate plans pertaining to child care that increase the quality of children's experience and reduce the impact of chaos on child development.

References

Ackerman, B. P., Brown, E. E., & Izard, C. E. (2004). The relations between persistent poverty and contextual risk and children's behavior in elementary school. *Developmental Psychology, 40*, 367–377.

Administration for Children and Families. (2002). *Making a difference in the lives of children and families: The impacts of Early Head Start programs on infants and toddlers and their families*. Washington, DC: U.S. Department of Health and Human Services.

Bakermans-Kranenburg, M. J., van IJzendoorn, M. H., & Bradley, R. H. (2005). Those who have receive: The Matthew-effect in early childhood intervention. *Review of Educational Research, 75*, 1–26.

Ball, T. M., Holberg, C., Aldoiun, M., Martinez, F., & Wright, A. (2002). Influence of attendance at day care on the common cold from birth through 13 years of age. *Archives of Pediatrics & Adolescent Medicine, 156,* 121–126.

Belsky, J., Vandell, D. L., Burchinal, M., Clarke-Stewart, K. A., McCartney, K., Owen, M., & the NICHD Early Child Care Research Network. (2007). Are there long term effects of early child care? *Child Development, 78,* 681–701.

Benasich, A. A., Brooks-Gunn, J., & Clewell, B. C. (1992). How do mothers benefit from early intervention programs? *Journal of Applied Developmental Psychology, 13,* 311–362.

Berghout, M., Godfrey, J., Larsen, L., Knudsen, L., Shelly, L., et al. (1996). Determinants of children's satisfaction with child care providers. *Early Child Development & Care, 115,* 19–36.

Blau, D. M. (2001). *The child care problem: An economic analysis.* New York: Russell Sage.

Bowman, B. T. (1997). Preschool as family support. In C. Dunst & M. Wolery (Eds.), *Advances in early education and day care* (pp. 157–172). Greenwich, CT: JAI Press.

Bradley, R. H. (2007). Parenting in the breach: How parents help children cope with developmentally challenging circumstances. *Parenting: Science & Practice, 7,* 99–148.

Bradley, R. H., & Corwyn, R. F. (2002). SES and child development. *Annual Review of Psychology, 53,* 371–399.

Bradley, R. H., & Vandell, D. L. (2007). Child care and the well-being of children. *Archives of Pediatrics & Adolescent Medicine, 161,* 669–676.

Britner, P. A., & Phillips, D. A. (1995). Predictors of parent and provider satisfaction with child day care dimensions: A comparison of center-based and family child care. *Child Welfare, 74,* 1135–1168.

Bronfenbrenner, U. (1974). *Is early education effective?* (Publication No. OHD76-30025). Washington, DC: Department of Health, Education and Welfare.

Bronfenbrenner, U., & Morris, P. A. (1998). The ecology of developmental processes. In W. Damon (Series Ed.) & R. M. Lerner (Vol. Ed.), *Handbook of child psychology: Vol. 1: Theoretical models of human development* (pp. 993–1028). New York: Wiley.

Burchinal, M. R., Peisner-Feinberg, E., Bryant, D. M., & Clifford, R. (2000). Children's social and cognitive development and child-care quality: Testing for differential associations related to poverty, gender, or ethnicity. *Applied Developmental Science, 4,* 149–165.

Castro, D. C., Bryant, D. M., Peisner-Feinberg, E. S., & Skinner, M. L. (2004). Parent involvement in Head Start programs: The role of parent, teacher, and classroom characteristics. *Early Childhood Research Quarterly, 19,* 413–430.

Caughy, M. O., DePietro, J. A., & Strobino, D. M. (1994). Day-care participation as a protective factor in the cognitive development of low-income children. *Child Development, 65,* 457–471.

Chan, J. B., & Sigafoos, J. (2001). Does respite care reduce parental stress in families with developmentally disabled children? *Child & Youth Care Forum, 30,* 253–263.

Clarke-Stewart, K. A. (1991). A home is not a school. In M. Lewis & S. Feinman (Eds.), *Social influences and socialization in infancy* (pp. 41–62). New York: Plenum Press.

Clarke-Stewart, K. A., Gruber, C. P., & Fitzgerald, L. M. (1994). *Children at home and in day care.* Hillsdale, NJ: Erlbaum.

Coleman, P. K., & Karraker, K. H. (1997). Self-efficacy and parenting quality: Findings and future applications. *Developmental Review, 18,* 47–85.

Conger, R. D., & Donnellan, M. B. (2007). An interactionist perspective on the socioeconomic context of human development. *Annual Review of Psychology, 58,* 175–199.

Corapci, F., & Wachs, T. D. (2002). Does parental mood or efficacy mediate the influence of environmental chaos on parenting behavior? *Merrill-Palmer Quarterly, 48,* 182–201.

Cost, Quality, and Outcomes Team. (1995). *Cost, quality, and outcomes in child care centers.* Denver, CO: University of Colorado at Denver.

Cryer, D., Tietze, W., & Wessels, H. (2002). Parents' perceptions of their children's child care: A cross-national comparison. *Early Childhood Research Quarterly, 17,* 259–277.

Dumas, J. E., Nissley, J., Nordstrom, A., Smith, E. P., Prinz, R. J., & Levine, D. W. (2005). Home chaos: Sociodemographic, parenting, interactional, and child correlates. *Journal of Clinical Child & Adolescent Psychology, 34,* 93–104.

Duncan, G. J., & Brooks-Gunn, J. (1997). *The consequences of growing up poor.* New York: Russell Sage Foundation.

Endsley, R. C., Minish, P. A., & Zhou, Q. (1993). Parent involvement and quality day care in proprietary centers. *Journal of Research in Childhood Education, 7,* 53–61.

Erwin, P. J., Sanson, D., Amos, D., & Bradley, B. S. (1993). Family day care and day care centers: Career, family and child differences and their implications. *Early Child Development & Care, 86,* 89–103.

Evans, G. W. (2004). The environment of childhood poverty. *American Psychologist, 59,* 77–92.

Feagans, L., & Manlove, E. (1994). Parents, infants, and day-care teachers: Interrelations and implications for better child care. *Journal of Applied Developmental Psychology, 15,* 585–602.

Fulmer, K. A. (1997). Parents' decision making strategies when selecting child care: Effects of parental awareness, experience, and education. *Child & Youth Care Forum, 26,* 391–409.

Gable, S., & Cole, K. (2000). Parents' child care arrangements and their ecological correlates. *Early Education & Development, 11,* 549–572.

Ghazvini, A. S., & Readdick, C. A. (1994). Parent-caregiver communication and a quality of care in diverse child care settings. *Early Childhood Research Quarterly, 9,* 207–222.

Goosens, F. A., & van IJzendoorn, M. H. (1990). Quality of infants' attachment to professional caregivers: Relation to infant-parent attachment and daycare characteristics. *Child Development, 61,* 832–837.

Gormley, W. T. (1995). *Everybody's children: Child care as public problem.* Washington, DC: Brookings Institute.

Han, W. (2004). Nonstandard work schedules and child care decisions: Evidence from the NICHD Study of Early Child Care. *Early Childhood Research Quarterly, 19,* 231–256.

Hand, K., & Wise, S. (2006). *Children from diverse cultural backgrounds in Australian child care service: What their carers say about partnerships with parents* (Research Paper No. 36). Melbourne: Australia Institute of Family Studies.

Haskett, M. E., & Kistner, J. A. (1991). Social interactions and peer perceptions of young physically abused children. *Child Development, 62,* 979–990.

Hayden, J., De Gioia, K., & Hadley, F. (2003, October). *Enhancing supports and networks for families with children 0-5.* Paper presented at the Families First in Practice Conference, Sydney, Australia.

Hofferth, S. L., Brayfield, A., Deich, S., & Holcomb, P. (1991). *The national child care survey, 1990.* Washington, DC: Urban Institute Press.

Hofferth, S. L., & Phillips, D. (1991). Child care policy research. *Journal of Social Issues, 47,* 1–13.

Hofferth, S. L., & Wissoker, D. A. (1992). Price, quality and income in child care choices. *Journal of Human Resources, 27,* 70–110.

Holloway, S. D., & Reichart-Erickson, M. (1989). Child-care quality, family structure, and maternal expectations: Relationship to preschool children's peer relations. *Journal of Applied Developmental Psychology, 10,* 281–298.

Horowitz, E. G. (1984). Parent-caregiver communication in two kinds of day-care settings. *Child & Youth Care Forum, 13,* 142–148.

Howes, C., & Eldridge, R. (1985). Responses of abused, neglected, and non-maltreated children to the behavior of their peers. *Journal of Applied Developmental Psychology, 6,* 261–270.

Hughes, P., & MacNaughton, G. (2002). Preparing early childhood professionals to work with parents: The challenges of diversity and dissensus. *Australian Journal of Early Childhood, 27*(2), 14–20.

Johnston-Brooks, C. H., Lewis, J., Evans, G. W., & Whalen, C. K. (1998). Chronic stress and illness in children: The role of allostatic load. *Psychosomatic Medicine, 60,* 597–603.

Knopf, J., & Swick, K. (2007). How parents feel about their child's teacher/school: Implications for early childhood professionals. *Early Childhood Education Journal, 34,* 291–296.

Kontos, S., & Dunn, S. (1989). Attitudes of caregivers, maternal experiences with day care, and children's development. *Journal of Applied Developmental Psychology, 10,* 37–51.

Lamb-Parker, F., Piotrkowski, C. S., Baker, A. J., Kessler-Sklar, S., Clark, B., & Peay, L. (2001). Understanding barriers to parent involvement in Head Start: A research-community partnership. *Early Childhood Research Quarterly, 16,* 35–51.

Leavitt, R. L. (1995). Parent-provider communication in family day care homes. *Child & Youth Care Forum, 24,* 231–245.

Leibowitz, A., Klerman, J. A., & Waite, L. J. (1992). Employment of new mothers and child-care choice: Differences by children's age. *Journal of Human Resources, 27,* 112–133.

Leslie, L. A., Ettenson, R., & Cumsille, P. (2000). Selecting a child care center: What really matters to parents? *Child & Youth Care Forum, 29*, 299–322.

Lewis, C., & Lamb, M. E. (2003). Fathers' influence on children's development: The evidence from two-parent families. *European Journal of Education, 18*, 211–228.

Lu, N., Samuels, M., Shi, L., Baker, S., Glover, S., & Sanders, J. (2004). Child day care risks of common infectious diseases revisited. *Child: Care, Health & Development, 30*, 361–368.

Marcon, R. A. (1999). Positive relationships between parent school involvement and public school inner-city preschooolers' development and academic performance. *School Psychology Review, 28*, 395–412.

Masten, A. S. (2001). Ordinary magic: Resilience processes in development. *American Psychologist, 56*, 227–238.

Matheny, A., Wachs, T. E., Ludwig, J., & Phillips, K. (1995). Bringing order out of chaos: Psychometric characteristics of the Louisville Chaos Scale. *Journal of Applied Developmental Psychology, 16*, 429–444.

Maxwell, L. E. (1996). Multiple effects of home and day care crowding. *Environment & Development, 28*, 494–511.

McBride, B. A., & Rane, T. R. (1997). Father/male involvement in early childhood programs: Issues and challenges. *Early Childhood Education Journal, 25*, 11–15.

McCartney, K. (2007). The family–child-care mesosystem. In K. A. Clarke-Stewart & J. Dunn (Eds.). *Families count: Effects on child and adolescent development* (pp. 155-175). New York: Cambridge University Press.

McCartney, K., Dearing, E., & Taylor, B. A. (2003, April). *Is higher-quality child care an intervention for children from low-income families?* Paper presented at the biennial meeting of the Society for Research in Child Development, Tampa, Florida.

McClow, C., & Gillespie, C. (1998). Parental reactions to the introduction of the Reggio Emilia approach in Head Start classrooms. *Early Childhood Education Journal, 26*, 131–136.

Melhuish, A., Mooney, A., Martin, S., & Lloyd, E. (1990). Type of childcare at 18 months: 1. Differences in interactional experience. *Journal of Child Psychology & Psychiatry & Allied Disciplines, 31*, 849–859.

Miedel, W. T., & Reynolds, A. J. (1999). Parent involvement in early intervention for disadvantaged children: Does it matter? Chicago Longitudinal Study. *Journal of School Psychology, 37*, 379–402.

Nafstad, P., Hagen, J., Oie, L., Magnus, P., & Jaakkola, J. (1999). Day care centers and respiratory health. *Pediatrics, 103*, 753–758.

National Institute of Child Health and Human Development Early Child Care Research Network. (1997a). The effects of infant child care on infant–mother attachment security: Results of the NICHD Study of Early Child Care. *Child Development, 68*, 860–879.

National Institute of Child Health and Human Development Early Child Care Research Network. (1997b). Familial factors associated with the characteristics of non-maternal care for infants. *Journal of Marriage & Family, 59*, 389–408.

National Institute of Child Health and Human Development Early Child Care Research Network. (1998a). Early child care and self-control, compliance, and problem behavior at twenty-four and thirty-six months. *Child Development, 69*, 1145–1170.

National Institute of Child Health and Human Development Early Child Care Research Network. (1998b). Relations between family predictors and child outcomes: Are they weaker for children in child care? *Developmental Psychology, 34*, 1119–1128.

National Institute of Child Health and Human Development Early Child Care Research Network. (2000a). Characteristics and quality of child care for toddlers and preschoolers. *Applied Developmental Science, 4*, 116–135.

National Institute of Child Health and Human Development Early Child Care Research Network. (2000b). The relation of child care to cognitive and language development. *Child Development, 71*, 960–980.

National Institute of Child Health and Human Development Early Child Care Research Network. (2001a). Child care and common communicable illnesses. *Archives of Pediatrics & Adolescent Medicine, 155*, 481–488.

National Institute of Child Health and Human Development Early Child Care Research Network. (2001b). Child care and family predictors of MacArthur preschool attachment and stability from infancy. *Developmental Psychology, 37*, 847–862.

National Institute of Child Health and Human Development Early Child Care Research Network. (2002). The interaction of child care and family risk in relation to child development at 24 and 36 months. *Applied Developmental Science, 6*, 144–156.

National Institute of Child Health and Human Development Early Child Care Research Network. (2003). Child care and common communicable illnesses in children aged 37 to 54 months. *Archives of Pediatrics & Adolescent Medicine, 157*, 196–200.

National Institute of Child Health and Human Development Early Child Care Research Network. (2005a). *Child care and child development: Results from the NICHD study of early child care and youth development.* New York: Guilford Press.

National Institute of Child Health and Human Development Early Child Care Research Network. (2005b). Early child care and children's development in the primary grades: Follow-up results from the NICHD Study of Early Child Care. *American Educational Research Journal, 42*, 537–570.

Nelson, F., & Garduque, L. (1991). The experience and perception of continuity between home and day care from the perspectives of child, mother and caregiver. *Early Child Development & Care, 68*, 99–111.

Overturf, J. J. (2005). *Who's minding the kids? Child care arrangements: Winter 2002* (Current Population Reports No. P70-101). Retrieved April 23, 2006, from the U. S. Bureau of the Census Web site: http://www.census.gov/prod/2005pubs/p70-101.pdf

Peisner-Feinberg, E., Burchinal, M., Clifford, R., Culkin, M., Howes, C., Kagan, S., & Yazejian, N. (2001). The relation of preschool child-care quality to children's cognitive and social developmental trajectories through second grade. *Child Development, 72*, 1534–1553.

Phillips, D., McCartney, K., & Sussman, A. (2006). Child care and early development. In K. McCartney & D. Phillips (Eds.), *Blackwell handbook of early childhood development* (pp. 471–489). Malden, MA: Blackwell Publishing.

Phillips, D. A., Voran, M., Kisker, E., Howes, C., & Whitebook, M. (1994). Child care of children in poverty: Opportunity or inequality? *Child Development, 65*, 472–492.

Powell, D. R. (1978). The interpersonal relationship between parents and caregivers in day care settings. *American Journal of Orthopsychiatry, 48*, 680–689.

Powell, D. R. (1990). The responsiveness of early childhood initiatives to families: Strategies and limitations. *Marriage and Family Review, 15*, 149–170.

Provost, M. A., Garron, D., & LaBarre, R. (1991). Social adjustment of young children of preschool age in relation to continuity of day care. *Canadian Journal of Behavioural Science, 32*, 183–194.

Pugello, E., & Kurtz-Costes, B. (1999). Why and how working women choose child care: A review with a focus on infancy. *Developmental Review, 19*, 31–96.

Radke-Yarrow, M. (1989). Developmental and contextual analysis of continuity. *Human Development, 32*, 204–209.

Rettig, M. A. (1998). Guidelines for beginning and maintaining a toy lending library. *Early Childhood Education Journal, 25*, 229–232.

Schliecker, E., White, D. R., & Jacobs, E. (1991). The role of day care quality in the prediction of children's vocabulary. *Canadian Journal of Behavioural Science, 23*, 12–24.

Segerstrom, S. C., & Miller, G. E. (2004). Psychological stress and the human immune system: A meta-analytic study of 30 years of inquiry. *Psychological Bulletin, 130*, 601–630.

Shpancer, N. (2002). The home-daycare link: Mapping children's new world order. *Early Childhood Research Quarterly, 17*, 374–392.

Stile, S., & Ortiz, R. (1999). A model for involvement of fathers in literacy development with young at-risk and exceptional children. *Early Childhood Education Journal, 26*, 221–224.

Swick, K. J. (2004). What parents seek in relations with early childhood family helpers. *Early Childhood Education Journal, 31*, 217–220.

Swick, K. J., & Bailey, L. B. (2004). Communicating effectively with parents and families who are homeless. *Early Childhood Education Journal, 32*, 211–215.

Van IJzendoorn, M. H., Tavecchio, L. W., Stams, G., Verhoerven, M., & Reling, E. (1998). Attunement between parents and professional caregivers: A comparison of childrearing attitudes in different child-care settings. *Journal of Marriage & Family, 60*, 771–781.

Votruba-Drzal, E., Coley, R. L., & Chase-Lansdale, P. L. (2004). Child care and low-income children's development: Direct and moderated effects. *Child Development, 75*, 296–312.

Wachs, T. D. (2000). *Necessary but not sufficient: The respective roles of single and multiple influences of individual development.* Washington, DC: American Psychological Association.

Waite, L. J., Leibowitz, A., & Witsberger, C. (1991). What parents pay for: Child-care characteristics, quality, and costs. *Journal of Social Issues, 47,* 33–48.

Winkelstein, E. (1987). Day care / family interaction and parental satisfaction. *Child Care Quarterly, 10,* 334–340.

10

Disorder, Turbulence, and Resources in Children's Homes and Neighborhoods

Jeanne Brooks-Gunn, Anna D. Johnson, and Tama Leventhal

Ecological systems theory highlights the role that physical and psychosocial components of all environmental levels play in healthy development (Bronfenbrenner, 1979, 2005; Wachs, 1989, 1992). In our opinion, more work has examined both the social and physical dimensions of the environment simultaneously when focusing on the macrosystem level than when focusing on the mesosystem level of the environment. Very little research has honed in on the intersection of the two systems with a focus on both social and physical components of the environment. Indeed, family as a context for children's development has been considered in numerous articles (Bornstein, 2002; Collins, Maccoby, Steinberg, Hetherington, & Bornstein, 2000; Maccoby & Martin, 1983). The neighborhood as a context has received increasing scrutiny in the past decade (Brooks-Gunn, Duncan, & Aber, 1997; Leventhal & Brooks-Gunn, 2000b; Sampson, Morenoff, & Gannon-Rowley, 2002). Can research on the two be brought together under a common rubric, in this case the theme of chaos? The goal of our chapter is to attempt such a synthesis.

This chapter, which has four main sections, focuses on the child's home and neighborhood. In the first section, a common language is proposed to categorize the physical and social dimensions of families and neighborhoods that might be indicative of chaos. In the second section, family-level research on the proposed categories is discussed. The third section does the same for neighborhood-level research. In the fourth section, we address future steps for research and we conclude with a brief discussion of prevention efforts.

We, the chapter authors, thank the Virginia and Leonard Marx Foundation, the National Institute for Child Health and Human Development, and the March of Dimes Foundation for their support. We also extend our gratitude to Rachel McKinnon and Anne Martin for help with manuscript preparation.

How Do Families and Neighborhoods Matter?

Measures of chaos are sometimes considered the consequence of low resources, thereby acting as a mediator between resources and child well-being (see chap. 14). For example, a parent whose low income is earned in a job with variable hours and shifts may find it difficult to organize the household. Furthermore, if the parent is also a single parent, he or she may have few adults who can help with household management. The direct consequence could be a household that is chaotic. Children in low-income families may be at a double disadvantage, in that they are both more likely to experience the direct effects of chaos, such as the temporal–spatial instability discussed by Hertzman (see chap. 8), as well as the indirect effects of chaos, such as interference with effective proximal processes such as parent–child interactions. Consequently, chaos in the home may be conceptualized as a link between resources and more direct parental behavior. Do parents in chaotic homes talk to their children less, supervise them less, or play with them less? However, some parenting behaviors, such as lack of a bedtime routine, might be considered measures of chaos. Given that parenting behavior could be a consequence of or an indicator of chaos, we turn to definitions in the next section.

Defining Chaotic Environments

We consider the following six constructs as dimensions of chaos: crowding and density, noise and confusion, clutter and messiness, fluidity and instability of residents, lack of predictability and routines, and low supervision and monitoring. Although all six constructs may be indicative of chaos in a child's life, we use the term *disorder* for the first three dimensions, and the term *turbulence* for the second three.

Crowding and Density

Crowding and density, often considered characteristic of disorder, are usually defined by how crowded a home is, as measured by the number of household members or by the number of people per room (Evans, 2006; Maxwell, 1996). Sometimes the proportion of children to adults within a household is considered. The question still remains as to what ratios or numbers constitute crowding or high density (Matheny, Wachs, Ludwig, & Phillips, 1995), and it is not known whether such variables are best conceptualized in linear or threshold terms.

From a neighborhood perspective, crowding is typically measured as the number of housing units per census block group or tract, or by the number of individuals or households per census block group or tract. The proportion of housing units with more than a certain number of household members has also been used. Density measures are typically not included in neighborhood studies (in part because such measures are correlated with concentrated disadvantage). As raised elsewhere in this volume (see chap. 11), defining neighborhoods by census block group or tracts may not be as meaningful as asking residents to identify the spatial parameters of their own neighborhood. In addition, the meaning of high density probably differs as a function of culture, class, income, and neighborhood. Comparisons across geographic units such as cities, states, and countries become difficult if influences of density are relative rather than absolute.

Noise and Confusion

Another dimension of disorder that is commonly studied has to do with noise and confusion. The Confusion, Hubbub, and Order Scale (CHAOS) has relevant items as well as items on crowding (Matheny et al., 1995). These authors provide information on the construct validity of their scale, which correlates well with direct observations of existing environmental conditions. Homes with especially high traffic patterns and high degrees of crowding had significantly higher scores on CHAOS than did homes with fewer visitors or residents. Given correlations with crowding, we wonder if the conceptual distinction between (a) density and crowding and noise and (b) confusion can be made at the measurement level with assessments of perceived chaos. One can surely envision a household that is crowded but has relatively little confusion. However, if in reality very few of those households exist, then in practical terms the two dimensions cannot be parsed.

With regard to neighborhoods, little has been done with respect to noise and confusion, at least at the census block group or tract level. Some research has focused on ambient noise in a particular neighborhood (see Evans, 2006). Other work has looked at specific neighborhoods, such as those under an airport's flight path or close to a highway (e.g., Hygge, Evans, & Bullinger, 2002).

Clutter and Messiness

Another dimension of disorder is clutter and messiness. In light of research on clutter, messiness, and dirtiness at both the family and neighborhood levels, we treat this dimension as separate from the others. The Home Observation for Measurement of the Environment (HOME) scale (Bradley et al., 1989) contains as part of its physical subscale a few items on how cluttered, dark, and dirty the house is (Caldwell & Bradley, 1984; Linver, Martin, & Brooks-Gunn, 2004).

The analogue for the neighborhood level might be what has been termed *physical disorder* or *broken windows*. Sociologists have been characterizing neighborhoods in terms of a neighborhood's clutter, trash, glass shards, graffiti, alcohol bottles, and drug paraphernalia since the Chicago School's observations over 50 years ago. A series of techniques has been developed for observing neighborhood clutter and cleanliness. One involves a walk around the proximity of the household unit to do physical ratings comparable to the HOME rating of the household. The Fragile Families and Child Well-Being Study, the Project on Human Development in Chicago Neighborhoods (PHDCN), and the Canadian National Longitudinal Survey of Children and Youth have all used this technique (Kohen, Brooks-Gunn, Leventhal, & Hertzman, 2002). Another approach involves observing features of neighborhoods or blocks from a car (Sampson & Raudenbush, 1999).

Fluidity and Instability of Residents

One aspect of turbulence has to do with entrances and exits of individuals in children's lives, or what we are terming *fluidity* or *instability* (Baydar, Greek, & Brooks-Gunn, 1997; Haveman, Wolfe, & Spaulding, 1991; Wu, 1996). These events have typically been studied at the family level, with the more general term *turbulence* sometimes being used to characterize them (Moore & Vandivere, 2000). In chapter 15 of this volume, Lustig highlights an extreme example of turbulence in the case of refugees. As Lustig argues, the physical upheaval and uncertainty associated with fleeing one's home can have a range of negative effects on child development, including interfering with healthy sexual, cognitive, and moral development. We use the term *turbulence* to include fluidity, lack of predictability, and lack of supervision, however. Though some events may be considered major crises, generally most instances of fluidity or instability represent temporary disturbances or changes that might be associated with changes in parenting and child well-being (Meadows, McLanahan, & Brooks-Gunn, 2008).

Fluidity or instability has been conceptualized as residential moves, as well as by household entrances and exits. At the household level, research has tended to consider residential moves separately from the entrances and exits of individuals, although fluidity implies that a series of events co-occur. Questions are being asked about the number of cumulative events as well as the timing of such events in a child's life. More is known about each event than the accumulation or timing of entrances and exits (Leventhal & Brooks-Gunn, 2000a).

The analogues at the neighborhood level are quite straightforward. They would include the proportion of single-parent families in a neighborhood, the proportion of neighbors who have lived in the neighborhood for the past 5 years, and the proportion of housing units that are owned rather than rented (ownership being associated with residential stability). Indeed, factor analyses of neighborhood characteristics (from census tract data) almost always identify a factor termed *Residential Stability* (Sampson, Raudenbush, & Earls, 1997; Xue, Leventhal, Brooks-Gunn, & Earls, 2005).

Measures of change (from one census period to another) might also be used as indicators of fluidity; these measures might include increases in the proportion of single-parent families, unemployed adults, rental as opposed to owner-occupied units, or proportions of poor families. Little research has been conducted on these types of indicators, although work is progressing on changes in neighborhood poverty and gentrification (Leventhal & Brooks-Gunn, 2007; Sampson & Morenoff, 2006).

Lack of Predictability and Routines

Yet another dimension of what we are calling *turbulence* considers lack of predictability and routines in a child's life. It is believed that routines or a socially ordered life is conducive to learning and regulation (Fiese, 2006; Johnson, Martin, Brooks-Gunn, & Petrill, 2008; Wachs, 2006). Indeed, a growing body of research is certainly suggestive of a positive association between absolute frequency of family routine activities and children's social development, as Fiese and Winter noted in chapter 4 of this volume. Scholars have developed a series of measures to tap routines, including the use of subscales or items on larger scales such as CHAOS (Matheny et al., 1995) and HOME (Bradley et al., 1989). Other attempts involve the development of separate scales (e.g., Family Routines Inventory; Boyce, Jensen, James, & Peacock, 1983; Jensen, James, Boyce, & Hartnett, 1983). Yet others focus on lists of activities, such as eating meals together, bedtime routines, the time that a child goes to bed and is awakened, and doing chores that are performed daily or near daily.

Perhaps the most popular single item used to assess routines has to do with the number of nights per week that a family eats dinner together. Because measures of household-level chaos or disruption of family routines have not yet been implemented in a comparable manner across the large, national studies of children (Brooks-Gunn, Fuligni, & Berlin, 2003), it is difficult to determine how the dimension of family routines maps onto or overlaps with the other dimensions of chaos that we discuss. However, as Fiese and Winter recognized (see chap. 4, this volume), enough research has been done to suggest that children are keenly aware of disruption or reduction of regular routines, as well as of the stress parents experience when trying to plan for the day and balance multiple obligations. Thus, we are confident that if increased efforts are made to assess the effects of family routines on child development in large, national studies, similar results will emerge.

On the neighborhood front, routines are not often measured, although they are discussed theoretically. Wilson (1987) in *The Truly Disadvantaged* considered the rhythm of daily life in neighborhoods where few adults are employed and individuals may be on the streets late at night. In such settings, the routine of rising and leaving for work may be absent, just as the routine of coming home and making dinner may be. The lack of routine, at a neighborhood level, could potentially influence individual family life. Studies in general do not include such measures (although census information allows for proxy measures such as the proportion of adults who are not employed, and neighborhood observation allows for proxy measures such as number and density of bars and presence of individuals loitering on street corners).

Supervision and Monitoring

A third aspect of turbulence has to do with parental supervision and monitoring of children. Measuring these constructs is difficult. Supervision subscales are common in variants of the HOME. Measures include how often parents who are working check in with their children after school (or whether children are required to contact their parents), how often (or for how long) a child of a particular age is allowed to be at a store or a mall without adult supervision, whether a child of a particular age is allowed to go to a movie without an adult, and so on. Time-use diaries can sometimes yield information about supervision, in that it is possible to calculate the proportion of time that a child, for example, is in a room or in a yard without an adult present (Fuligni & Brooks-Gunn, 2002). Another literature focuses on the interconnectedness of parents to their children's friends or to the parents of their children's friends (Fletcher, Darling, Steinberg, & Dornbusch, 1995). It is presumed that more supervision is possible if a parent is connected to the appropriate child or parent networks. A few studies ask both parents and children for this information to examine the concordance between the two. And sometimes children are asked how often their parents do not know where they are. With the advent of cell phones and text messaging, more detailed questions on monitoring can now be asked (Brooks-Gunn & Donahue, 2008).

The supervision and monitoring of children at the neighborhood level has certainly been measured. Perhaps the most well-known attempt involves *collective efficacy*, defined as social cohesion and commitment to the common good among residents of a community. Collective efficacy has been measured by neighbors' responses to questions about their willingness to intervene if they were to witness a violent exchange, for example (Sampson et al., 1997).

Family-Level Disorder and Turbulence: Associations With Child Well-Being

Research on chaos at the family level demonstrates links to parenting behavior and child well-being. In general, research has been done on each of the six dimensions outlined in the previous section. The literature is reviewed in two sections: one on disorder (crowding, noise, clutter) and one on turbulence (fluidity, lack of routines, and low supervision).

Disorder

Quite a bit of evidence is accumulating on links between the CHAOS Scale and child well-being, in terms of emotional regulation and behavior problems as well as academic achievement and cognitive and linguistic abilities. We review literature on children here although links have been found for adolescents (see, e.g., Evans, Gonnella, Marcynyszyn, Gentile, & Salpekar, 2005). The CHAOS Scale includes items on crowding and on noise and confusion. Both are associated with children's social, emotional, and cognitive outcomes (Evans et al., 2005; Maxwell & Evans, 2000). However, research has not always distinguished among these dimensions. Our third dimension of disorder—clutter and messiness—is discussed later in this section.

High CHAOS scores have been linked to 4- to 6-year-olds' problem behavior, as reported by mothers, even after controlling for parenting quality (Coldwell, Pike, & Dunn, 2006). Another study of preschool-age and school-age children found that externalizing behavior and attention problems were associated with high levels of chaos in the home, as measured by the CHAOS scale (Dumas et al., 2005). Similar findings were reported in a sample of 4-year-old twins (Pike, Iervolino, Eley, Price, & Plomin, 2006).

Links have been found with linguistic and cognitive measures as well. Early studies were more likely to look at these indicators of well-being rather than at socioemotional markers (Gottfried & Gottfried, 1984; Wachs, 1979). The twin study mentioned earlier saw an association between high levels of CHAOS and expressive vocabulary and nonverbal cognitive abilities (Pike et al., 2006), even after controlling for income.

In a study using an older sample of twins in which only a subset of the CHAOS items were included, links between two aspects of CHAOS (household calm and order) and cognitive development were found (Hart, Petrill, Deater Deckard, & Thompson, 2007). In our own work, we have examined early reading achievement (Johnson, Martin, Brooks-Gunn, & Petrill, 2008) in this same sample of twins. After factor analysis of the CHAOS Scale, two components emerged. The first factor was indicative of household confusion and consisted of items asking about household routines and parental perceptions of confusion in the home. The other factor represented household noise and included items asking about the degree of calm and quiet in the household. We found that early reading was linked to household confusion but not to household noise, perhaps because confusion and noise are somewhat distinct constructs (an area of chaos research that certainly calls for more attention). We hypothesized that different aspects of chaos might operate in different ways. Although prior research strongly suggests that noise does interfere with the development of early reading skills, perhaps in this sample of middle-class twins noise may not be as critical for early reading as the presence of routines in the household. It is also possible that noise may be more endemic in households with twins, so that our noise measure did not discriminate among households in this particular sample. A third possibility is that perhaps our subscale is not picking up routines but is representative of a less cluttered household, a topic to which we turn now.

Clutter and dirtiness are represented in the HOME scale of the physical environment. However, few studies have focused on this particular dimension specifically (rather than as part of the physical environment). One interesting analysis did examine clutter and dirtiness, using a unique feature of the Panel Study of Income Dynamics (PSID; Hill, 1992). Between 1968 and 1972, field staff rated each home on cleanliness (a 5-point scale), which showed variability (although the mean scores in each year were quite high, at 4.13) and exhibited high interyear reliability (alpha of 0.86; Dunifon, Duncan, & Brooks-Gunn, 2001). Though the field staff were instructed to code for cleanliness, we believe that the measure taps both clutter and dirtiness. We examined connections between cleanliness of the homes in the late 1960s to labor force outcomes between 1994 and 1996 for the offspring (i.e., Generation 1's household was rated and Generation 2's outcomes were examined). Even in the presence of controls for a vast array of possible covariates (Generation 1's education, occupation, race, number of siblings, urban or rural location, region, poverty, and so on), the association between cleanliness of the child-rearing household and wages as an adult persisted. This association was slightly reduced, although still statistically significant, when analyses controlled for possible psychological characteristics and variables that might be related to housework burden from 1968 to 1972.

Analyses were also conducted for the cleanliness of Generation 2 households, with respect to outcomes in their offspring (Generation 3), focusing on completed schooling and hourly earnings (Dunifon et al., 2001). Again, even in the presence of extensive controls, there was a strong and significant association between the cleanliness of Generation 2's households and Generation 3's years of completed schooling. We argue that children benefit from growing up in a clean and noncluttered household, with the benefits being seen 25 years later and across generations (a "grandparent effect" of a clean home).

Of course, the question is why would a clean home be associated with wages of the children and schooling of the grandchildren a quarter of a century later? Though part of the link is accounted for by other characteristics of the home and the parents (grandparents), the association does not disappear. It is possible that our measure is picking up other aspects of the home environment, if other measures of turbulence and disorder are correlated with cleanliness and clutter. Regrettably, other measures were not available in the PSID. Indeed, it is imperative for the next generation of studies to examine the intercorrelations of the various indicators of chaos. Another explanation might be that parents who keep a clean home are more likely to exhibit personality characteristics such as conscientiousness. In other words, parents who are conscientious (as conceptualized as one of the Big Five personality characteristics) may be more likely to have a clutter-free home, as well as to have more predictable household routines and supervision of children. Research has not proceeded in this promising direction yet.

Turbulence

As stated earlier, most of the literature on turbulence in the home has focused on the fluidity or instability of household residents, with entrances and exits from the household typically considered separately; few studies include multiple indicators of changes in the household (Adam, 2004). In other words, separations from the caregiver, exit of a husband or boyfriend, entrance of a new husband or boyfriend, the birth of siblings, and residential moves are not included in any one analysis. In addition, little is known about the accumulation of these household events (Ackerman, Kogos, Youngstrom, Schoff, & Izard, 1999).

In our research, we have attempted to model associations between all of these events and child well-being, rather than looking at each separately (Leventhal & Brooks-Gunn, 2000a). In addition, we have examined the accumulation of household events. Using data from the Infant Health and Development Program, we examined household events (maternal separations, moves in and out of male partners, birth of a sibling, and residential moves) that occurred during the first 8 years of life. IQ scores at ages 3, 5, and 8 and achievement test scores at age 8 were the outcomes of interest. All analyses controlled for maternal age, ethnicity, education, and verbal ability; household income; and child birth weight and neonatal health. Findings from this research illustrate several points. First, even with an extensive number of covariates considered, cumulative household events were associated with child IQ and achievement. Second, these links were stronger at age 3 than at age 8. And third, income (or resources) accounted for about one third to one half of the association. Further analyses of the timing of such household events suggested that the number of events occurring in the first 3 years of life and the 6th to 8th year of life were equally likely to be associated with age 8 IQ and reading achievement; however, events in the 4th and 5th years of life were not associated with age 8 outcomes. Again, adding income into the equation reduced but did not totally account for the links between events and age 8 test scores.

What we want to stress through highlighting our study is the importance of considering multiple instances of fluidity or instability in the household, rather than focusing on one at a time, as well as controlling for the possibility that associations are due to the presence of low resources (such as low income or parental education) rather than the coming and going of various household members and residences. We are still left, however, in need of an explanation for why these household events are linked to child outcomes. Here is where the examination of other indicators of chaos would be useful. For example, perhaps what characterizes families with high fluidity or instability is an environment where routines and supervision are low, or perhaps households with high fluidity are more likely to be crowded, noisy, or cluttered. Of course, other aspects of parenting (such as high levels of punitive behavior or detachment and low levels of teaching, language use, or sensitivity) might also be salient. The research based on routines and supervision is much more limited than that based on household changes. As stated earlier, it is not even known how much overlap exists between the three dimensions of disorder discussed in an earlier section and these two dimensions of turbulence. We look forward to the next generation of studies, such as Early Head Start and the Early Childhood Longitudinal studies, which are gathering data on routines and, to a lesser extent, supervision.

Neighborhood-Level Disorder and Turbulence: Associations With Child Well-Being

The literature on neighborhoods and chaos is much less well developed that that on families. In addition, almost nothing is known about the interaction between family and neighborhood chaos. As a consequence, the following review is brief.

Disorder

Neighborhood indicators of disorder, as presented earlier, include crowding or density of families, individuals, or housing units. A question raised by Dunn et al. in chapter 11 of this volume is relevant here: Is neighborhood social and physical disorder interchangeable with neighborhood-level chaos? We agree that the answer is no, and that more research needs to be done to further tease out the aspects of neighborhood chaos that overlap with neighborhood disorder, as well as the aspects that are distinct. We contend that, as mentioned earlier, one dimension of neighborhood disorder that may overlap with chaos but is not itself chaos is neighborhood density. Almost nothing is known about neighborhood density, over and above neighborhood-level indicators of disadvantage. In census tract analyses, density typically is not identified as a factor separate from concentrated disadvantage. Some studies have attempted to define a neighborhood structural variable that includes proportion of children or youth in a neighborhood; however, this is not the same as density.

Another dimension of disorder is noise. A research base does exist (see Evans, 2006). Suffice it to say here that residence in noisy urban neighborhoods is associated with poorer attention, memory, reading, and achievement (Evans, 2006; Evans & Hygge, 2007). Findings from a natural experiment suggest that the installation of noise-abating materials on the noisy side of a public school building reduced reading difficulties in second-, fourth-, and sixth-grade children (Bronzaft, 1981). Neighborhood noise stemming from air-traffic patterns has also been implicated in achievement (Hygge et al., 2002).

The final dimension discussed here, cleanliness, has been studied under the rubric of physical disorder. In general, studies using the observed disorder measures have examined links to juvenile delinquency and crime, rather than child well-being (Sampson & Raudenbush, 1999). As Dunn et al. argue (see chap. 11, this volume), much more work has been done to develop a connection between disorder and crime in neighborhoods than has been done to examine a connection between neighborhood disorder and child health and development (studies to this end of adults, and to a lesser extent adolescents, have relied almost exclusively on self-report neighborhood and outcome data and, therefore, suffer from problems of shared method variance). Thus, we feel that the question of whether physical disorder has an effect over and above concentrated disadvantage has not yet been well explored.

Turbulence

Fluidity or instability has been examined at the neighborhood level in terms of residential stability. Again, direct links with child outcomes have focused on adolescent delinquency and crime (e.g., Sampson et al., 1997). For young children, residential instability is associated with child maltreatment (Coulton, Crampton, Irwin, Spilsbury, & Korbin, 2007).

Neighborhood routines have been discussed theoretically but not empirically. Research on supervision and monitoring does exist, via the construct of collective efficacy. High collective efficacy is associated with lower juvenile delinquency and crime, even when controlling for structural features of the neighborhood (concentrated disadvantage, residential stability, and racial–ethnic composition of the neighborhood; Sampson et al., 1997). Links are also found for children's internalizing problems and school achievement (Leventhal, Xue, & Brooks-Gunn, 2006; Xue et al., 2005). Furthermore, differences in collective efficacy account in large part for the racial differences in the timing of sexual initiation (Browning, Leventhal, & Brooks-Gunn, 2004).

One interesting question is how family-level and neighborhood-level indicators of turbulence may interact to influence child well-being. We have conducted illustrative analyses of the joint effects of family-level and neighborhood-level supervision and monitoring on youth sexual behavior using the PHDCN nested data (Browning, Leventhal, & Brooks-Gunn, 2005). Three measures of parental control were calculated: place monitoring (i.e., amount of time allowed to be in public places without adults), peer monitoring (i.e., contact with the youth's friends), and family emotional and social support (i.e., family members having confidence and being available). Moderated analyses indicated that collective efficacy was associated with later sexual onset only when place monitoring was taken into account; collective efficacy was significant when youth experienced low levels of parental place monitoring. These associations were found when controlling for a host of individual-level and neighborhood-level characteristics. This last study is an example of what is needed in the field—that is, an examination of chaos (in this case, low supervision and monitoring) across environments. Few examples currently exist.

Implications for Research and Prevention

With regard to what has been covered in this chapter, the implications for research and prevention overlap in that a prevention agenda, if properly evaluated, will provide insights into the extent to which disorder and turbulence play a significant role in children's lives. That is, experimental evaluations would provide evidence for both the likelihood that change in disorder and turbulence will occur and the likelihood that such changes will affect child well-being. This approach is in keeping with our belief that nothing is as theoretical as a well-designed intervention (Brooks-Gunn, 2004). We outline our key points for future research directions and prevention efforts in the following sections.

Research

A theme that runs throughout our chapter is the call for more rigorous methodological approaches to the study of chaos and its effect on children. An experimental study in which children are randomly assigned to programs aimed at reducing characteristics of chaos in the home and neighborhood would greatly improve researchers' ability to draw conclusions about when, how, and to what extent different dimensions and indictors of chaos impact different child outcomes.

The consistent use of a set of chaos measures across large national studies of children would also offer great insight into how much the various dimensions of chaos overlap, in what ways they overlap, and on whether the overlap depends on sample characteristics. For those observational studies that have included measures of chaos, selection bias has been a problem. Longitudinal models, fixed-effect approaches (individual, family, and neighborhood), nested designs, regression discontinuity analyses, and natural experiments (exogenous shocks) can do much to address such biases.

Further research also needs to be conducted on the mechanisms through which disorder and turbulence influence child well-being. Also, a focus on multiple levels of the environment is needed, especially in studies where more than one environmental level is assessed. In particular, more work is needed with respect to other meso- and macrosystem links, particularly between home and school and between school and neighborhood. Although some research has been done to explore associations between several dimensions of chaos in the school and child outcomes, more attention must be paid to the interaction of school-level chaos with chaos at other environmental levels. Using our classification of aspects of chaos as belonging under disorder or turbulence, we can draw some preliminary parallels between home and neighborhood chaos and child development, and school chaos and child development.

Under indicators of disorder, and consistent with findings on the home and neighborhood levels, research has identified negative links between crowding and noise in schools and child outcomes. In particular, off-task behavior was observed more often among kindergarten children in classrooms with greater density (Krantz, 1974). Other studies have found that teachers in crowded classrooms tend to spend more time disciplining and less time actually teaching their students (Ehrenberg, Brewer, Gamoran, & Willms, 2001). Similarly, high degrees of noise in schools negatively affect teachers' abilities to teach effectively; teachers in noisy schools are more likely to be tired, annoyed, and less patient than are teachers of similar backgrounds who teach in quieter schools (Evans & Hygge, 2007). To date, research on associations between clutter and messiness in schools, and child well-being, is extremely limited. As Maxwell pointed out in chapter 6 of this volume, visual complexity and overstimulation stemming from cluttered classroom walls may disrupt the learning of elementary-school and adolescent children. However, this hypothesis needs further testing.

Regarding dimensions of turbulence, our other indicator of chaos, prior research has focused on examining connections between child well-being and school fluidity or instability. Again, findings are consistent with research on home- and neighborhood-level turbulence. For example, one study found a weak negative association between school transfers and school performance among elementary-school-age children (Alexander, Entwisle, & Dauber, 1996), and a different study on adolescents found evidence of a negative association between school transitions and social development (Seidman, Allen, Aber, Mitchell, & Feinman, 1994). Instability in both the school and the home appears to be most detrimental to adolescents' school performance compared with instability in either domain alone (Pribesh & Downey, 1999), though other research suggests that school changes, especially in the later years, have more adverse effects on educational outcomes than do residential moves (Swanson & Schneider, 1999). Beyond the research discussed by Corapci in chapter 5 of this volume on daily instability and arrangement multiplicity of child care, we found no quantitative studies that examine links between lack of predictability and routines in schools and child development (although case studies indicate that this aspect of school turbulence is likely related to student outcomes). Also, work has yet to be done to examine whether there are associations between a lack of supervision and monitoring in school and child outcomes.

Prevention

Much less time has been spent on prevention than has been spent on research. A few promising approaches are discussed here. In all cases, evidence needs to focus on whether the environment can be altered and on whether such changes result in changes in child well-being.

In general, prevention efforts might be designed to alter the level of disorder in the home or neighborhood. Several efforts have attempted to alter disorder in the home. These include experiments to improve housecleaning techniques in residences with high levels of allergens, to reduce the incidence of asthma. Overall, these approaches have not been effective; however, it is not known whether the lack of results is due to the compliance of caregivers with the household regime or whether reducing allergens in the home is more difficult than anticipated (Richardson, 2005; Tortolero, 2007). With regard to noise and confusion, interventions have attempted to reduce the number of hours that televisions are on in households (Brooks-Gunn & Donahue, 2008), although success is quite variable across experiments.

With respect to disorder in the neighborhood, the best examples are in the realm of public health, including reducing lead in gasoline (likely differentially benefiting children who lived in congested neighborhoods; Berney, 1993) as well as cleaning up water supplies on Indian reservations, which altered the incidence of illness (Watson, 2005). Regression discontinuity designs could also be used to investigate changes in child well-being when airport or highway traffic patterns are altered (itself a form of prevention).

Some prevention efforts have focused on reducing turbulence in homes, at least in terms of fluidity and instability of household members. The federal government has mounted and is evaluating the efficacy of psychological approaches to strengthening marriages (Dion, 2005). This initiative is based on small-scale experiments demonstrating the benefits of relationship training and counseling in preventing divorce (Markman et al., 2004). Lowering of the breakup rate could influence children in a variety of ways, from the presence of a father in the home, to the increase in routines and predictability, to the reduction in residential moves.

On the neighborhood level, programs to reduce turbulence are also possible. Homeownership in poor neighborhoods is enhanced by Home Builders and Habitat for Humanity programs, although whether such programs alter child well-being has not been evaluated. Employment programs (such as the Enterprise Zones) aim to increase the number of employed adults in a neighborhood; if Wilson (1987) is right, such increases might alter the routines in a neighborhood as well as the turn-over in residents (because of the ability to pay rent). Such initiatives could possibly influence child well-being (although such possibilities have not been assessed). Reduction of neighborhood disorder and increase in safety have the potential to reduce anxiety among children (Xue et al., 2005). Such programs have been initiated in large cities, although they are not experiments and impacts on children are not examined.

In conclusion, we wish to underline our overarching point: If we want to demonstrate that chaos impacts children's lives, then we must be able to show that a change in chaos is linked to a change in outcomes and that it is possible to alter the level and type of chaos in families and neighborhoods. We hope that this volume will move chaos theory to the forefront of research and prevention efforts on behalf of vulnerable children. We also hope that it will propel scholars, including ourselves, to link their work more closely to the construct of chaos.

References

Ackerman, B. P., Kogos, J., Youngstrom, E., Schoff, K., & Izard, C. (1999). Family instability and the problem behaviors of children from economically disadvantaged families. *Developmental Psychology, 35*, 258–268.

Adam, E. K. (2004). Beyond quality: Parental and residential stability and children's adjustment. *Current Directions in Psychological Science, 13*, 210–213.

Alexander, K. L., Entwisle, D. R., & Dauber, S. L. (1996). Children in motion: School transfers and elementary school performance. *The Journal of Educational Research, 90*, 3–12.

Baydar, N., Greek, A., & Brooks-Gunn, J. (1997). A longitudinal study of the effects of the birth of a sibling during the first six years of life. *Journal of Marriage and the Family, 59*, 939–956.

Berney, B. (1993). Round and round it goes: The epidemiology of childhood lead poisoning, 1950-1990. *The Milbank Quarterly, 17*, 3–39.

Bornstein, M. (Ed.). (2002). *Handbook of parenting* (2nd ed.). Hillsdale, NJ: Erlbaum.

Boyce, W. T., Jensen, E. W., James, S. A., & Peacock, J. L. (1983). The Family Routines Inventory: Theoretical origins. *Social Science and Medicine, 17*, 193–200.

Bradley, R. H., Caldwell, B. M., Rock, S. L., Barnard, K., Gray, C., Hammond, M., et al. (1989). Home environment and cognitive development in the first 3 years of life: A collaborative study involving six sites and three ethnic groups in North America. *Developmental Psychology, 25*, 217–235.

Bronfenbrenner, U. (1979). *The ecology of human development.* Cambridge, MA: Harvard University Press.

Bronfenbrenner, U. (2005). *Making human beings human.* Los Angeles, CA: Sage.

Bronzaft, A. L. (1981). The effect of a noise abatement program on reading ability. *Journal of Environmental Psychology, 1*, 215–222.

Brooks-Gunn, J. (2004). Don't throw out the baby with the bathwater: Incorporating behavioral research into evaluations. *Social Policy Report, 18*(2), 14–15.

Brooks-Gunn, J., & Donahue, E. (Eds.). (2008). Electronic media in the lives of children and youth. *The Future of Children, 18*(1), 3–10.

Brooks-Gunn, J., Duncan, G. J., & Aber, J. L. (Eds.). (1997). *Neighborhood poverty: I. Context and consequences for children.* New York: Russell Sage.

Brooks-Gunn, J., Fuligni, A. S., & Berlin, L. J. (2003). *Early child development in the 21st century: Profiles of current research initiatives.* New York: Teachers College Press.

Browning, C., Leventhal, T., & Brooks-Gunn, J. (2004). Neighborhood context and racial differences in early adolescent sexual activity. *Demography, 41*, 697–720.

Browning, C., Leventhal, T., & Brooks-Gunn, J. (2005). Sexual initiation in early adolescence: The nexus of parental and community control. *American Sociological Review, 70*, 758–778.

Caldwell, B., & Bradley, R. H. (1984). *Home Observation for Measurement of the Environment.* Little Rock, AR: University of Arkansas.

Coldwell, J., Pike, A., & Dunn, J. (2006). Household chaos: Links with parenting and child behaviour. *Journal of Child Psychology and Psychiatry, 47*, 1116–1122.

Collins, W. A., Maccoby, E. E., Steinberg, L., Hetherington, M. E., & Bornstein, M. H. (2000). Contemporary research on parenting: The case for nature and nurture. *American Psychologist, 55*, 218–232.

Coulton, C. J., Crampton, D. S., Irwin, M., Spilsbury, J. C., & Korbin, J. E. (2007). How neighborhoods influence child maltreatment: A review of the literature and alternative pathways. *Child Abuse & Neglect, 31*, 1117–1142.

Dion, R. M. (2005). Healthy marriage programs: Learning what works. The Future of Children, 15(2), 139–156.

Dumas, J. E., Nissley, J., Nordstrom, A., Phillips Smith, E., Prinz, R. J., & Levine, D. J. (2005). Home chaos: Sociodemographic, parenting, interactional, and child correlates. *Journal of Clinical Child and Adolescent Psychology, 34*, 93–104.

Dunifon, R., Duncan, G. J., & Brooks-Gunn, J. (2001). As ye sweep, so shall ye reap. *The American Economic Review, 91*, 150–154.

Ehrenberg, R. G., Brewer, D. J., Gamoran, A., & Willms, J. D. (2001). Class size and student achievement. *Psychological Science in the Public Interest, 2*, 1–30.

Evans, G. W. (2006). Child development and the physical environment. *Annual Review of Psychology, 57,* 423–451.
Evans, G. W., Gonnella, C., Marcynyszyn, L. A., Gentile, L., & Salpekar, N. (2005). The role of chaos in poverty and children's socioemotional adjustment. *Psychological Science, 16,* 560–565.
Evans, G. W., & Hygge, S. (2007). Noise and performance in children and adults. In L. Luxon & D. Prasher (Eds.), *Noise and its effects* (pp. 549–566). London: Wiley.
Fiese, B. H. (2006). *Family routines and rituals.* New Haven, CT: Yale University Press.
Fletcher, A. C., Darling, N. E., Steinberg, L., & Dornbusch, S. M. (1995). The company they keep: Relation of adolescents' adjustment and behaviour to their friends' perceptions of authoritative parenting in the social network. *Developmental Psychology, 31,* 300–310.
Fuligni, A. S., & Brooks-Gunn, J. (2002). Meeting the challenge of new parenthood: Responsibilities, advice and perceptions. In N. Halfon, K. T. McLearn, & M. A. Schuster (Eds.), *Child rearing in America: Challenges facing parents with young children* (pp. 83–116). New York: Cambridge University.
Gottfried, A. E., & Gottfried, A. W. (1984). Home environment and cognitive development in young children of middle socio-economic status families. In A. Gottfried (Ed.), *Home environment and early cognitive development* (pp. 57–116). Orlando, FL: Academic Press.
Hart, S. A., Petrill, S. A., Deater Deckard, K., & Thompson, L. A. (2007). SES and CHAOS as environmental mediators of cognitive ability: A longitudinal genetic analysis. *Intelligence, 33,* 233–242.
Haveman, R., Wolfe, B., & Spaulding, J. (1991). Childhood events and circumstances influencing high school completion. *Demography, 28,* 133–157.
Hill, M. (1992). *The Panel Study of Income Dynamics.* Newbury Park, CA: Sage.
Hygge, S., Evans, G. W., & Bullinger, M. (2002). A prospective study of some effects of aircraft noise on cognitive performance in schoolchildren. *Psychological Science, 13,* 469–474.
Jensen, E. W., James, S. A., Boyce, W. T., & Hartnett, S. A. (1983). The Family Routines Inventory: Development and validation. *Social Science and Medicine, 17,* 201–211.
Johnson, A. D., Martin, A., Brooks-Gunn, J., & Petrill, S. A. (2008). Order in the house! Associations among household chaos, the home literacy environment, maternal reading ability, and children's early reading. *Merrill-Palmer Quarterly, 54,* 445–472.
Kohen, D. E., Brooks-Gunn, J., Leventhal, T., & Hertzman, C. (2002). Neighborhood income and physical and social disorder in Canada: Associations with young children's competencies. *Child Development, 73,* 1845–1860.
Krantz, P. J. 1974. *Ecological arrangements in the classroom.* Unpublished doctoral dissertation. Lawrence: University of Kansas.
Leventhal, T., & Brooks-Gunn, J. (2000a). *"Entrances" and "exists" in children's lives: Associations between household events and test scores.* Unpublished manuscript.
Leventhal, T., & Brooks-Gunn, J. (2000b). The neighborhoods they live in: The effects of neighborhood residence upon child and adolescent outcomes. *Psychological Bulletin, 126,* 309–337.
Leventhal, T., & Brooks-Gunn, J. (2007). *Changes in neighborhood poverty from 1990 to 2000 and changes in children's behavior problems over six years.* Manuscript submitted for publication.
Leventhal, T., Xue, Y., & Brooks-Gunn, J. (2006). Immigrant differences in school-age children's verbal trajectories: A look at four racial/ethnic groups. *Child Development, 77,* 1359–1374.
Linver, M. R., Martin, A., & Brooks-Gunn, J. (2004). Measuring infants' home environment: The IT-HOME for infants between birth and 12 months in four national data sets. *Parenting Science and Practice, 4,* 115–137.
Maccoby, E., & Martin, J. (1983). Socialization in the context of the family: Parent-child interaction. In E. M. Hetherington (Ed.), *Handbook of child psychology: Vol. 4. Socialization, personality and social development* (pp. 1–101). New York: Plenum Press.
Markman, H. J., Whitton, S. W., Kline, G. H., Stanley, S. M., Thompson, H., St. Peters, M., et al. (2004). Use of an empirically based marriage education program by religious organizations: Results of a dissemination trial. *Family Relations, 53,* 504–512.
Matheny, A. P., Wachs, T. D., Ludwig, J. L., & Phillips, K. (1995). Bringing order out of chaos: Psychometric characteristics of the Confusion, Hubbub, and Order Scale. *Journal of Applied Developmental Psychology, 16,* 429–444.

Maxwell, L. E. (1996). Multiple effects of home and day care crowding. *Environment and Behavior, 28*, 494–511.

Maxwell, L. E., & Evans, G. W. (2000). The effects of noise on pre-school children's pre reading skills. *Journal of Environmental Psychology, 20*, 91–97.

Meadows, S. O., McLanahan, S. S., & Brooks Gunn, J. (2008). Stability and change in family structure and maternal health trajectories. *American Sociological Review, 73*, 314–334.

Moore, K. A., & Vandivere, S. (2000). *Stressful family lives: Child and parent well-being* (Assessing the New Federalism Series B, No. B-17). Washington, DC: Urban Institute Press.

Pike, A., Iervolino, A. C., Eley, T. C., Price, T. S., & Plomin, R. (2006). Environmental risk and young children's cognitive and behavioral development. *International Journal of Behavioral Development, 30*, 55–66.

Pribesh, S., & Downey, D. B. (1999). Why are residential and school moves associated with poor school performance? *Demography, 36*, 521–534.

Richardson, G. (2005) How is the indoor environment related to asthma?: Literature review. *Journal of Advanced Nursing, 52*, 328–339.

Sampson, R. J., & Morenoff, J. (2006). Durable inequality: Spatial dynamics, social processes, and the persistence of poverty in Chicago Neighborhoods. In S. Bowles, S. Durlauf, & K. Hoff (Eds.), *Poverty traps* (pp. 176–203). Princeton, NJ: Princeton University Press.

Sampson, R. J., Morenoff, J. D., & Gannon-Rowley, T. (2002). Assessing 'neighborhood effects': Social processes and new directions in research. *Annual Review of Sociology, 28*, 443–478.

Sampson, R. J., & Raudenbush, S. W. (1999). Systematic social observation of public spaces: A new look at disorder in urban neighborhoods. *The American Journal of Sociology, 105*, 603–651.

Sampson, R. J., Raudenbush, S., & Earls, F. (1997, August 15). Neighborhoods and violent crime: A multilevel study of collective efficacy. *Science, 277*, 918–924.

Seidman, E., Allen, L., Aber, J. L., Mitchell, C., & Feinman, J. (1994). The impact of school transitions in early adolescence on the self-system and perceived social context of poor urban youth. *Child Development, 65*, 507–522.

Swanson, C., & Schneider, B. (1999). Students on the move: Residential mobility and educational mobility in American's schools. *Sociology of Education, 72*, 54–67.

Tortolero, S. R. (2007). Prevalence of asthma symptoms in a screened school-age population of poor children in Houston, Texas (1997-1998). *Pediatric Asthma, Allergy & Immunology, 20*, 11–18.

Wachs, T. D. (1979). Proximal experience and early cognitive-intellectual development: The physical environment. *Merrill-Palmer Quarterly, 25*, 3–42.

Wachs, T. D. (1989). The nature of the physical microenvironment: An expanded classification system. *Merrill-Palmer Quarterly, 35*, 399–402.

Wachs, T. D. (1992). *The nature of nurture*. Newbury Park, CA: Sage.

Watson, T. (2005). Public health investments and the infant mortality gap: Evidence from federal sanitation interventions on U.S. Indian reservations. *Journal of Public Economics, 90*, 1537–1560.

Wilson, W. J. (1987). *The truly disadvantaged: The inner-city, the underclass, and public policy*. Chicago: University of Chicago Press.

Wu, L. L. (1996). Effects of family instability, income, and income instability on the risk of a premarital birth. *American Sociological Review, 61*, 386–406.

Xue, Y., Leventhal, T., Brooks-Gunn, J., & Earls, F. (2005). Neighborhood residence and mental health problems of 5- to 11-year-olds. *Archives of General Psychiatry, 62*, 1–10.

Part IV

Chaos at the Exosystem Level

11

Neighborhood Chaos and Children's Development: Questions and Contradictions

James R. Dunn, Nicole J. Schaefer-McDaniel, and Jason T. Ramsay

Individuals and families who live in a neighborhood collectively create a social context that influences the developing child, and this influence has an additive effect above the effects of family and individual characteristics (Sampson, Morenoff, & Earls, 1999). Although Bronfenbrenner (1977) originally suggested that neighborhoods function as exosystems, current neighborhood researchers argue they are a part of the microsystem because residents, particularly children and young people, spend a significant time outside exploring their neighborhoods (Chawla, 2002; Hart, 1979; Lynch, 1977; Proshansky & Fabian, 1987). Children are the primary consumers of the neighborhood (Holaday, Swan, & Turner-Henson, 1997), and it is more or less their "turf" (Burton & Price-Spratlen, 1999, p. 78).

In this chapter, we provide a broad overview of the existing literature on the relationship between neighborhoods and child development, and we explore how chaos at the neighborhood level might influence children's lives and well-being.

Conceptualizing the Ecology of Child Development

According to Bronfenbrenner (1977, 1979, 1986), several propositions about the influence of the environment on child development are required to elaborate a theory of neighborhood influences on child development.

> Proposition I: Experience (including both objective and subjective experience) is a key element of the ecological model of development.
> Proposition II: Children develop healthily and competently when they benefit from repeated, complex, and bidirectional interactions with other people (particularly the primary caretaker) and objects in their everyday life. Healthy interactions are characterized as positive in nature and respectful of the child's person and emotions.

Proposition III: Development occurs as a joint function of the characteristics of (a) the developing person, (b) the environment—both immediate and remote—in which the processes are taking place, (c) the nature of the developmental outcomes under consideration, and (d) the continuities and changes occurring in the environment over time, over the life course, and over the historical period in which the person has lived.

Proposition IV: Mutual emotional attachment between the child and the child's primary caregiver motivates the child's interest in the immediate physical, social, and symbolic environment inviting exploration, manipulation, elaboration, and imagination.

In addition to establishing a broad outline of key principles of child development and how the environment may influence these processes, it is necessary to establish the key elements of child development so that they can be measured and the relative influence of neighborhood factors can be investigated for each element. Following the work of Kagan (1992), Doherty (1997), and Janus and Offord (2007), we adopt the position that the key elements of child development are social development, emotional development, cognitive development, language development, and healthy physical development.

Defining Neighborhoods

A central issue facing neighborhood research is how to conceptualize, spatially define, and measure neighborhoods. *Neighborhoods* have been conceptualized as local social contexts in which individuals and families interact and engage with local institutions (Aber, Gephart, Brooks-Gunn, Connell, & Spencer, 1997). Research has not yet determined an optimal strategy to define meaningful geographical neighborhood units that conform to actual patterns of social interaction for analysis (Sampson, Morenoff, & Gannon Rowley, 2002). We do not yet know the exact mechanisms and pathways through which neighborhoods influence children and the scale at which they operate. Moreover, neighborhood-based mechanisms may not function at a single scale: Networks among parents may have a different geographic scale than do the influence of local institutions or facilities such as libraries and parks. Here we employ a recent definition of neighborhoods as a "bundle of spatially based attributes associated with residences, sometimes in conjunction with other land uses" (Galster, 2001, p. 2113).

Most studies typically use census tracts or other administrative boundaries to define neighborhood units, which may not conform to the "actual" neighborhoods perceived or experienced by the people who live there (Diez Roux, 2001). Some researchers have attempted to define more meaningful neighborhood boundaries using people's perceptions of the spatial boundaries of their neighborhood (Coulton, Korbin, Chan, & Su, 2001; Hume, Salmon, & Ball, 2005). This strategy is not a panacea, however, because it omits the plethora of processes, structural and social, operating at the neighborhood level, which a person may not experience every day, but which, over time, may impinge significantly on developmental trajectories. For example, areas that respondents would not consider part of their neighborhood despite physical proximity to their neighborhood (e.g., a street known to be violent or frequented by drug dealers) may have an impact on the developmental processes despite not being a place where children go to play. Further, disagreement among individuals in a local area about neighborhood boundaries may make deciding on particular boundaries difficult.

Though census tracts may not coincide with other competing neighborhood boundary definitions, recent research suggests that these spatial areas may be the best approximations of actual neighborhoods. In a multilevel analysis of neighborhood socioeconomic status (SES) influences on health, N. Ross, Tremblay, and Graham (2004) found that empirical results for census tracts were remarkably similar to "natural" neighborhoods; the latter are defined through a consultative process with a variety of community groups and citizens. Another advantage of using census tracts is the availability of census data and their standardized measurement.

Evidence of Effects of Neighborhoods on Child Development: A Summary

Most neighborhood effects studies have focused on the outcomes of older children, especially adolescents. This research has generally examined the role of neighborhood factors on high school completion (Brooks-Gunn, Duncan, Klebanov, & Sealand, 1993; Halpern-Felsher et al., 1997), academic achievement (Entwisle, Alexander, & Olson, 1994; Garner & Raudenbush, 1991), behavior problems and criminal activity (Ennett, Fewelling, Lindrooth, & Norton, 1997; Peeples & Loeber, 1994), sexuality and teenage pregnancy (Billy, Brewster, & Grady, 1994; Billy & Moore, 1992; Brewster, 1994a, 1994b; Ku, Sonenstein, & Pleck, 1993), and smoking (Billy & Moore, 1992; Briggs, 1997a, 1997b). This research has also relied mainly on census data to measure neighborhood characteristics and has found modest but statistically significant relations between increasing neighborhood disadvantage and problem behaviors among adolescents and youth.

Despite the strong focus on teenagers, a smaller body of research has demonstrated that the health and well-being of young children (0 to 6 years old) are also influenced by the neighborhood characteristics of where they live. This research has found neighborhood effects on both cognitive (Brooks-Gunn et al., 1993; Chase-Lansdale & Gordon, 1996; Chase-Lansdale, Gordon, Brooks-Gunn, & Klebanov, 1997; Duncan, Brooks-Gunn, & Klebanov, 1994; Klebanov, Brooks-Gunn, McCarton, & McCormick, 1998; McCulloch & Joshi, 2001) and behavioral–emotional outcomes for young children (Brooks-Gunn et al., 1993; Chase-Lansdale & Gordon, 1996; Chase-Lansdale et al., 1997; Duncan et al., 1994; Kalff et al., 2001, Kohen, Brooks-Gunn, Leventhal, & Hertzman, 2002). These studies suggest that residing in a disadvantaged neighborhood is associated with poorer developmental outcomes even after adjusting for individual and family characteristics. Further, stronger neighborhood effects have typically been reported for cognitive outcomes than for behavioral and social outcomes of young children.

Studies exploring the relationship between neighborhoods and cognitive, behavioral, and emotional child development typically use data from national longitudinal surveys of a nationally representative sample of children and youth. In the United States, neighborhood effects studies have used data from the National Longitudinal Survey of Youth (NLSY; Chase-Lansdale & Gordon, 1996; Chase-Lansdale et al., 1997) and the Infant Health and Development Program (IHDP; Brooks-Gunn et al., 1993; Duncan et al., 1994; Klebanov et al., 1998). The National Child Development Study (NCDS) has been used in Britain (McCulloch & Joshi, 2001) and the National Longitudinal Survey of Children and Youth (NLSCY) has been used in Canada (Boyle & Lipman, 2002; Kohen, Hertzman, & Brooks-Gunn, 1998; Kohen, Hertzman, & Wiens, 1998; Kohen et al., 2002).

These studies have found an association between neighborhood deprivation and reduced cognitive development and achievement of young children when controlling for individual factors. For example, McCulloch and Joshi (2001) found a significant association between neighborhood deprivation and lower cognitive test scores of children ages 4 to 5 using data from the NCDS. On the basis of data from the IHDP, neighborhood affluence was positively associated with the IQ of children age 3 (Brooks-Gunn et al., 1993) and children age 5 (Duncan et al., 1994). Chase-Lansdale et al. (1997), using the NLSY, found that high neighborhood SES was positively associated with reading achievement and the Peabody Picture Vocabulary Test—Revised (PPVT–R), a commonly used measure of cognitive ability, for children ages 5 and 6. Also using the NLSY, researchers have found a positive association between neighborhood SES and verbal IQ of 5-year-olds (Chase-Lansdale & Gordon, 1996). Last, using NLSCY data, Kohen, Hertzman, and Brooks-Gunn (1998) found a positive relation between higher neighborhood income and toddlers' PPVT–R scores.

In summary, these studies, using different neighborhood units, data sets, child outcome measures, and neighborhood characteristics, have consistently demonstrated that neighborhood SES can influence young children's emotional and behavioral outcomes, independently of their family background. Unfortunately, most of these studies estimating neighborhood effects on child development are not able to use hierarchical statistical techniques required for a more reliable estimation of neighborhood effects and may consequently underestimate neighborhood effects (Duncan & Raudenbush, 1999). Another shortcoming of neighborhood studies such as these is the absence of consideration for the mechanisms that "cause" these effects to occur. However, a small but growing body of literature has sought to develop better theory to explain relationships between neighborhood characteristics and child health and well-being (Furstenberg & Hughes, 1997).

For example, affluent neighborhoods are thought to establish shared norms regarding child rearing among adults and a willingness of adults to intervene on behalf of children, and to consist of adults who have resources (e.g., time, money) to organize and establish adequate institutional assets such as schools, libraries, and parks to promote child development (Jencks & Mayer, 1990; Sampson et al., 1999; W. J. Wilson, 1987). Concentrated neighborhood poverty, in contrast, is theorized to produce neighborhood environments that are disadvantageous for children (Jencks & Mayer, 1990; Sampson et al., 1999; W. J.

Wilson, 1987). In particular, impoverished neighborhoods are believed to lack the resources required to establish an environment beneficial for child development (Sampson et al., 1999).

In the following section, we examine the existing research on neighborhoods and child development. Our purpose is to determine how well the concept of chaos at the neighborhood level integrates with existing theory on neighborhoods and child development. In particular, we examine the connections between chaos and (a) neighborhood SES, (b) neighborhood social and physical disorder, and (c) neighborhood social capital.

Conceptualizing Chaos at the Neighborhood Level

Our goal is a conceptual one: to develop an integrate set of concepts regarding the impact of neighborhood chaos on child development across disciplines. We use this conceptual framework for critically reflecting on gaps in current research and suggesting fruitful avenues of future research for the continued examination of chaos and its impact on child development.

Chaos as Neighborhood Poverty or Disadvantage?

The neighborhood literature is limited in its measurement of the setting because researchers rely largely on crude census data to provide information about neighborhoods. In practice, researchers select variables on the basis of theoretical importance in the relevant literature and enter them into models individually or combine them into an index. The latter technique is particularly advantageous for overcoming problems of multicollinearity but problematic because it does not allow for an examination of specific neighborhood socioeconomic characteristics that are most important (Small & Newman, 2001). Despite a significant variation in census variables selected in neighborhood effects studies, they most commonly represent the neighborhood dimensions of SES, ethnic diversity, family composition, and residential stability.

This approach then raises the question of whether one can equate chaos at the neighborhood level with disadvantage or low SES as measured in the census. This proposition would then assume that lower SES neighborhoods are inherently more chaotic than higher income or wealthier neighborhoods. One might argue that lower income neighborhoods are more troubled by disruption or unexpected turbulence (see chap. 10, this volume) than are higher income neighborhoods, where life might proceed more peacefully. For example, consider the turbulent lives of residents in poor neighborhoods such as the South Bronx in New York City (Kozol, 1996, 2001) or Chicago (Kotlowitz, 1991), where residents are often witnesses to neighborhood violence, gang activity, and social isolation and abandonment. However, classifying and thinking about chaos in such a dichotomous way (i.e., chaos = poverty; no chaos = wealth) essentially ignores the foundation of Bronfenbrenner's ecological theory. Such a definition of chaos is based on static (and problematic) classifications derived from the census report and therefore ignores his focus and emphasis on the lived experience of people in their settings. Further, in its true definition, *chaos* refers to sudden, unexpected, and unintended disruptions, but poverty levels, particularly at the neighborhood level, can be relatively stable circumstances that may nevertheless undermine development. So, rather than relying on socioeconomic indicators to classify a neighborhood as chaotic, a better method would be to focus on the residents' lived experiences in the setting.

Chaos as Neighborhood Disorder?

Neighborhood disorder generally refers to physical incivilities such as obvious abandonment and deterioration of housing or commercial units, but the term can also apply to social or behavioral incivilities such as vandalism or gang activity. Physical incivilities set the stage for social incivilities. Neighborhoods typically reach a state of physical disorder because of a lack of fiscal resources for maintenance of amenities and generalized disinvestment within the neighborhood. That in itself is connected to the presence of low-income census tracts and a low-income tax base. Without the ability to generate capital to improve physical order (e.g., renovating derelict buildings), there is a strong likelihood that these neighborhoods will not attract businesses or other investments. This situation can increase the proliferation of further physical and social incivilities and especially can create a crime-receptive environment, which further perpetuates physical and social incivilities. In an environment with a high prevalence of disorder, the perception of social trust is very low and the fear of violent crime high. Such an environment can constrain children's (and parents') perceptions of the types of activities that are possible within their surroundings. It both increases instability in their lives through perceived threat and unexpected alterations in daily routines and, in many cases, produces an overly cautious and excessively routinized or controlled life experience.

The roots of scholarly research on neighborhood disorder can be found in research in the sociology of crime. Both social and physical disorder have been demonstrated to be related to a number of different outcomes (J. Q. Wilson & Kelling, 1982). In social epidemiology and child development studies, the same measurement tools have been used to study the relationship between neighborhood disorder and health (e.g., Cohen et al., 2000) and child development outcomes (Calvert, 2002). However, the theoretical logic of the connection between neighborhood disorder and crime is much better developed than that of the connection between neighborhood characteristics and health or child development. Part of the problem is that the outcome measures used in health and child development neighborhood studies are too broadly focused; effects of neighborhoods on health and child development are likely specific to focused outcomes such as physical well-being and cognitive development, but less specific to emotional and social development. At the root of the problem in such studies lies a lack of theoretical development in terms of how neighborhood characteristics influence children's development. For example, researchers rarely consider a theoretical rationale for why (or whether) the presence of abandoned buildings in a neighborhood relates to children's academic achievement or development of social competencies.

Further, although neighborhood disorder assesses components of neighborhood life that go beyond the simple dichotomy of affluent versus poor neighborhood, the measurement of disorder remains problematic. Even one of the more celebrated innovations in the measurement of neighborhood, systematic social observations (SSO), is not without its problems. SSO is a methodology in which outsiders observe and rate a neighborhood in terms of social and physical dimensions (Sampson & Raudenbush, 2004). Although this methodology is better than the use of static census data to measure neighborhoods, it has a number of limitations associated with unexamined issues related to repeat measurement, geographical scale, and the influence of observer characteristics on ratings (Schaefer-McDaniel, 2007). Bronfenbrenner might also criticize the fact that an outsider was describing and assessing chaos in a given neighborhood rather than the residents who live there and experience it every day.

A small but growing body of research recognizes the limitation of relying on outsiders to measure neighborhood characteristics that are best described by people who live there. Therefore, a small number of researchers are currently surveying residents about their views of neighborhood disorder (e.g., C. E. Ross & Mirowsky, 1999, 2001). Such a methodology is better if one equates neighborhood disorder with chaos, but we still have reservations in terms of whether these two concepts are the same. The term *disorder* generally refers to and is defined in terms of extreme circumstances such as devastating physical abandonment or deterioration whereas *chaos* can apply to more nuanced and less extreme situations and conditions. Thus, the term *disorder* comprises elements of chaos (e.g., noise, violence) but acts more as an umbrella term for anything undesirable in a neighborhood.

The other potential problem with disorder is that if one accepts its overlap with chaos and agrees that both chaos and disorder are undesirable attributes of a neighborhood, is the opposite—order—what is really wanted? Can there be too much order as well? If so, then what state should be desired for children's environments? Taken to an extreme, efforts to increase order in developmental environments for children may also have important shortcomings, which would have important policy implications.

In *The Uses of Disorder* Sennett (1970) suggested that the myth of the "purified community," which is arguably dominant in North American society, is the outcome of the same motives that cause racial segregation in U.S. cities, and particularly the exclusion of poor, minority families from "better" neighborhoods. If this myth produces a state of order for those who remain, is it desirable, or even acceptable? Sennett was clear on the issue. In his chapter "Ordinary Lives in Disorder," he introduced the idea of an anarchic community as a growth opportunity for both individuals and groups:

> This book contrasts a society that is with a society that could be. On the one hand, there exists a life in which the institutions of the affluent city are used to lock men [sic] into adolescence even when physically adult. On the other hand, there is a possibility that affluence and the structures of a dense, disorganized city could encourage men to become more sensitive to each other as they become fully grown. I believe the society that could be is not a utopian ideal; it is a better arrangement of social materials, which as organized today are suffocating people. (Sennett, 1970, p. 189)

Sennett's views call into question the compatibility of chaos with the disorder–order continuum. Consider the example of a gated community. There is little disorder, and a great deal of order, by design. But is there chaos? Governed by rules about how many people can live there, a gated community is indeed a place that would not have a high density of people or a high level of crowding. In addition, gated communities are not likely to be noisy or cluttered and messy. They are likely to be quite stable in terms of who lives there, as are most homeownership communities. At the neighborhood level, the predictability of routines is probably less important than the predictability of family routines, but in a controlled environment like a gated community, few things in the neighborhood would interrupt those family routines. The final dimension, low supervision and monitoring in children, could occur in a gated community because parents believe that the "gates provide the supervision" (thanks to Gary Evans for this phrasing) and protect their children from harm.

The opposite of chaos may not be disorder per se. The opposite of a chaotic environment would be one that is characterized as "dynamic." Drawing on Bronfenbrenner's ecological model, we propose the following characteristics of a dynamic neighborhood:

Proposition I: A range of feelings, thoughts, emotions and motivations are experienced. The environment provides a safe, generally accepting forum in which to reflect on such feelings and motivations.

Proposition II: Interaction between the child and the environment, especially significant others, is active and progressively more complex over extended periods. This implies that the transit of people into and from the child's life is not rapid, abrupt, or excessively discontinuous.

Proposition III: The neighborhood allows for bidirectional engagement between children and adults. This interaction is a form of mutual coregulation that is viewed as positive by both the child and adult. Neighborhood factors could impinge on coregulation through several avenues: A relatively poor neighborhood may place stresses on adults and especially parents that limit the levels of mutual coregulation that can occur. Coregulation may occur within the family, but not within other areas. A prerequisite for the development of mutual coregulation is contact between the child and adults in the child's life. Excessive transit of people in the child's life or lack of opportunities to spend time together limit the ability of mutual coregulation to evolve.

Proposition IV: The neighborhood could limit the activities that the child participates in, especially complex activities provided by recreation groups, reading clubs, and other groups that enable the child to develop the capacity to engage in more complex forms of mutual coregulation with parents, peers, teachers, and significant others in the child's realm. A neighborhood high in chaos could curtail this possibility because of perceptions of safety or lack of social cohesion or organization. Factors at the neighborhood level could limit the ability of children and adults to form mutual attachments through lack of opportunity, time, or space for these relationships to develop.

Proposition V: All of the factors operating at the neighborhood level that are related to poor developmental outcomes, such as poverty, lack of social cohesion, perceived mistrust, lack of basic levels of physical and social order, and lack of social capital, work against the neighborhood, giving rise to opportunities for mutual coregulation.

We believe mutual coregulation is closely related to the development of self-regulation in children. Neighborhoods that are chaotic and disordered are likely to have poor cohesion, lack of trust, low income, and little social capital. The existence of such factors may hamper not only bidirectional influences between children and peers and between children and adults but also by extension the development of self-regulation. The existing literature on social capital and child development, however, is not sufficiently theorized to allow a connection to be made between chaos and Bronfenbrenner's propositions.

Chaos as Lack of Social Capital?

Social capital generally refers to qualities that lie in social relationships and networks such as trust and reciprocity. Key problems with the concept are its analytical vagueness, the underdevelopment of theoretical accounts of its links to health and child development, and the relative lack of analysis that situates social capital in the context of real events and settings, rather than abstracting from daily life via survey questions and opinion polls.

A good example of a study that does situate social capital in the context of real events at the neighborhood level appears in the book *Heat Wave: A Social Autopsy of a Disaster in Chicago* (Klinenberg, 2002). Klinenberg contrasted two neighborhoods in Chicago with similar SES that had radically different mortality in the 1996 heat wave: The first neighborhood was an African American one with a high level of gang and drug activity where elderly people, too afraid to go outside and lacking social connections to ask others to come and help them, died alone in their apartments from extreme heat stroke. He contrasted this with a Hispanic neighborhood where the streets were safer, street life was vibrant, and social networks were denser, meaning that individuals and local community organizations knew to check up on people who were vulnerable to the heat wave. Although this summary is a gross oversimplification of the analysis, it does illustrate how a textured, context-bound account of real-life events are important to building an understanding of the influence of neighborhood chaos on child development and the shortcomings of an approach focused just on survey question or systematic social observations that abstract these neighborhood characteristics from their context.

Klinenberg (2002) essentially illustrated that chaotic neighborhoods, namely, those low in social capital, had poorer outcomes during the heat wave compared with nonchaotic neighborhoods where residents took care of each other. However, simply equating chaos with an absence of social capital raises problems. What are the material preconditions for creating social capital? Is sentencing a large proportion of single parents to poverty via public policy an impediment to social capital? To then suggest it is simply the poor's unwillingness or lack of interest in participating in formal groups and organizations that prevents social capital from being created is a classic example of "blame the victim." Moreover, another problem with social capital is that some examples of particularly strong social capital can be found in certain organizations and participation that progressive social policy would not want to encourage; the Nazi Party, the Ku Klux Klan, and neighborhood gangs are commonly cited.

Connecting the Literature on Neighborhoods to Chaos and Child Health

Where is the chaos? In the child, the family, or the neighborhood? None of the above, and all three at the same time? We argue that chaos resides in the patterns of activity that are woven through the nested ecological contexts of everyday life, rather than in specific attributes of the neighborhood, family, or child. Chaos, we argue further, is an emergent property of the interactions among the constituents of the ecological niche, including the child's own qualities and personality. Developmental science excels at measuring the chaos "within" the child, in many individual differences from stress response to personality structures. Social epidemiology and social geography excel at measuring the "chaos" present in the physical and social environment. However, it is our assertion that a complete understanding of chaos as process must take into account the complex interaction between the constituent components.

Chaos may not be reducible to its constituent components in a reductionist approach favored by developmental science, but Gibson's (1982) concept of affordance provides a means of identifying and measuring contributors to chaos and also preserves the interactive nature of the phenomenon, which is an inherent feature of Bronfenbrenner's ecological model. "Roughly, the affordances of things are what they furnish, for good or ill, that is, what they *afford* the observer" (Gibson, 1982, p. 401). Affordances then are inherent in objects, making perception an active rather than passive activity. Affordances are thus defined as "action possibilities" present, either explicitly or latently, in the environment (Gibson, 1982). In essence, affordances are present in the environment, regardless of whether the person is aware of them, but measurable only in relation to the subject's capabilities. The concept of affordance allows for individual difference to come into play at the child and family levels because an affordance is a perceived action possibility, and perception is an important component in understanding how people experience life in a particular neighborhood and therefore how perception constrains their actions. Similarly, the roots of chaos cannot be reduced to measures of the individual entities that constitute the system; they must be examined as a dynamic phenomenon, as process. Chaos depends at least partly on contingent circumstances: A particular activity may be chaotic for one family or child but not for another.

Unlike neighborhood SES, disorder, and social capital, chaos is a dynamic concept that is used to characterize patterns of activity and the interaction between a child and his or her environment. Chaos, therefore, rather than being a discrete attribute of a neighborhood, is best characterized as an emergent attribute of the pattern of activities of a child and his or her interaction with other forces, social and institutional practices, and other specific conditions within an ecological niche.

However, before thought is given to how chaos should be defined and, more generally, what aspects of a neighborhood should be of concern in research on child development, it is necessary first to be clear on the states of being or conditions of living, at the neighborhood level, that are most beneficial to child development. This clarification should be done with respect to specific dimensions of child development.

The propositions put forth by Bronfenbrenner that were listed at the beginning of this chapter are an excellent place to start. In summary, they suggest the need for attention to be paid to both subjective and objective experience. We suggest a slight modification: that the distinction between chaos mediated by perception and chaos not mediated by perception is a better distinction. Noise, crowding, lack of routine activity, and uncontrolled stimuli may be so familiar to one's experience that their negative effects are experienced irrespective of any perceptual processes that they may go through. There is ample evidence of this in the research on work conditions and health.

Bronfenbrenner's theory also suggests that processes of experience and interaction with the environment over time are critical to development, as is increasing complexity of stimuli as the child's development permits. Finally, Bronfenbrenner suggested that attachment to caregivers, notably the parent(s), is critical to all types of development at all stages. In our view, a critical state of being should be considered the bedrock for all development, as drawn from the social theory of Anthony Giddens (1991).

Giddens (1991) claimed that there are "existential questions," which are "queries about basic dimensions of human existence, in respect of human life as well as the material world" that people "answer" in the course of the routinized activity of their day-to-day life (p. 243). The ongoing ability to answer such questions is a fundamental part of individuals' ability to maintain a sense of *ontological security*, which Giddens defined as a "sense of continuity and order in events, including those not directly within the perceptual environment of the individual" (1991, p. 243). Ontological security has direct links to Bronfenbrenner's preliminary definitions of chaos; where chaos is an overriding element of both objective and subjective experience, then the possibility exists that ontological security can be undermined.

Ontological security is arguably at once a precursor to a number of dimensions of child development and an outcome of successful development, in a cumulative, iterative way, over time. In the case of physical well-being, such as the existence of a safe neighborhood environment with opportunities for safe, exploratory physical play, including increasing elements of complexity in play structures and the environment more generally, the ontological security of a child will influence his or her confidence to initiate physically demanding play in the environment, and the environment, in turn, can provide a reflection that reinforces and builds on that confidence to produce further development in the child's physical development and well-being. Many of the roots of ontological security, such as those that would be measured as *emotional maturity* and *social knowledge and competence*, will be primarily the outcome of more proximal processes (the family) initially, but again, the neighborhood environment may provide resources (or affordances) that mitigate shortcomings of the family (e.g., mentors, service providers, other families) or may provide stressors that magnify the shortcomings of the family of origin.

For language and cognitive development, what the neighborhood environment can provide to boost ontological security may be limited to a neighborhood-level analogue to attachment as described by Bronfenbrenner. This analogue may come in the form of teachers, other parents, other adults living in the neighborhood, older children, and supportive community services. The resources at the neighborhood level would also need to have *substantive content* that is supportive of development of cognitive skills (e.g., school curriculum, local norms supportive of learning).

Another important element of Giddens's formulation, instrumental to the maintenance of ontological security, is the individual's ability to maintain a coherent sense of self-identity. *Self-identity*, according to Giddens (1991), is "the self as reflexively understood by the individual in terms of his or her biography" (p. 244). This biography, in turn, is structured according to an individual's narrative of the self, which is "the story or stories by means of which self-identity is understood, both by the individual concerned and by others" (p. 243). This biography also provides an important means to understanding how the experience of neighborhood environments in childhood and at other critical periods of development in the life course may enhance or inhibit further development. In particular, the phenomenon by which an individual internalizes the identity of a group to which he or she belongs coalesces into an identity that in turn shapes the individuals' own self-perception of his or her limits and expectations for the future. This identity may act in a positive direction (e.g., poor minority kids who grow up in mixed-income neighborhoods tend to have better outcomes than poor, minority kids who grow up in poor, minority neighborhoods), or it may limit future opportunity.

Future Research Directions

In this section, we conclude our discussion of conceptualizing the idea of chaos to the neighborhood level by recommending directions for future research. First, researchers need to give careful thought and attention to clarifying the definition of chaos at the neighborhood level, including its relationship to poverty and disadvantage, disorder, and social capital. We further recommend that such research should take residents' subjective and objective experiences into account. In other words, we suggest developing narrative accounts of chaos at the neighborhood level that show how it affects, or has affected, real-life events. A good illustration of this kind of method is seen in the book *Heat Wave: A Social Autopsy of Disaster in Chicago* (Klinenberg, 2002).

Second, we urge investigators to consider the influential roles of culture and personal history when defining the concept. Chaos is a highly subjective phenomenon that can be experienced similarly or differently depending on the person's personality and/or temperament, the place or context, and the person's history and experiences. For example, consider a neighborhood in which people hang out on the street, drink together, talk, and argue. A resident who has lived in that neighborhood for many years might not consider this behavior as a source of chaos and might join in on the discussions. However, a new resident might be intimidated by such a situation and consequently view it as chaotic because he or she has not experienced this occurrence before and feels uncomfortable. A person's perception and evaluation of whether a situation constitutes chaos or an ordinary everyday occurrence can also be influenced by personal characteristics (e.g., racial status, appearance) and social relations. Recognizing (and incorporating) these issues will get to the heart of Bronfenbrenner's theory of ecological development.

We further believe that the groundwork for defining and developing the measurement of chaos at the neighborhood level must be largely qualitative in nature (Burke et al., 2005; O'Campo, Burke, Peak, McDonnell, & Gielen, 2005) and participant observation (Small, 2006) might be particularly appropriate. Only following the successful development of a definition and measure of chaos at the neighborhood level do we urge researchers to examine the relationship between chaos and child development more closely. In this research agenda, we particularly recommend that investigators consider this relationship with reference to specific aspects of child development and specific age groups of children. All too often, researchers apply analytical frameworks without a theoretical foundation, for example, assuming that neighborhood effects must occur for all areas of child development. Oliver, Dunn, Kohen, and Hertzman (2007) demonstrated, however, that this is not the case. Their study revealed little effect of neighborhood characteristics on social competence, knowledge, and emotional maturity in kindergarten children in Vancouver, Canada, and a much larger effect of neighborhood on physical health and well-being and language and cognitive development. Further, they found an especially large effect of neighborhood factors on communication skills and general knowledge. These results should come as no surprise, as it is quite logical that social competence and emotional maturity would be more dependent on family characteristics than on neighborhood characteristics. In language and cognitive ability and physical health and well-being, it also makes sense that there is more of an effect of the neighborhood (although by no means does the effect of the neighborhood outweigh the effects of family characteristics). The language and cognitive effects at the neighborhood level are likely due to peer effects (interacting with peers who have large vocabularies, use complex sentence structures, and have good abstract reasoning may have an effect on children who do not get such exposures in the home). For physical health and well-being, neighborhood effects may well be a result of the norms around physical activity as well as the availability of safe places for unstructured play and programs, leagues, and classes for structured play (that are accessible to the local population, in all the dimensions of access).

It is also worth mentioning briefly that a full account of mediating and moderating factors in the relationship between neighborhood chaos and child development is necessary. It is especially important for the development of policy options. If, for example, some aspect of chaos cannot be stopped, it is possible that its effects may be ameliorated by helping the population subgroups most vulnerable to it or by changing some other factor that would mitigate its effects. Similarly, it is important to investigate whether there are critical periods for exposure to chaos that may affect its manifestation in child development.

Future research should also explore how chaos at the neighborhood level affects a child's family processes, proximal development, and development of resiliency. For example, if chaos is a qualitative entity that is the emergent function of the interaction of a number of neighborhood-level variables with family- and personal-level variables, then by extension, some disorder in the form of changes in the environment that challenge the individual to adapt is crucial to development. The key, however, is that the environment affords not only this opportunity but also its resolution in the form of a return to equilibrium and an established routine. Chaotic neighborhoods are ones that are a constant challenge and either never allow a full return to equilibrium or cause the person to adapt by creating a false and rigidly enforced artificial equilibrium. As Sennett's research indicates, the gated community does not provide enough manageable discontinuity in experience to challenge the child's ability to return to equilibrium. Any experience outside of this narrow range of experience, therefore, may appear markedly chaotic to the child. In other words, efforts to protect children from chaos may undermine their ability to deal with it when it will almost inevitably occur in their lives.

Though the discussion so far has been concerned with chaos and levels of disruption in a neighborhood, we would also like to see researchers consider optimal states of life in neighborhoods. This would contribute to an understanding of what healthy neighborhoods (systems) should look like to promote and maximize positive child development and potential. In this chapter, ontological security has been put forth as a foundation for successful development, but other conceptions should be explored as well. As researchers, we are well placed to develop a metaphor for the type of community we should be seeking to build as a means to successful child development. The admonition that we should reduce and avoid chaos is not enough, no matter how convincing our scientific evidence of their undesirable effects.

Finally, although this chapter focused on one level of the dynamic systems affecting child development, it is critical to synthesize findings across all systems for a holistic view of the production of child development when considering policy options and designing potential interventions. In the example of chaos, it is essential to explore this phenomenon in every setting of a child's life, including home, day care, school, or job, and neighborhood, as well as the further removed contexts of the exo- and macrosystems. This broad examination is particularly important because children may experience disruption in more than one context and subsequently experience a cumulative effect, which may be more damaging for a child's development.

References

Aber, J., Gephart, M., Brooks-Gunn, J., Connell, J., & Spencer, M. (1997). Neighborhood, family, and individual processes as they influence child and adolescent outcomes. In J. Brooks-Gunn,

G. Duncan, & J. Aber (Eds.), *Neighborhood poverty: Vol. 1. Contexts and consequences for development* (pp. 41–61). New York: Russell Sage Foundation.

Billy J. O., Brewster, K. L., & Grady, W. (1994). Contextual effects on the sexual behaviour of adolescent women. *Journal of Marriage and the Family, 56*, 387–404.

Billy, J. O., & Moore, D. E. (1992). A multilevel analysis of marital and nonmarital fertility in the US. *Social Forces, 70*, 977–1001.

Boyle, M., & Lipman, E. (2002). Do places matter? Socioeconomic disadvantage and behavioural problems of children in Canada. *Journal of Consulting and Clinical Psychology, 70*, 378–389.

Brewster K. L. (1994a). Neighborhood context and the transition to sexual activity among Black women. *Demography, 31*, 603–614.

Brewster, K. L. (1994b). Race differences in sexual activity among adolescent women: The role of neighborhood characteristics. *American Sociological Review, 59*, 408–424.

Briggs, X. S. (1997a). Moving up versus moving out: Neighborhood effects in housing mobility programs. *Housing Policy Debate, 8*, 195–234.

Briggs, X. S. (1997b). *Yonkers revisited: The early impacts of scattered-site public housing on families and neighborhoods.* New York: New York Teachers College.

Bronfenbrenner, U. (1977). Toward an experimental ecology of human development. *American Psychologist, 32*, 513–531.

Bronfenbrenner, U. (1979). *The ecology of human development: Experiments by nature and design.* Cambridge, MA: Harvard University Press.

Bronfenbrenner, U. (1986). Ecology of the family as a context for human development: Research perspectives. Developmental Psychology, 22, 723–742.

Brooks-Gunn, J., Duncan, G. J., & Aber, J. L. (1997). *Neighborhood poverty: Context and consequences for children.* New York: Russell Sage Foundation.

Brooks-Gunn, J., Duncan, G., Klebanov, P., & Sealand, N. (1993). Do neighborhoods influence child and adolescent development? *American Journal of Sociology, 99*, 353–395.

Burke, J. G., O'Campo, P., Peak, G. L., Gielen, A. C., McDonnell, K. A., & Trochim, W. (2005). An introduction to concept mapping as a participatory public health research method. *Qualitative Health Research Methods, 15*, 1392–1410.

Burton, L. M., & Price-Spratlen, T. (1999). Through the eyes of children: An ethnographic perspective on neighborhoods and child development. In A. S. Masten (Ed.), *Cultural processes in child development* (pp. 77–96). Mahwah, NJ: Erlbaum.

Calvert, W. J. (2002). Neighborhood disorder, individual protective factors, and the risk of adolescent delinquency. *ABNF Journal, 13*(6), 127–135.

Chawla, L. (2002). *Growing up in an urbanizing world.* London: Earthscan Unicef.

Chase-Lansdale, L., & Gordon, R. (1996). Economic hardship and the development of five- and six-year-olds: Neighborhood and regional perspectives. *Child Development, 67*, 3338–3367.

Chase-Lansdale, L., Gordon, R., Brooks-Gunn, J., & Klebanov, P. (1997). Neighborhood and family influences on the intellectual and behavioral competence of preschool and early school-age children. In J. Brooks-Gunn, G. Duncan, & J. Aber (Eds.), *Neighborhood poverty: Vol. 1. Contexts and consequences for development* (pp. 79–118). New York: Russell Sage Foundation.

Cohen, D., Spear, S., Scribner, R., Kissinger, P., Mason, K., & Wildgen, J. (2000). "Broken windows" and the risk of gonorrhoea. *American Journal of Public Health, 90*, 230–236.

Coulton, C. J., Korbin, J., Chan, T., & Su, M. (2001). Mapping residents' perceptions of neighborhood boundaries: A methodological note. *American Journal of Community Psychology, 29*, 371–383.

Diez Roux, A.V. (2001). Investigating neighborhood and area effects on health [Commentary]. *American Journal of Public Health, 91*, 1783–1788.

Doherty, G. (1997). *Zero to six: The basis for school readiness* (Working Paper R-97-8E). Ottawa, ON: Human Resources Development Canada.

Duncan, G., Brooks-Gunn, J., & Klebanov, P. (1994). Economic deprivation and early childhood development. *Child Development, 65*, 296–318.

Duncan, G., & Raudenbush, S. (1999). Assessing the effects of context in studies of child and youth development. *Educational Psychologist, 34*, 29–41.

Ennett, S. T., Fewelling, R. L., Lindrooth, R. C., & Norton, E. C. (1997). School and neighborhood characteristics associated with school rates of alcohol, cigarette, and marijuana use. *Journal of Health and Social Behaviour, 38*, 55–71.

Entwisle, D. R., Alexander, K. L, & Olson, L. S. (1994). The gender-gap in math—Its possible origins in neighborhood effects. *American Sociological Review, 59*, 822–838.

Furstenberg, F., & Hughes, M. E. (1997). The influence of neighborhoods on children's development: A theoretical perspective and research agenda. In J. Brooks-Gunn, G. Duncan, & J. Aber (Eds.), *Neighborhood poverty: Vol. 1. Contexts and consequences for development* (pp. 23–47). New York: Russell Sage Foundation.

Galster, G. (2001). On the nature of the neighborhood. *Urban Studies, 38*, 2111–2124.

Garner, C. L., & Raudenbush, S. W. (1991). Neighborhood effects on educational attainment: A multilevel analyses. Sociology of Education, *64*, 251–262.

Gibson, J. J. (1982). Notes on affordances. In E. Reed & R. Jones (Eds.), *Reasons for realism* (pp. 401–418). Hillsdale, NJ: Erlbaum.

Giddens, A. (1991). *Modernity and self-identity. Self and society in the late modern age.* Palo Alto, CA: Stanford University Press.

Halpern-Felsher, B. L., Connell, J. P., Spencer, M. B., Aber, J. L., Duncan, G. P., Clifford, E., et al. (1997). Neighborhood and family factors predicting educational risk and attainment in African American and white children and adolescents. In J. Brooks-Gunn, G. Duncan, & J. Aber (Eds.), *Neighborhood poverty: Vol. 1. Contexts and consequences for development* (pp. 146–173). New York: Russell Sage Foundation.

Hart, R. (1979). *Children's experience of place.* New York: Irvington Publishers.

Holaday, B., Swan, J. H., & Turner-Henson, A. (1997). Images of the neighborhood and activity patterns of chronically ill schoolage children. *Environment & Behavior, 29*, 348–373.

Hume, C., Salmon, J., & Ball, K. (2005). Children's perceptions of their home and neighborhood environments, and their association with objectively measured physical activity: A qualitative and quantitative study. *Health Education Research, 20*, 1–13.

Janus, M., & Offord, D. (2007). Psychometric properties of the Early Development Instrument (EDI): A teacher-completed measure of children's readiness to learn at school entry. *Canadian Journal of Behavioural Science, 39*, 1–22.

Jencks, C., & Mayer, S. (1990). The social consequences of growing up in a poor neighborhood. In L. Lynn & M. McGeary (Eds.), *Inner-city poverty in the United States* (pp. 111–189). Washington, DC: National Academy Press.

Kagan, S. L. (1992). Readiness past, present and future: Shaping the agenda. *Young Children, 48*(1), 48–53.

Kalff, A., Kroes, M., Vles, J., Hendriksken, J., Feron, F., Steyaert, J., et al. (2001). Neighborhood level and individual level SES effects on child problem behaviour: A multilevel analysis. *Journal of Epidemiology and Community Health, 55*, 246–250.

Klebanov P., Brooks-Gunn, J., McCarton, C., & McCormick, M. (1998). The contribution of neighborhood and family income to developmental test scores over the first three years of life. *Child Development, 69*, 1420–1436.

Klinenberg, E. (2002). *Heat wave: A social autopsy of disaster in Chicago.* Chicago, IL: University of Chicago Press.

Kohen, D. E., Brooks-Gunn, J., Leventhal, T., & Hertzman, C. (2002). Neighborhood income and physical and social disorder in Canada: Associations with young children's competencies. *Child Development, 73*, 1844–1860.

Kohen, D., Hertzman, C., & Brooks-Gunn, J. (1998). *Neighborhood influences on children's school readiness* (Working Paper W-98-15E). Ottawa, ON: Human Resources Development Canada.

Kohen, D. E., Hertzman, C., & Wiens, M. (1998). *Environmental changes and children's competencies* (Working Paper W-98-25E). Ottawa, ON: Human Resources Development Canada.

Kotlowitz, A. (1991). *There are no children here: The story of two boys growing up in the other America.* New York: Anchor Books.

Kozol, J. (1996). *Amazing grace: The lives of children and the conscience of a nation.* New York: HarperCollins Press.

Kozol, J. (2001). *Ordinary resurrections: Children in the years of hope.* New York: Crown Publishers.

Ku, L., Sonenstein, F., & Pleck, J. H. (1993). Neighborhood, family, and work: Influences on the premarital behaviors of adolescents. *Social Forces, 72*, 497–503.

Lynch, K. (1977). *Growing up in cities.* Cambridge, MA: MIT Press.

McCulloch, A., & Joshi, H. (2001). Neighborhood and family influences on the cognitive ability of children in the British National Child Development Study. *Social Science and Medicine, 53*, 579–591.

O'Campo, P., Burke, J., Peak, G. L., McDonnell, K. A., & Gielen, A. C. (2005). Uncovering the neighborhood influences on intimate partner violence using concept mapping. *Journal of Epidemiology and Community Health, 59*, 603–608.

Oliver, L., Dunn, J. R., Kohen, D., & Hertzman, C. (2007). Do neighborhoods influence the readiness to learn of kindergarten children in Vancouver? A multilevel analysis of neighborhood effects. *Environment & Planning A, 39*, 848–868.

Peeples, F., & Loeber, R. (1994). Do individual factors and neighborhood context explain ethnic differences in juvenile delinquency? *Quantitative Criminology, 10*, 141–157.

Proshansky, H. M., & Fabian, A. K. (1987). The development of place identity in the child. In C. S. Weinstein & T. G. David (Eds.), *Spaces for children* (pp. 21–40). New York: Plenum Press.

Ross, C. E., & Mirowsky, J. (1999). Disorder and decay: The concept and measurement of perceived neighborhood disorder. *Urban Affairs Review, 34*, 412–432.

Ross, C. E., & Mirowsky, J. (2001). Neighborhood disadvantage, disorder and health. *Journal of Health and Social Behavior, 42*, 258–276.

Ross, N., Tremblay, S., & Graham, K. (2004). Neighborhood influences on health in Montreal, Canada. *Social Science and Medicine, 59*, 1485–1494.

Sampson, R. J., Morenoff, J. D., & Earls, F. (1999). Beyond social capital: Spatial dynamics of collective efficacy for children. *American Sociological Review, 64*, 633–660.

Sampson, R. J., Morenoff, J. D., & Gannon Rowley, T. (2002). Assessing "neighborhood effects": Social processes and new directions in research. *Annual Review of Sociology, 28*, 443–478.

Sampson, R. J., & Raudenbush, S. W. (2004). Seeing disorder: Neighborhood stigma and the social construction of "broken windows." Social Psychology Quarterly, *67*, 319–342.

Schaefer-McDaniel, N. (2007). *Children talk about their New York City neighborhoods: The role of subjective and objective neighborhood evaluations in understanding child health*. Saarbrücken, Germany: VDM Verlag Dr. Müller.

Sennett, R. (1970). *The uses of disorder*. New York: Norton.

Small, M. L. (2006). Neighborhood institutions as resource brokers: Childcare centers, interorganizational ties, and resource access among the poor. *Social Problems, 53*, 274–292.

Small, M. L., & Newman, K. (2001). Urban poverty after *The Truly Disadvantaged*: The rediscovery of the family, the neighborhood, and culture. *Annual Review of Sociology, 27*, 23–45.

Wilson, J. Q., & Kelling, G. (1982, March). The police and neighborhood safety: Broken windows. *The Atlantic Monthly, 127*, 29–38.

Wilson, W. J. (1987). *The declining significance of race: The truly disadvantaged—The inner city, the underclass, and public policy*. Chicago: University of Chicago Press.

12

Parent Employment and Chaos in the Family

Rena Repetti and Shu-wen Wang

Urie Bronfenbrenner's (1989) bioecological model recognizes that experiences outside of the home can be important to child development. Such experiences include those in environments that the child may never encounter directly, such as the parent's workplace. There is no question that parent employment is almost always a net benefit to the family. Jobs generally bring income, health care, routines, sources of social support, and enhanced well-being, all of which reduce or protect against chaotic elements in a child's life. However, parents' experiences in the workplace are not always uniformly advantageous; this chapter focuses on employment experiences that can add to, or exacerbate, chaos in the home. We begin with research that examines how psychosocial characteristics of the workplace, particularly stressors, can affect family relationships. Job-related time commitments that can detract from a healthy family environment are considered next, followed by child-care arrangements and certain job benefits, particularly paid leave, all of which are critical elements in the broader context that shape how workplace experiences extend into the home. Finally, we turn to the sometimes devastating impact that parental job loss and long-term unemployment can have on families. Our chapter concludes with a conceptual model that integrates key findings and promising directions for future research.

Psychosocial Characteristics of the Workplace

The impact that jobs can have on physical and mental health (Kuper & Marmot, 2003; Repetti & Mittmann, 2004) is thought to also shape employees' interactions with family members (Crouter & Bumpus, 2001; Perry-Jenkins, Repetti, &

We, the chapter authors, thank the Alfred P. Sloan Foundation and the University of California, Los Angeles, Center on Everyday Lives of Families for supporting our work. Shu-wen Wang's work on this chapter was also supported by a predoctoral fellowship from the American Psychological Association Minority Fellowship Program. We appreciate suggestions from participants in the First Bronfenbrenner Conference on the Ecology of Human Development, especially Feyza Corapci, John Eckenrode, Gary Evans, Barbara Fiese, Taryn Morrissey, and Henry Ricciuti. Lisa Flook, Darby Saxbe, Jacqui Sperling, and Rich Slatcher also provided helpful comments on an earlier draft of this chapter.

Crouter, 2000). The research reviewed next points to the potential for work experiences to introduce or exacerbate chaos in the family environment.

Overload and Social Stressors

Work overload occurs when heavy demands are placed on workers, that is, when the number of tasks and responsibilities, and the pace at which they must be completed, are elevated. Research suggests that work overload can have an impact on the family via its influence on an employed parent's energy, mood, and behavior. One pattern of findings suggests that heavy demands for attention and energy in the workplace drain parents' resources and detract from their monitoring and supervision of children. For example, more time pressure at work has been associated with less allocation of time to parenting and with less knowledge of children's whereabouts and activities (Bumpus, Crouter, & McHale, 1999; Greenberger, O'Neil, & Nagel, 1994).

Repetti (1992) proposed that social withdrawal—a reduction in social interactions and emotional responsiveness with family members that helps an individual recuperate from high-stress workdays—is one pathway through which job stressors impact the parent–child relationship. In one study, mothers were more withdrawn from their preschoolers on days when they reported greater workload (Repetti & Wood, 1997a). Furthermore, mothers who reported experiencing more psychological distress were the most vulnerable to this daily effect. In a daily report study of male air traffic controllers, fathers were less behaviorally and emotionally involved with their children on higher workload days (Repetti, 1994). Though social withdrawal may be an adaptive short-term coping response, over time, chronic withdrawal may damage feelings of closeness and gnaw at family cohesion (Repetti & Wood, 1997b).

The perception of work overload has also been linked with greater tension and conflict and less warmth and acceptance in parents' relationships with adolescents (Galambos, Sears, Almeida, & Kolaric, 1995). In the mother–preschooler study mentioned earlier, reports of aversive or impatient parent–child interactions increased on higher workload days for mothers who endorsed more depressive symptoms (Repetti & Wood, 1997a).

Negative interactions with supervisors and coworkers can lead to feelings of frustration, anger, and discouragement that carry over from the workplace into the home. In the mother–preschooler study, mothers were more withdrawn on days of high interpersonal stress at work, an effect that was especially pronounced for the women who reported more Type A behaviors (Repetti & Wood, 1997a). Researchers have also identified a "negative emotion spillover" response, a short-term process in which negative emotions generated at work are expressed at home, increasing the likelihood of conflict in the family (e.g., Piotrkowski, 1979). In the air traffic controller study mentioned earlier, working fathers reported greater use of discipline and more negative emotion expression (e.g., anger) with their children on days that were more interpersonally stressful at work (Repetti, 1994). Even interactions with infants appear to be influenced by social stressors at work. Costigan, Cox, and Cauce (2003) found that a more negative interpersonal workplace atmosphere experienced by mothers predicted the quality of parenting behaviors displayed by both mothers and fathers 3 months later; negative (e.g., intrusive) behaviors increased and positive (e.g., stimulating and sensitive) behaviors decreased.

Another way that stress can be introduced from the parent's workplace into a child's life is through its impact on the marital relationship. Daily report studies have demonstrated a same-day linkage between stressors at work and both social withdrawal and tension in marital interactions (Bolger, DeLongis, Kessler, & Wethington, 1989; Repetti, 1989; Schulz, Cowan, Cowan, & Brennan, 2004; Story & Repetti, 2006). Studies on paths of stress and emotion flow suggest that tension in the marital relationship influences parents' interactions with their children (Almeida, Wethington, & Chandler, 1999). Therefore, any impact that work stressors have on the marital dyad may also be echoed in the parent–child dyad.

Autonomy and Complexity

Two job characteristics—control or autonomy and cognitive complexity—are thought to influence child-rearing values and socialization practices. Melvin Kohn, who studied under Urie Bronfenbrenner, theorized that work environments influence personality tendencies, which then shape parenting values and behaviors (Kohn, 1969; Kohn & Schooler, 1982). Because autonomy and complexity are more stable characteristics of a job than are the stressors discussed earlier, they are more closely tied to the skills and personal qualities that are required of the worker and are therefore more closely linked with social class. Kohn and Schooler (1982) found that middle-class jobs involving more autonomy and complex tasks socialized workers to be more oriented toward self-directedness, whereas working-class jobs socialized workers toward greater conformity and obedience to rules and supervisors. Their conclusion was that the occupation's socialization of workers' values translated into qualitatively different parenting practices.

This line of inquiry has continued, with studies suggesting that greater job autonomy and complexity is associated with a host of more positive and effective parenting practices, including an emphasis on internalization of social norms by children instead of the use of direct parental control (Parcel & Menaghan, 1993), less restrictive parental control (Mason, Cauce, Gonzales, Hiraga, & Grove, 1994), greater parental acceptance of children and less authoritarian parenting (Grimm-Thomas & Perry-Jenkins, 1994), more use of inductive reasoning with children (Whitbeck et al., 1997), and less harsh parenting (Greenberger et al., 1994; Whitbeck et al., 1997). In addition, parents who report greater job autonomy and work complexity have been found to be more warm and responsive (Greenberger et al., 1994) and to provide home environments with more emotional support and intellectual stimulation (Cooksey, Menaghan, & Jekielek, 1997; Menaghan & Parcel, 1991; Parcel & Menaghan, 1994).

Thus, research suggests that the degree of autonomy and complexity in parents' jobs is associated with child-rearing values and behaviors that are, to some degree, consistent with the behaviors needed for success at work. However, selection is a powerful force in this process. An individual's intellectual capacity, educational background, personality, and values limit the range of his or her potential jobs or careers. Because these individual variables also shape parenting, it is critical that the causal paths connecting social class, job selection factors, and job characteristics be carefully considered in any effort to understand how occupations are linked to parent socialization practices (see, e.g., Greenberger et al., 1994). Unfortunately, teasing apart the unique effects of job characteristics from the characteristics of the individuals that select into these jobs is a research goal that has rarely been met.[1]

Few studies have uncovered significant associations between parent job stressors and child developmental outcomes and, when these associations have been found, the effect sizes have been small (see Repetti, 2005). However, evidence for an impact of job stressors on the quality of family relationships has been much more consistent. When overload and social stressors deplete an employee's physical, cognitive, and emotional stores, the effects at home can range from less parental attention and monitoring and greater social withdrawal to more interpersonal conflict. In addition, child-rearing values and behaviors may be shaped to some degree by the autonomy and complexity that parents experience at work. Of course, there is no simple equation to characterize how parents' responses to experiences at work shape their behavior at home. A host of individual and family variables—such as socioeconomic status, personality traits, psychological distress, and family conflict—influence work–family dynamics.

Time at Work

Jobs and families have been described as "greedy institutions" that require much, including commitments of time, from their members (Coser, 1974). An economic analysis points to the family's fixed budget of time; by working more hours in the labor market, the family has fewer hours to spend together, and vice versa (Leibowitz, 2005). In a Dutch sample, fathers who spent more time at work spent less time in activities with their adolescent children, although no such effect was observed for mothers (Dubas & Gerris, 2002). On the one hand, more time in the labor market usually means that more goods can be purchased for the family, including basics such as better housing and food. On the other hand, unpaid labor in the home, such as the care of children, shopping, and meal preparation, also benefits the family. The question here is this: What kinds of time commitments at work detract from a healthy family environment?

Long Hours

Time demands at work are increasing; since 1970, the number of Americans working more than 50 hours per week has gradually increased, with jobs that entail long hours more likely held by fathers and by employees with higher

---

[1]Brooks-Gunn et al. (see chap. 10, this volume) provided an interesting discussion of how parallel selection factors challenge causal modeling in neighborhood effects research.

educations (Barnett, 2006).² Across the globe, working parents lament the loss of time with their children resulting from demands on the job (Heymann, Simmons, & Earle, 2005). A recent report from the Pew Research Center showed that among working mothers in the United States, only 21% felt that full-time work was the ideal situation for them, down from 32% who expressed that opinion 10 years ago (Taylor, Funk, & Clark, 2007).

Although longer work hours can be stressful, researchers have not found a reliable association between work hours, per se, and individual well-being outcomes. In fact, the length of the workweek is often associated with positive individual outcomes, such as better physical and mental health. That may be because more work hours are associated with increased income, more opportunities for advancement, greater job security, and greater access to health insurance and other benefits. Moreover, according to the "healthy worker effect," parents who feel most able to manage both job and family are the most likely to work longer hours, whereas parents in poor health reduce their work hours. When time spent commuting is considered, however, the story may be different. For example, in a study of rail commuters, Evans and Wener (2006) found that longer commutes were associated with elevations in cortisol, a stress hormone.

The association between time spent in work-related activities and family outcomes is complex. A daily report study found that when husbands spent more hours working (this measure included both paid and unpaid work), both they and their wives described less warmth in their marital interactions that day (Doumas, Margolin, & John, 2003). In another study, longer driving commutes predicted more self-reported negative mood when employees arrived home (Novaco, Kliewer, & Broquet, 1991). Excessive time spent commuting on public transportation interferes with the establishment of family routines, especially among single mothers (Roy, Tubbs, & Burton, 2004). Despite the fact that longer work hours are associated with scheduling difficulties at home and perceptions of work–family strain, in two-parent families they are generally not linked to more marital strain, less marital companionship, or a lower quality of the home environment provided to children (Barnett, 2006).³

In dual-earner families, more maternal work hours are associated with greater father involvement in household labor and child care. For example, Crouter, Bumpas, Maguire, and McHale (1999) found that when wives worked longer hours, fathers were significantly better informed about their children's daily activities, whereabouts, and companions. In a study of seasonal changes in work hours, Crouter and McHale (1993) showed that fathers whose wives were employed during the school year but not in the summer had better knowledge of their children during school months, a pronounced drop in knowledge over the summer, and then a sharp increase when their wives returned to work the following school year. Increases in paternal involvement and knowledge

---

²See chapter 2 of this volume for a longer historical perspective on working time and American families.
³The probability of divorce does increase as wives work more hours. However, the direction of causality is not clear because women anticipating a divorce may seek to increase their financial security by working more hours, and time spent at work may be a way to avoid the kind of severe marital difficulties that precede a divorce (Barnett, 2006).

may explain Barnett's (2006) finding that, in a sample of dual-earner families, fathers rated the quality of their experiences in the parental role more positively when their wives worked longer hours.[4]

The best predictor of family outcomes is not the objective number of work hours but the subjective experience of those hours, such as the fit between parents' work schedules and their families' needs, and feelings of role overload (Barnett, 2006). This pattern is consistent with research reviewed earlier connecting perceptions of work overload with disruptions in marital and parent–child interactions.

Nonstandard Hours

Barnett's (2006) review of the literature shows that the distribution of parents' work hours over the day is more important for families than is the absolute number of hours that parents work. Among shift workers, night, evening, and rotating shifts are far more disruptive than is the day shift. In general, people who are more socially and economically disadvantaged are more likely to work nonstandard schedules (weekends or nonday hours), although education does not show a linear association with nonstandard hours. Shift work is mostly found in relatively low-paying service sector jobs (e.g., cashiers, truck drivers, sales workers, waiters), and future job growth in the United States is projected to be disproportionately high in these occupations. Single mothers are also more likely than married mothers to work nonstandard schedules. In two-parent, dual-earner families, working nonstandard hours is more common among younger, less educated parents and among those with more children. Younger child age is also associated with greater likelihood that parents work nonstandard schedules (Barnett, 2006).

One reason mothers and fathers choose nonstandard schedules is to maximize the amount of time their children are cared for by a parent. For example, mothers may try to synchronize employment hours with children's school schedules and, in two-parent homes, "tag team" work hours with a spouse (Bianchi & Raley, 2005). There do appear to be some immediate benefits to this approach. Research indicates that when wives work nonday shifts, fathers spend more time on homework and child care, know more about their children's activities, and receive more disclosures from their children (Barnett & Gareis, 2007; Staines & Pleck, 1983). However, some of these short-term advantages to shift work may be offset by long-term disadvantages. For instance, dual-earner couples in which wives work nonday shifts report more overall work–family conflict (Staines & Pleck, 1983). When both parents work full-time, night, evening, and rotating shifts are far more disruptive of family routines and relationships than is the day shift. Most men working nonstandard shifts report a decrease in their functioning in the family (Barnett, 2006).[5] For poor single mothers, the situation

---

[4]It is interesting to note that wives who worked longer hours described their own parent role more positively, but they also reported more work–family conflict and more psychological distress (Barnett, 2006).

[5]There is also evidence of the negative effects of shift work on men's individual health outcomes (especially true for night and rotating shifts). The symptoms include sleep disturbance, fatigue,

can be extreme, with some women taking on responsibilities for full-time caregiving during the day and full-time employment at night (Roy et al., 2004).

Among two-parent families, it may be more important to consider the degree of overlap in parents' work schedules than the impact of one spouse's shift schedule alone. The most obvious consequence when one parent's time at work does not overlap with the school and work schedules of the rest of the family is the limited time that the members spend together. Families with school-age children tend to live in sync with the rest of the world but out of sync with that parent, making it difficult to develop or maintain family routines and rituals (Crouter & McHale, 2005). As Fiese and Winter (see chap. 4, this volume) pointed out, structure and routine are valuable resources used by families in child rearing. When a parent works a nonstandard shift, even the family dinner—one of the most common family rituals—may be a rare event (Presser, 2005). Shift work also reduces couple time at home, so that parents have less time for their own relationship building. Perhaps it is not surprising that among dual-earner couples there is greater marital dissatisfaction when either spouse works a nonday shift and in families with children the odds of marital disruption are 2.5 times higher when wives work nights (Barnett, 2006).

When shift schedules are voluntarily chosen, the likelihood of any negative outcome is reduced (Barnett, 2006; Staines & Pleck, 1983). However, most people do not voluntarily choose nonstandard schedules; the seniority that often confers greater schedule control is typically not achieved until a time when children are past school-age. Perry-Jenkins (2005) found, for example, that the often random nature and timing of overtime work for the working-class parents in her study caused difficulties, such as the need to arrange child care at the last minute. As she pointed out, the level of control over scheduling extra hours at work that is possible for most professionals may explain why long hours do not have uniform associations with family functioning.

Long work hours can add stress to, and reduce time spent with, the family, but it is the subjective experience of those hours and the way that they are distributed over the day, rather than the mere number of hours spent on the job, that appears to be most important. Nonstandard shift schedules, particularly when they are not voluntarily chosen, can add chaos to family life.

Arranging Time and Care for Children and for Family Health

Family-friendly employer benefits, supportive federal policies, and reliable child care constitute critical components of the larger work–family context. In the United States, some employer-provided benefits that can help workers manage their responsibilities at home, such as paid holidays, vacation, and sick leave, are common for full-time workers. Other paid benefits, such as paid personal leave, assistance for child care, and on-site or off-site child care, are rare. Conversely, unpaid family leave has been much more common, especially after the late 1990s with the enactment of federal legislation (Ruhm, 2005). Workers at

---

gastrointestinal ailments, trouble concentrating, headaches, and substance use or abuse (cigarettes, caffeine, and other stimulants), as well as heart disease and other illnesses (Barnett, 2006).

smaller companies as well as part-time and less-skilled workers have access to fewer employer-provided protections and opportunities in the United States (Henly & Lambert, 2005; Ruhm, 2005).[6]

Ruhm's (2005) review shows that, compared with European countries, the U.S. labor market is one with long work hours, short vacations, limited availability of parental leave, and little support for child care. He further describes economic analyses indicating that many Americans willingly trade off more time at work and less workplace flexibility for increased wages. In other words, employers provide and the government mandates few family-friendly benefits because the public's behavior and choices reveal a preference for higher wages.[7] U.S. government policies are aimed at maintaining high employment rates and maximizing individual earnings as well as national income. As Ruhm (2005) concluded, within this context, balancing increasing job responsibilities with the needs of families through employer actions, or government-mandated policies, such as those that would increase paid leave and child-care options, is expensive and controversial.

Paid and Unpaid Leaves

The Family Medical Leave Act (FMLA) requires public agencies and large employers to offer 12 weeks of unpaid leave for employees to care for babies, sick relatives, or their own health problems. Workers may be required to first use accrued sick leave or vacation time; however, health insurance continues during the leave (Ruhm, 2005). There are often hidden costs for using up all paid leaves prior to the return to work, a strategy that is often voluntarily adopted by mothers to maintain income during a maternity leave. One major cost, after the mother returns to work, is the accumulation of unexcused absences (e.g., to care for a sick baby or keep appointments with the pediatrician) that result in written warnings and other penalties, such as pay docking (Perry-Jenkins, 2005). The Project on Global Working Families, which interviewed working parents in six countries (Mexico, Botswana, Vietnam, the United States, Russia, and Honduras), found that in all the countries that were studied, when children are sick, poor parents find it difficult to make arrangements for their care, such as getting leave time from their jobs (Heymann et al., 2005).

Unfortunately, poverty and illness go together. One study found that it was quite common for poor working mothers in the United States to care for children or elderly relatives with chronic illnesses. These women often lost or resigned from full-time jobs because of the hospitalizations, frequent medical

---

[6]However, even when family-friendly policies are available to employees (e.g., allowing employees to pay for child care with pretax dollars), they may not be used to maximum benefit, at least not by working-class parents (Perry-Jenkins, 2005).

[7]Employers who voluntarily offer extensive family-friendly benefits that are not required by the government may attract a disproportionate number of employees who expect to take advantage of the costly benefits. Because these benefits are likely to be financed by lower wages, individuals who do not expect to use them will avoid those companies. In that scenario, the family-friendly benefits are even more expensive because they are spread over a high concentration of people who use them rather than being pooled with many employees with lower expected benefit use (Ruhm, 2005).

appointments, and other heavy caregiving demands that they faced (Burton, Lein, & Kolak, 2005). It is interesting that the most common reason for taking FMLA leave is for one's own health problems, not for care of newborns or newly adopted children (Ruhm, 2005).[8] However, employed mothers in low-wage jobs often ignore their own health symptoms and postpone regular checkups and treatment in order to keep up with their work and family responsibilities (Burton et al., 2005).

Child Care

The cost of child care is a great challenge for many working families. Employed single parents have a particularly high cost burden because a larger percentage of their total family income is spent on child care, even though single parents use the cheapest sources of care. Free child care (usually provided by grandparents and other relatives) is used disproportionately by workers with the fewest resources in terms of earnings and education (Ruhm, 2005). Some research suggests that working-class families in the United States have lower quality care arrangements than do either poor or high-income families. They earn too much to receive child-care subsidies or support but too little to afford the cost of high-quality child care (Perry-Jenkins, 2005).[9]

Unpredictable and nonstandard work schedules present the greatest difficulties. Even though many parents choose shift work as a way of avoiding nonparental care for their children, the majority of low-skilled single mothers in the United States cite labor market rather than child-care reasons for working nonstandard hours (Henly & Lambert, 2005). Unfortunately, it is rare for providers to offer child care at times that can accommodate nonstandard work schedules, such as weekends, late evenings, or early mornings (Henly & Lambert, 2005). Most U.S. parents working nonstandard hours rely on informal care, especially relatives, and multiple arrangements. Many low-wage workers must also cope regularly with unpredictable schedules, which means that parents often must patch together a plan both within and outside of the formal child-care sector at the last minute. Even when provisions can be made, the process is hectic, stressful, and sometimes inadequate (Henly & Lambert, 2005).[10]

It is common for low-wage parents, especially single mothers, shift workers, and workers with some unpredictability in their hours, to use a collection of multiple formal and informal arrangements for child care; this usually consists of a primary arrangement and then one or more auxiliary caregivers who supplement that care on either a routine or a "just in case" basis (Henly & Lambert, 2005). Depending on their ages, children may also spend some time alone at

---

[8]One reason is that many women eligible for maternity leave under the FMLA already had rights to job absences under other workplace benefits and state and federal laws (Ruhm, 2005).

[9]See chapter 9 of this volume for a discussion of how family and child-care variables can function synergistically to influence child development.

[10]Preliminary research in The Netherlands has not uncovered any negative effects on child development associated with greater use of flexible options in child-care centers, such as extended hours and evening meals, when they are available (De Schipper, Tavecchio, Van IJzendoorn, & Linting, 2003).

home or in the care of siblings. The international study mentioned earlier found that serious accidents can occur under these circumstances. Moreover, among poor and single-parent households in the third-world countries that were studied, older children, especially girls, were often taken out of school to provide care for younger siblings or cousins (Heymann et al., 2005).

Although child-care quantity and quality do not appear to be important mediators of the effects of parental work on child developmental outcomes (Korenman & Kaestner, 2005), that does not mean that families do not benefit when children are in reliable, safe, high-quality care. A parent's peace of mind, predictable schedules, and child comfort make for a better family climate, whereas shifting and unpredictable child-care arrangements contribute to a chaotic environment for children (Henly & Lambert, 2005).

Job Loss and Unemployment

Unstable employment is a major source of chaos for the family. The loss of a job is not a single stressful event; it leads to a cascade of secondary stressors, most resulting from the sudden change in financial circumstance, such as moving to less expensive housing or deferring payment on household bills (Price, Friedland, & Vinokur, 1998). Everyone in the household is exposed to these downstream effects, so it is not surprising that secondary stressors also affect the spouse's mental health and degrade the quality of the couple's relationship (Howe, Levy, & Caplan, 2004). Children also react to the changes in the family. For example, parental job loss predicts initiation of smoking in early adolescence, an effect that is observed even after controlling for socioeconomic status, educational, and parenting characteristics (Unger, Hamilton, & Sussman, 2004).

The impact of job loss is much more destructive when it leads to a prolonged period of unemployment. Research in Scandinavian countries indicates that a decline in the functioning and well-being of the unemployed parent can have profound effects on the family. A birth cohort study found that unemployed parents are at increased risk of drug abuse, violence, and out-of-home placements for their children (Christoffersen, 2000). Of course, unemployment does not always have such devastating effects. However, whatever the repercussions for the family, they are magnified when there is only one parent in the home and, unfortunately, unemployment rates tend to be higher among single parents (Strom, 2001). In two-parent homes, the negative effects on well-being spread to the other parent, and long-term unemployment in either spouse increases the risk of divorce, particularly for couples with limited financial resources (Strom, 2003).

Parental unemployment also has direct effects on children. Mothers' long-term unemployment has been associated with an increased risk of serious child abuse, even after controlling for the other adverse family factors linked to unemployment (Christoffersen, 2000). The outcomes for offspring include more psychological and physical health problems and serious accidents (Strom, 2001, 2003), as well as higher rates of hospitalization, psychiatric problems, suicide attempts, and prison sentences (Christoffersen, 2000). Research conducted in Slovakia found that paternal unemployment lasting more than 1 year was linked to lower self-rated health among adolescents; this was true even after controlling for the father's education, perceived financial stress, and family affluence (Sleskova et al., 2006).

Selection factors are a very important consideration in any analysis of the effects of unemployment. Moreover, characteristics that increase a parent's chance of becoming unemployed, such as mental health problems and drug abuse, are also aggravated during the period of unemployment. In addition, couples often share risk factors for becoming unemployed, such as poor educational background and regional employment rates, resulting in a phenomenon sometimes called "couple unemployment" whereby spouses of the unemployed are at higher risk for unemployment themselves. But selection factors do not explain all of the effects, especially during periods of high unemployment. In fact, most of the findings cited earlier controlled for the effects of adverse selection on unemployment, with the exception of one critical variable. Of all the risk factors for, and mediators of, job loss and unemployment, low income is by far the most significant. Jobs that pay poorly are less stable, and economic hardship is the dominant mediating force for all of the major effects of unemployment on families (Strom, 2003).

Not only do adverse selection factors get exacerbated and act as mediators of the negative effects of unemployment, but the picture is further complicated by processes of reverse causality. For instance, job loss is often due to absenteeism and, among employed parents, especially single mothers with young children, unreliable child-care arrangements and health problems are the two primary reasons for missed days from work (Holzer, 2005). Thus, chaos in the family increases risk of parental job loss, which not only increases economic hardship but also exacerbates other existing strains in the family. On the other hand, benefits such as health care, child-care assistance, and flexible working schedules can help parents be more stable and productive employees (Holzer, 2005).

Integration of Research Findings

The research record is consistent with Bronfenbrenner's ecological model in which the parent's workplace acts as an exosystem, a setting that children do not directly experience but that nonetheless contributes to their development. Our review reveals different microsystems—work, day care, home—acting together and influencing each other, and highlights the kinds of parental work experiences that can contribute to a chaotic atmosphere in families. Parents' jobs pose threats to the family when they are unstable and when they are characterized by an overload of responsibilities and demands, distressing interpersonal relationships, nonstandard and unpredictable schedules, or insufficient paid leave time. Inextricable from this mix is the availability of quality child care; the effects of risky job characteristics are greatly exacerbated when child care does not meet both the needs of the family and the demands of the workplace. However, the financial resources that parental employment brings to the family provide protection against chaos, as is underscored by our discussion of the repercussions of economic hardship resulting from job loss and long-term employment.

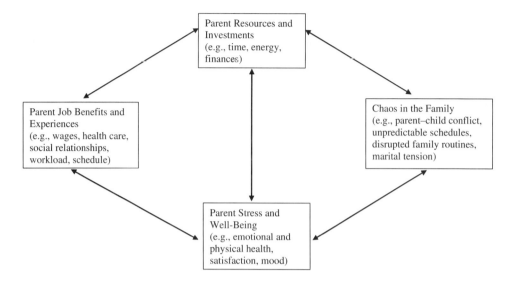

Figure 12.1. Model of the effects of parent employment on chaos in the family.

A simple model that integrates key findings reviewed here is depicted in Figure 12.1. It includes the impact that jobs have on the resources that parents bring to the family, as well as their emotional and physical well-being. On the one hand, when a job leads to a reduction in parental investments and an increase in parent stress, there is a potential addition or exacerbation of chaotic elements in the home. On the other hand, most jobs benefit families; as a job increases the resources that a parent can invest in the family or enhances a parent's well-being, there is a potential reduction in chaos or a strengthening of the protections against chaos (cf. van Steenbergen, Ellemers, & Mooijaart, 2007).

Investments include the parent's devotion of time, energy, attention, and financial resources to the family. In addition to wages, jobs usually bring other resources that can be invested in the family, such as access to health care, a broader social network, knowledge, and values. According to the research literature reviewed here, a parent's time spent at work detracts from family life when it interferes with the family's ability to maintain structure and routines. Interference from work occurs when (a) work hours are distributed in an awkward pattern across the week, such as in nonstandard work shifts; (b) work schedules are unpredictable and uncontrollable; and (c) parents cannot arrange reliable, high-quality child care, either because the costs are prohibitive or the care that is available does not meet the family's needs. Workers in low-wage jobs and single mothers are disproportionately represented in all three categories.

The second mediator in Figure 12.1 is the parent's level of stress and well-being, which can have direct effects on the family's social environment as well as indirect effects through their impact on parent investments. An example of an indirect effect is when stress generated at work detracts from a parent's energy, patience, and cognitive stores, leaving less of each to invest at home. Job stress can also have a more direct impact, as the evidence of negative emotion spillover indicates. Jobs can also enhance parent well-being in many ways, such as when employees derive satisfaction, pride, hope, and a sense of control from their jobs, or when medical benefits mean improved health. Under these circumstances, positive effects on the family are mediated through the parent's mood and physical well-being. Of course the model presented in Figure 12.1 is highly simplified. As discussed next, future research will suggest more complex models with variables that act as moderators and selection factors.

Research Agenda

Our review suggests several promising directions for future research. First, to observe how parental employment contributes to chaos in the family, researchers should focus on unstable jobs, the cumulative effects of different risky job characteristics, and mismatches between the demands of a job and the family's resources and needs. Second, because the effects of parental employment are not uniform, investigators must carefully consider the role of moderator variables, particularly social class, the number and ages of children in the household, and a variety of social and psychological attributes of parents, children, and families. These variables help to define the particulars of families' lives, and it is those details, in combination with certain aspects of the parents' jobs, that determine how the parental workplace adds to, or detracts from, chaos in children's lives.

Recognizing the roles of group differences and moderators means choosing an appropriate sampling strategy in research. One method is to use a random sample with the expectation that both the sample size and the variability in family and job characteristics will be sufficient to detect interaction effects. However, low-income households are typically underrepresented in studies of work–family issues, so that the full range of job and family characteristics needed to identify moderator effects is not always present. Another approach focuses on more homogeneous subgroups to uncover associations that might otherwise be too small to observe in a random sample. For example, investigators can address questions about the impact of different kinds of work schedules by focusing exclusively on single-mother families, or on two-earner families, or on families with children within a particular age range (infants, or school-age children, or teens). The varying child-care needs and practices of these different family types are integral to understanding how work hours affect family routines. In the absence of very large samples that can distinguish among the variety of consequences of work hours that would likely emerge in diverse subgroups, it is more productive to focus on subpopulations in which the effects might be more consistent and therefore easier to detect.

Our third suggestion, therefore, is for more detailed, descriptive investigations of groups of families who share certain features. The complexity of the work–family interface really comes to life in research studies that provide a close-up view of families as they cope with job demands (e.g., Perry-Jenkins' 2005 study of working-class families, Burton et al.'s 2005 study of low-income families, Henly and Lambert's 2005 investigation of low-skilled and hourly wage job holders). The particulars and nuances of families' lives tend to get buried in the general trends that are the focus of large-scale investigations. Our experience with this type of research has been through the Alfred P. Sloan Foundation Center on Everyday Lives of Families at the University of California at Los Angeles. This interdisciplinary research center created a digital video archive of the daily lives of middle-class, dual-earner families over the course of an average week. With these data we observe naturalistic social interaction and the regulation of emotion within families (Ochs, Graesch, Mittmann, Bradbury, & Repetti, 2006). For example, among the fathers, more socially distressing days at work were followed by higher levels of cortisol at home in the evening (Saxbe, Repetti, & Nishina, 2008). Other analyses indicate that these fathers are often ignored by their children when they return home from work (Campos, Graesch, Repetti, Bradbury, & Ochs, in press). Our videotapes show parents sometimes struggling to cope with their children's busy schedules, echoing increases in the number of Americans who report "always feeling rushed" in national surveys (Bianchi & Raley, 2005; see also chap. 4, this volume). This situation was highlighted by Bronfenbrenner in his rather alarming statement that "the hectic pace of modern life poses a threat to our children second only to poverty and unemployment" (Father of Hum Ec Dies, 2005, ¶ 13).

A focus on the particulars of family life also makes clear that jobs are not randomly distributed in the population. Our fourth suggestion is that researchers take selection factors into account when addressing questions about the effects of work on family life. The same variables that are associated with risk for chaos in the family—for example, parent education, personality, and level of functioning—also help to determine the kinds of jobs that parents hold and help to shape their experiences at work. Many of those risk factors are, in turn, exacerbated by stressors at work or the loss of a job creating a complex cycle of influence. We believe that ignoring the role of selection factors and the dynamic ways in which they are entangled with and influence family processes will only limit an understanding of how parents' jobs influence their families.

Finally, we encourage a search for general factors outside of the home that impede parents' functioning and interfere with family life. Although our chapter focuses on the workplace, other than the financial benefits that jobs bring, we do not believe there is anything unique about them in this analysis. Parents need to somehow cope, perhaps through a period of social withdrawal, in response to any set of daily responsibilities that overload them with too many tasks, or that sap their energy and patience, or that crowd their minds with worries. There would be a similar residue of emotional reaction to any kind of social interaction that entails conflict or otherwise generates feelings of anger and frustration, whether those relationships are with family members, neighbors, or coworkers and supervisors. Any rigid, awkward, or unpredictable schedule that is imposed on parents and makes it difficult to be with, and care for, their children will disrupt family routines and relationships; it does not matter whether that schedule is determined by college classes, by the needs of an elderly or sick relative, or by a job or career. When it comes to managing the care of children under these circumstances, ready and reliable assistance from others—which can come from a spouse, a grandparent or other relative, or someone who is paid to provide that assistance—is essential. And when parents are sick, or their children have special needs, such as when they are infants or require medical care, mothers and fathers need a break from their other demanding responsibilities. That can mean a semester off from school, or an adult sibling taking over the care of a debilitated elderly parent, or paid leave from a job. Instead of an approach that focuses exclusively on the workplace, we believe an approach that identifies general processes in the exosystem that can impact microsystem chaos is ultimately more consistent with, and will help to advance, Urie Bronfenbrenner's goal of a general theory of bioecological development.

References

Almeida, D. M., Wethington, E., & Chandler, A. L. (1999). Daily transmission of tensions between marital dyads and parent-child dyads. *Journal of Marriage and the Family, 61*, 49–61.

Barnett, R. C. (2006). Relationship of the number and distribution of work hours to health and quality-of-life (QOL) outcomes. In P. L. Perrewé & D. C. Ganster (Eds.), *Research in occupational stress and well being: Vol. 5. Employee health and coping methodologies* (pp. 99–138). New York: Elsevier.

Barnett, R. C., & Gareis, K. C. (2007). Shift work, parenting behaviors, and children's socioemotional well-being. *Journal of Family Issues, 28*, 727–748.

Bianchi, S. M., & Raley, S. B. (2005). Time allocation in families. In S. M. Bianchi, L. M. Casper, & R. B. King (Eds.), *Work, family, health, and well-being* (pp. 21–42). Mahwah, NJ: Erlbaum.

Bolger, N., DeLongis, A., Kessler, R. C., & Wethington, E. (1989). The contagion of stress across multiple roles. *Journal of Marriage and the Family, 51*, 175–183.

Bronfenbrenner, U. (1989). Ecological systems theory. *Annals of Child Development, 6*, 187–249.

Bumpus, M. F., Crouter, A. C., & McHale, S. M. (1999). Work demands of dual-earner couples: Implications for parents' knowledge about children's daily lives in middle childhood. *Journal of Marriage and the Family, 61*, 465–475.

Burton, L. M., Lein, L., & Kolak, A. (2005). Health and mothers' employment in low-income families. In S. M. Bianchi, L. M. Casper, & R. B. King (Eds.), *Work, family, health, and well-being* (pp. 493–509). Mahwah, NJ: Erlbaum.

Campos, B., Graesch, A. P., Repetti, R., Bradbury, T., & Ochs, E. (in press). Opportunity for interaction? A naturalistic observation study of dual-earner families after work and school. *Journal of Family Psychology*.

Christoffersen, M. N. (2000). Growing up with unemployment: A study of parental unemployment and children's risk of abuse and neglect based on national longitudinal 1973 birth cohorts in Denmark. *Childhood, 7*, 421–438.

Cooksey, E. C., Menaghan, E. G., & Jekielek, S. M. (1997). Life-course effects of work and family circumstances on children. *Social Forces, 77*, 637–667.

Coser, L. (1974). *Greedy institutions: Patterns of undivided commitment*. New York: Free Press.

Costigan, C. L., Cox, M. J., & Cauce, A. M. (2003). Work-parenting linkages among dual-earner couples at the transition to parenthood. *Journal of Family Psychology, 17*, 397–408.

Crouter, A. C., & Bumpus, M. F. (2001). Linking parents' work stress to children's and adolescents' psychological adjustment. *Current Directions in Psychological Science, 10*, 156–159.

Crouter, A. C., Bumpas, M. F., Maguire, M. C., & McHale, S. M. (1999). Linking parents' work pressure and adolescents' well-being: Insights into dynamics in dual-earner families. *Developmental Psychology, 35*, 1453–1461.

Crouter, A. C., & McHale, S. M. (1993). Temporal rhythms in family life: Seasonal variation in the relation between parental work and family processes. *Developmental Psychology, 29*, 198–205.

Crouter, A. C., & McHale, S. M. (2005). Work, family, and children's time: Implication for youth. In S. M. Bianchi, L. M. Casper, & R. B. King (Eds.), *Work, family, health, and well-being* (pp. 49–66). Mahwah, NJ: Erlbaum.

De Schipper, J. C., Tavecchio, L. W. C., Van IJzendoorn, M. H., & Linting, M. (2003). The relation of flexible child care to quality of center day care and children's socio-emotional functioning: A survey and observational study. *Infant Behavior and Development, 26*, 300–325.

Doumas, D., Margolin, G., & John, R. S. (2003). The relationship between daily marital interaction, work, and health promoting behaviors in dual-earner couples: An extension of the work-family spillover model. *Journal of Family Issues, 24*, 3–20.

Dubas, J. S., & Gerris, J. R. M. (2002). Longitudinal changes in the time parents spend in activities with their adolescent children as a function of child age, pubertal status, and gender. *Journal of Family Psychology, 16*, 415–427.

Evans, G. W., & Wener, R. E. (2006). Rail commuting duration and passenger stress. *Health Psychology, 25*, 408–412.

Father of hum ec dies. (2005, September 26). *The Cornell Daily Sun.* Available from http://cornellsun.com/node/15373

Galambos, N. L., Sears, H. A., Almeida, D. M., & Kolaric, G. C. (1995). Parents' work overload and problem behavior in young adolescents. *Journal of Research on Adolescence, 5*, 201–223.

Greenberger, E., O'Neil, R., & Nagel, S. K. (1994). Linking workplace and homeplace: Relations between the nature of adults' work and their parenting behaviors. *Developmental Psychology, 30*, 990–1002.

Grimm-Thomas, K., & Perry-Jenkins, M. (1994). All in a day's work: Job experiences, self-esteem, and fathering in working-class families. *Family Relations, 43*, 174–181.

Henly, J. R., & Lambert, S. (2005). Nonstandard work and child-care needs of low-income parents. In S. M. Bianchi, L. M. Casper, & R. B. King (Eds.), *Work, family, health, and well-being* (pp. 473–492). Mahwah, NJ: Erlbaum.

Heymann, J., Simmons, S., & Earle, A. (2005). Global transformation in work and family. In S. M. Bianchi, L. M. Casper, & R. B. King (Eds.), *Work, family, health, and well-being* (pp. 511–528). Mahwah, NJ: Erlbaum.

Holzer, H. J. (2005). Work and family life: The perspective of employers. In S. M. Bianchi, L. M. Casper, & R. B. King (Eds.), *Work, family, health, and well-being* (pp. 83–95). Mahwah, NJ: Erlbaum.

Howe, G. W., Levy, M. L., & Caplan, R. D. (2004). Job loss and depressive symptoms in couples: Common stressors, stress transmission, or relationship disruption? *Journal of Family Psychology, 18*, 639–650.

Kohn, M. L. (1969). *Class and conformity: A study in values.* Homewood, IL: Dorsey Press.

Kohn, M. L., & Schooler, C. (1982). Job conditions and personality: A longitudinal assessment of their reciprocal effects. *American Journal of Sociology, 87*, 1257–1286.

Korenman, S., & Kaestner, R. (2005). Work-family mismatch and child health and well-being: A review of the economics research. In S. M. Bianchi, L. M. Casper, & R. B. King (Eds.), *Work, family, health, and well-being* (pp. 297–312). Mahwah, NJ: Erlbaum.

Kuper, H., & Marmot, M. (2003). Job strain, job demands, decision latitude, and risk of coronary heart disease within the Whitehall II study. *Journal of Epidemiology and Community Health, 57*, 147–153.

Leibowitz, A. A. (2005). An economic perspective on work, family, and well-being. In S. M. Bianchi, L. M. Casper, & R. B. King (Eds.), *Work, family, health, and well-being* (pp. 187–200). Mahwah, NJ: Erlbaum.

Mason, C. A., Cauce, A. M., Gonzales, N., Hiraga, Y., & Grove, K. (1994). An ecological model of externalizing behaviors in African-American adolescents: No family is an island. *Journal of Research on Adolescence, 4*, 639–655.

Menaghan, E. G., & Parcel, T. L. (1991). Determining children's home environments: The impact of maternal characteristics and current occupational and family conditions. *Journal of Marriage and the Family, 53*, 417–431.

Novaco, R. W., Kliewer, W., & Broquet, A. (1991). Home environmental consequences of commute travel impedance. *American Journal of Community Psychology, 19*, 881–909.

Ochs, E., Graesch, A., Mittmann, A., Bradbury, T., & Repetti, R. L. (2006). Video ethnography and ethnoarchaeological tracking. In M. Catsouphes, E. E. Kossek, & S. Sweet (Eds.), *Handbook of work and family: Multi-disciplinary perspectives and approaches* (pp. 387–409). Mahwah, NJ: Erlbaum.

Parcel, T. L., & Menaghan, E. G. (1993). Family social capital and children's behavior problems. *Social Psychology Quarterly, 56*, 120–135.

Parcel, T. L., & Menaghan. E. G. (1994). Early parental work, family social capital, and early childhood outcomes. *American Journal of Sociology, 99*, 972–1009.

Perry-Jenkins, M. (2005). Work in the working class: Challenges facing families. In S. M. Bianchi, L. M. Casper, & R. B. King (Eds.), *Work, family, health, and well-being* (pp. 453–472). Mahwah, NJ: Erlbaum.

Perry-Jenkins, M., Repetti, R. L., & Crouter, A. C. (2000). Work and family in the 1990s. *Journal of Marriage and the Family, 62,* 981–998.

Piotrkowski, C. S. (1979). *Work and the family system.* New York: Free Press.

Presser, H. B. (2005). Embracing complexity: Work schedules and family life in a 24/7 economy. In S. M. Bianchi, L. M. Casper, & R. B. King (Eds.), *Work, family, health, and well-being* (pp. 43–48). Mahwah, NJ: Erlbaum.

Price, R. H., Friedland, D. S., & Vinokur, A. D. (1998). Job loss: Hard times and eroded identity. In J. H. Harvey (Ed.), *Perspectives on loss: A sourcebook. Death, dying, and bereavement* (pp. 303–316). Philadelphia: Brunner/Mazel.

Repetti, R. L. (1989). Effects of daily workload on subsequent behavior during marital interaction: The roles of social withdrawal and spouse support. *Journal of Personality and Social Psychology, 57,* 651–659.

Repetti, R. L. (1992). Social withdrawal as a short-term coping response to daily stressors. In H. S. Friedman (Ed.), *Hostility, coping, and health* (pp. 151–165). Washington, DC: American Psychological Association.

Repetti, R. L. (1994). Short-term and long-term processes linking job stressors to father-child interaction. *Social Development, 3,* 1–15.

Repetti, R. L. (2005). A psychological perspective on the health and well-being consequences of parental employment. In S. M. Bianchi, L. M. Casper, & R. B. King (Eds.), *Work, family, health and well-being* (pp. 245–258). Mahwah, NJ: Erlbaum.

Repetti, R. L., & Mittmann, A. (2004). Workplace stress. In A. Christensen, R. Martin, & J. Smyth (Eds.), *Encyclopedia of health psychology* (pp. 342–344). New York: Kluwer Academic/Plenum Publishers.

Repetti, R. L., & Wood, J. (1997a). Effects of daily stress at work on mothers' interactions with preschoolers. *Journal of Family Psychology, 11,* 90–108.

Repetti, R. L., & Wood, J. (1997b). Families accommodating to chronic stress: Unintended and unnoticed processes. In B. H. Gottlieb (Ed.), *Coping with chronic stress* (pp. 191–220). New York: Plenum Press.

Roy, K. M., Tubbs, C. Y., & Burton, L. M. (2004). Don't have no time: Daily rhythms and the organization of time for low-income families. *Family Relations, 53,* 168–178.

Ruhm, C. J. (2005). How well do government and employer policies support working parents? In S. M. Bianchi, L. M. Casper, & R. B. King (Eds.), *Work, family, health, and well-being* (pp. 313–325). Mahwah, NJ: Erlbaum.

Saxbe, D., Repetti, R. L., & Nishina, A. (2008). Marital satisfaction, recovery from work, and diurnal cortisol among men and women. *Health Psychology, 27,* 15–25.

Schulz, M. S., Cowan, P. A., Cowan, C. P., & Brennan, R. T. (2004). Coming home upset: Gender, marital satisfaction and the daily spillover of workday experience into couple interactions. *Journal of Family Psychology, 18,* 250–263.

Sleskova, M., Salonna, F., Geckova, A. M., Nagyova, I., Stewart, R. E., van Dijk, J. P., & Groothoff, J. W. (2006). Does parental unemployment affect adolescents' health? *Journal of Adolescent Health, 38,* 527–535.

Staines, G. L., & Pleck, J. H. (1983). *The impact of work schedules on the family.* Ann Arbor, MI: Institute for Social Research, University of Michigan.

Story, L. B., & Repetti, R. L. (2006). Daily occupational stressors and marital behavior. *Journal of Family Psychology, 20,* 690–700.

Strom, S. (2001). Keep out of the reach of children: Parental unemployment and children's accident risks in Sweden 1991-1993. *International Journal of Social Welfare, 11,* 40–52.

Strom, S. (2003). Unemployment and families: A review of research. *Social Service Review, 77,* 399–430.

Taylor, P., Funk, C., & Clark, A. (2007). *From 1997 to 2007: Fewer mothers prefer full-time work. A social & demographic trends report.* Washington, DC: Pew Research Center.

Unger, J. B., Hamilton, J. E., & Sussman, S. (2004). A family member's job loss as a risk factor for smoking among adolescents. *Health Psychology, 23,* 308–313.

van Steenbergen, E. F., Ellemers, N., & Mooijaart, A. (2007). How work and family can facilitate each other: Distinct types of work–family facilitation and outcomes for women and men. *Journal of Occupational Health Psychology, 12,* 279–300.

Whitbeck, L. B., Simons, R. L., Conger, R. D., Wickrama, K. A. S., Ackley, K. A., & Elder, G. H., Jr. (1997). The effects of parents' working conditions and family economic hardship on parenting behaviors and children's self-efficacy. *Social Psychology Quarterly, 60,* 291–303.

Part V

Chaos at the Macrosystem Level

13

Well-Being, Chaos, and Culture: Sustaining a Meaningful Daily Routine

Thomas S. Weisner

The experience of persisting chaotic household, family, and community life is troubling and undesirable for parents and children. A central reason why this is so goes beyond the problems created by noise, toxic air and water, danger, and the other indicators of chaotic settings. Chaotic environments make it very difficult to accomplish a task that is recognizable most everywhere in the world—the project of sustaining a daily routine of life, for oneself and one's family and household, that is reasonably predictable, helps meet values and goals that give meaning to life, fits with the resources available, and keeps conflicts and disagreement relatively low. Chaotic settings do not provide much opportunity for the experience of *well-being*, which is engaged participation in the daily activities of a cultural community that that community deems desirable. Therefore, chaos matters because it can so often make well-being and meaningful cultural engagement very difficult. It disrupts the ability to participate in cultural communities that give meaning to life through the experience of well-being, in addition to the other deleterious effects it has. In this chapter, I contrast chaos with what I think is its opposite—well-being—and suggest a framework for describing chaos and well-being that could be applicable across cultural communities.

It also is important to clearly frame one's conception of chaos so that plural pathways to good development around the world are not mistaken for pathways characterized by deficits and chaos. Some developmental contexts would be universally considered chaotic and undesirable for parents and children. Other social settings might seem to be chaotic, or at least developmentally inappropriate, but in fact they are not because they represent valued developmental pathways for some communities. Many other circumstances are a combination: settings that have elements indexical of chaotic and deleterious environments, yet with compensatory beliefs, values, goals, and practices that in local context are

---

The research reported in this chapter was supported by grants from the National Institute of Child Health and Human Development: HD19124 and HD11944, Ecocultural Contexts of Families With Children With MR/DD; HD36038 (A. C. Huston, principal investigator [PI]), Income and Employment Effects on Children and Families; and HD004612 (T. S. Weisner, PI core), Field Work Training and Qualitative Data. The Semel Institute (Center for Culture and Health), University of California, Los Angeles, also provided support.

not experienced as chaotic, or at least not to the same extent as would be expected.

In this chapter, I first identify features that are good indicators of chaos and deleterious conditions most anywhere, and some others that may appear so but are likely not in other communities. I then outline an ecocultural conceptualization of chaos that is useful for understanding chaos in context. I consider the opposite of chaos not as the absence of chaotic conditions but as the presence of well-being, and offer a definition of well-being that also suggests ways to understand chaos. Finally, I present two examples of research studies on well-being to illustrate this approach.

Some Chaotic Environmental Features Are Likely to Be Deleterious Anywhere

Strong evidence suggests that some known environmental features are not going to be good for children and families most anywhere. Several chapters in this volume identify such bioecological and psychosocial risk conditions. Lustig (see chap. 15) describes the war and refugee experience of mortal danger, deprivation, upheaval, fear, uncertainty, and loss, which are defining features of a chaotic context and experience. Chronically inadequate nutrition, routinely unresponsive social stimulation, and little or no physical protection from threat and attack coming from both inside and outside the home are all clearly deleterious. High exposure to toxic substances and environments, as well as low-quality and dangerous housing, schools, day care, and other neighborhood settings, also contributes to high-risk environments (Evans, 2006). Low income and persistent poverty are associated with chaotic indices such as household crowding, high noise levels, disrupted household routines and rituals, residential and school relocation, and parental partner instability. However, poverty or low socioeconomic status (SES) is not simply a marker of chaos (see chap. 14, this volume; see also Evans, Gonnella, Marcynyszyn, Gentile, & Selpekar, 2005).

Chronic and persistent anger, conflict, and violence in close relationships are not good experiences for children, families, or communities anywhere. Cold, unsupportive, and neglectful caretaking and relationships are similarly unhealthy, inhibiting self-regulation and social engagement (Repetti, Taylor, & Seeman, 2002)—if such relationships are experienced by community members that way. However, such care can have compensatory benefits for households facing harsh and dangerous environments by possibly helping children survive in very difficult circumstances (e.g., Goldstein, 2003). Rapid, unpredictable, or unwanted changes in caretakers for children, especially when others familiar to a child are unavailable to buffer change, are not good for children or families. A persistent sense of threat or danger undermining social connections and security does not promote good health and is associated with physiological stress (Taylor, Repetti, & Seeman, 1997). Relative social and economic status inequality, especially where compensatory social communities and social supports are not available, are difficult barriers to overcome. These circumstances are all too often part of chaotic situations.

Chaos in contemporary circumstances in the United States has been defined to include situations with little structure or stable routines: noise and crowding in an overly fast-paced world allowing fewer hours for household members to be together, fewer family rituals and ceremonies, and the experience of life as hectic and of things being "out of control" (see chap. 1). These conditions could affect children in various ways. Social exchanges are unpredictable, brief, and few. A sense of competency and effectance may be lower because of the uncertainty of the environment within which one is attempting to be effective. Environmental chaos (noise, density and crowding, lack of structure) can have similar deleterious effects in non-Western, developing societies as well, though there are mediating influences and different cultural understandings of these environmental features.

Contextual Understanding of Environmental Circumstances

Strong evidence from the ethnographic record and contemporary studies of children and families indicates that there are many different, pluralistic ways of raising children (LeVine, 2007). These pathways at times may appear chaotic, deleterious, or unacceptable by Eurocentric standards, but they are nonetheless morally and developmentally appropriate in other local contexts. Also, strong evidence shows that these alternative pathways can and do produce youth and adults who are socially competent and capable of contributing to their community and becoming parents themselves, as is true in the U.S. context (DeLoache & Gottlieb, 2000; Harkness & Super, 1996; Kagitçibasi, 1996; LeVine, Miller, & West, 1988; LeVine & New, 2008; Shweder, Minow, & Markus, 2002; Weisner, 2001).

An example of a very widespread practice and model of good child care in much of the world is having multiple caretakers for children. Does socially distributed multiple caretaking of children lead to relational insecurity, emotional loss, confusion, and anxiety, or does it develop a strong sense of empathy, nurturance, social responsibility, social intelligence, and social competence in children? Are children and parents in these socially distributed care settings in fact encouraging affiliative rather than egoistic or individualistic styles of competence and achievement that have important adaptive advantages? Many communities practice one or another form of child lending, fosterage, or kin adoption of children. Is the use of multiple caretakers a sign of deficient parenting? In such communities socially distributed care, defined as the placing of children in settings where others can assist in care and where children will learn to adapt, is a positive goal. This is an expectable, normative, taken-for-granted cultural pathway for children and parents (Gottlieb, 2004; Isiugo-Abanihe, 1985; Keller, 2007; Serpell, 1993; Seymour, 1999; Verhoef, 2005; Weisner, 1996).

Attachment and the development of "security" is another example of good child care in other parts of the world. Is there a single, optimal maternal–child attachment style, or are there plural cultural models of close relationships (between parents, siblings, kin, peers, authorities, romantic partners, and others)? Do these varied cultural models of ways to socialize trust lead to deficits, stress, and insecurity in children, or do they result in flexible, interdependent relational security (Harwood, Miller, & Irizarry, 1995; LeVine & Norman, 2001; Weisner, 2005)? "Authoritative" parenting is similarly widely cited in the United States as a putative candidate for an optimal parenting style for children (Baumrind, 1989). Are parenting practices described as "stern," "harsh," or "strict" by Euro-American researchers and on most parenting scales actually negative and deleterious for children, or are structure, respect, clear control, training, and the importance of understanding hierarchy and authority common and effective parenting models in the United States and in much of the world (e.g., Chao, 1994; Darling & Steinberg, 1993; Lareau, 2003)?

Placing children and parents into separate, less crowded individual rooms, spaces, and beds provides another example. Is separate sleeping desirable and optimal for the development of autonomy and independence, providing a developmentally essential sense of personal space and security in less chaotic spaces? Does this practice encourage individuation and self-regulation and discourage an inappropriate dependency on the mother while strengthening the parent–child relationship? Or do the various cosleeping arrangements for children that are common around the world, particularly crowded, dense, bed sharing among children, show not only that cosleeping is morally appropriate in a community emphasizing interdependence and "symbiotic harmony" as an overarching developmental goal but also that there are no negative consequences related to the practice (McKenna & McDade, 2005; Okami, Weisner, & Olmstead, 2002; Rothbaum, Pott, Azuma, Miyake, & Weisz, 2000; Shweder, Balle-Jensen, & Goldstein, 1995)?

These examples suggest that what might appear developmentally deleterious, dangerous, and chaotic using Eurocentric criteria may not be so to the same degree, or even at all, when considered from the viewpoints of communities with other standards for good parenting and goals for good development. As useful as many current environmental and social–relational indices of chaos or deleterious parenting practices are, middle- and upper-middle-class Euro-American families, children, and their neighborhoods often seem to rise to the top of the positive end of parenting practices, home environments, and child developmental scales. It is likely that some of these measures of parenting, home environments, and chaos implicitly have Euro-American development goals and pathways embedded in them and thus have been defined as more positive. Lichter and Wethington (see chap. 2, this volume), for example, demonstrated considerable change in health, poverty, mortality, and household conditions in the United States over the past few generations, usually in the direction of reduced chaos, poverty and mortality, and improved health. Furthermore, although minority and immigrant children are especially likely candidates for experiencing chaos in the home, the meanings of chaos or of alternative expressions of instability may vary substantially across cultural and racial or ethnic groups. Lichter and Wethington suggested that studies of growing inequality in environments and opportunity require understanding different forms of chaotic situations. Wachs and Corapci (2003) called for better theory regarding the kinds of cultural differences that would moderate or mediate environmental chaos. Strong evidence for the impacts of chaos on children and families will require integrated measures of objective conditions, subjective experience, and features of the cultural learning environment relevant to candidate indicators of chaos within an environment.

Well-Being and Chaos in an Ecocultural Context

In addition to being sure to conceptualize and assess what might be chaos or deleterious practices in cultural contexts, researchers need to offer a positive contrast to what is chaotic. Order and predictability alone, if defined as the opposite of chaos, are not sufficient; one might ask, predictability of what environment and which practices? A more useful contrast to chaos is well-being. Well-being incorporates the meaning and the sociocultural components of what are desirable everyday routines and activities into an understanding of what chaos, as the opposite of well-being, might be like.

Well-being is the engaged participation of a child or parent in the everyday routines and activities deemed desirable by a cultural community, and the psychological experiences produced as a result of such engagement. Chaos is, in part, the absence of such meaningful cultural engagement.

This conception of well-being can apply cross-culturally and prove useful in understanding what is and is not chaos in the environments of children and families elsewhere in the world (Weisner, 2009). Like chaos, the conditions that promote well-being are likely to have both universal and local aspects. There are already examples of such work in research on quality of life, well-being, and family and child assessment (Bornstein, Davidson, Keyes, & Moore, 2003; Brown, 1997). Earls and Carlson (1995), for example, called for the incorporation of the experiences of parents and children in local cultural context into family and child well-being work. Ben-Arieh and Goerge (2005) also included strong assessments of local cultural context in their global assessments of well-being. Vleminckx and Smeeding (2001) showed such variations across economically developed nations. Chaos also could be that set of specific contextual factors that in turn produces low well-being and disengagement (see chap. 1, this volume). Both causal paths (absence of well-being leads to the experience of chaos, or chaotic contexts lead to low well-being and then disengagement) have empirical support. Regardless of the causal sequence, the broader point is that meaningful cultural engagement is negatively linked to chaos.

Sustaining a Meaningful Routine of Life: A Universal Developmental Project for Providing Well-Being

Well-being is a family and ecocultural project. Parents everywhere have a common project: to construct a social ecology that balances what they want for themselves and their family with what is possible given their circumstances. This project involves sustaining a daily routine of life. Sustainability is a holistic conceptualization of how families are doing with respect to this project. Ecological–cultural (ecocultural) theory suggests that sustaining a daily routine is a universal adaptive problem for all families (Whiting & Edwards, 1988). Ecocultural family theory extends Super and Harkness's (1986) notion of a developmental niche for the child to the study of the family and ecological context (Weisner, 2002).

The unit of analysis for studying both sustainability and well-being is the everyday activities and practices in the context of the cultural learning environment of that community (Whiting, 1980). Activities and practices can be described by common dimensions with features that recur and structure life in that environment: a *script* for normative conduct; *goals and values* organizing the meaning and direction of the activity; a *task* and functional goals of the activity; *people* and their relationships that are present in the activity; the motivations and feelings that people have in the setting that influence their *engagement* in it; and the *resources* needed to constitute the activity and make it happen (Cole, 1996; Gallimore, Goldenberg, & Weisner, 1993; Rogoff, 2003).

Sustainability is an ongoing project, not a one-time end point; it is contextual, embedded in an everyday routine of life, and thus part of some local social context defined by appropriate action, speech, and morality. Sustainability includes both the local resources and ecology of the family and community and the goals, values, and meanings that the community affords and people bring to their practices.

Sustainability of a child's daily routine of activities should enhance well-being because it defines a higher quality family daily routine (which consists of activities prevalent in a community). Barring pathology and more universally chaotic circumstances that can and do prevail around the world, well-being is the engaged participation by children and families in such meaningful activities. Well-being and sustainability are contextual universals. The features of sustainability should enhance well-being in any community, but the local meanings and contexts of that community are a part of measuring and understanding well-being. Contextual analysis is not the same as relativism, or the view that practices can be understood only by using local community standards.

Sustainability links the psychological experience of well-being to those environmental features thwarting sustainability, which are likely candidates for understanding chaos in any community. For example, socially distributed care of children, such as child fosterage or adoption, might be experienced as a high-stress, chaotic family environment in one context in which it is not normative and valued, but is a much less chaotic situation, even a desired pathway, in another community. Describing the cultural learning environment carefully would go a long way toward the evidence needed to ascertain the extent to which an environmental feature that might be detrimental or chaotic in a local community in fact is, and how it affects families and children.

Sustainability as a Comparative Standard for Well-Being

Sustaining a daily routine involves four processes (Weisner et al., 2005): (a) fitting the routine to family resources, (b) balancing varied family interests and conflicts, (c) ensuring meaningfulness of family activities with respect to goals and values, and (d) providing stability and predictability of the daily routine. Routines that have better resource fit, less conflict, more balance, more meaningfulness, and enough predictability are posited to be better for families, and thus provide greater well-being for those participating in them. If these features are largely absent most of the time, these could be indicators of chaos.

Chaotic conditions are those that significantly interfere with the developmental project of sustaining a meaningful daily routine. To be chaotic, though, not only would these conditions have to be quite extreme and disruptive, but they also would have to persist across time in a child's development and show connections to one or more of the features that would make a family cultural routine of life reasonably meaningful and sustainable. Those candidate features of chaos would include an inability to fit the family routine into the resources available; continual conflict, violence, or threat; lack of fit with goals and values; and persistent unpredictability and instability. Hertzman (see chap. 8) emphasized the many pathways through which temporal and spatial stability can matter to family process and child development. Chaos and thwarting of sustainability are associated; however, the causal pathways can go in multiple directions, including the expectation that some features, such as resource scarcity itself, are antecedents that increase the risk of chaos.

Imagine developmental pathways as consisting of everyday activities (getting ready for bed, sleeping, having breakfast, going to church, sitting in classrooms, going to an after-school program, visiting relatives, playing video games, doing homework, hanging with friends, going to the mall, dating, "partying," watching TV). Those activities and their cultural and ecological contexts are the "stepping stones" children traverse as they move along a pathway through the day and the day's routine. These activities make up the life pathways that they engage in each day. Children not only actively and joyously engage in those activities but also resist and transform them as active agents.

Urie Bronfenbrenner's process, person, context, and time (PPCT) framework describes participants enmeshed in an active, dynamic socioecological system that is analogous to the pathways model of everyday activities (Bronfenbrenner, 2005). The PPCT framework includes cultural scripts for processes (such as interactional processes or processes of resource allocation), cultural models for understanding the conception of a person (child and youth agency, the self, identity), cultural learning environments for children (context creation, engagement, and meaning), and cultural transmission of these models and scripts across generations. I think Bronfenbrenner would appreciate the importance of the socially situated approach to mind. He commented that mental processes (perception, cognition, motivation, emotion, memory) "are about *something* ... [and] much of that content is in the outside world. ... In humans, the content turns out, early on, to be mainly about people, objects, and symbols" (p. 177). The PPCT framework seems more useful, at least operationally, than the perhaps more familiar nested hierarchical model of microsystem, mesosystem, exosystem, and macrosystem. In that model, culture, for example, is placed in the macrosystem, far from the everyday settings in which cultural models, scripts, values, and goals are actually lived out (Bronfenbrenner, 1979; Weisner, 2008).

Sustainability of family routines differs from stress, coping, and similar challenges, though those familiar constructs are also certainly involved in sustaining a family daily routine. The stress and coping models begin with an unusual, difficult perturbation that is a challenge or (potential) threat, and then responses of individuals or families to those perturbations are examined. Resilience or adaptation is the successful response to such threats. Sustainability, however, captures another, more enduring project: juggling ongoing demands in the service of meeting long-term shared and personal goals and developmental projects (Gallimore, Bernheimer, & Weisner, 1999). Sustainability focuses on the everyday accommodations made in a local context that keep life going and keep alive the daily routines expectable and meaningful in that community. Fiese and Winter (see chap. 4) reviewed the evidence supporting chaos as a family-wide construct, closely tracking its impacts on child socioemotional functioning within meaningful family routines.

Assessing Sustainability of the Daily Routine: The Ecocultural Family Interview

Sustainability of the daily routine can be reliably described and assessed over time. The Ecocultural Family Interview (EFI) provides the frame for such a conversation. It begins with an orienting question: "Walk me through your day. From the time you get up to the time you go to bed, what are your activities throughout the day? What does it take to make those activities happen?" The EFI focuses on the activities that make up the day, their importance and salience to parents, and the features that constitute the daily routine (Weisner, 2009; Weisner et al., 2005). The EFI is a conversation, not a question–response format interview. This kind of conversation has long been useful in qualitative and ethnographic research. The data from EFI-type conversations (along with fieldnotes and direct observations of daily routines) can be indexed, rated, and coded for quantitative analysis and comparison.

For example, the EFI has been used to better understand the process of family accommodation to children with developmental disabilities (Gallimore, Coots, Weisner, Garnier, & Guthrie, 1996; Nihira, Weisner, & Bernheimer, 1994). Our research team encouraged parents to "tell their story" about their child and about how they were or were not adapting their family routine, and in response to what or whom (Bernheimer & Weisner, 2007). There were no false negatives; if parents did not bring up topics, we used probes that carefully covered standard ecocultural domains we knew from theory and prior research would likely matter (e.g., income or equity resources, work or career, supports, child services, siblings, work schedules, goals, religious beliefs, kin and friendship connections, transportation, couple supports).

Overall ratings of sustainability in this study are related to family composition (lower ratings for single parents) and SES and family income (higher ratings for higher SES and income). Resources are not a proxy for achieving a sustainable daily routine, however, because there is considerable variation in income and SES. Higher levels of family sustainability are associated with high social and interpersonal kin and nonkin social and institutional connectedness, and with lower family workloads in caring for the child with a disability. However, none of these relationships are linear; how families integrate and balance work and make use of connections is important for sustainability.

Sustainability starts with a holistic appraisal of the family's goals, the context of their daily routine, and the varied features that seem to sustain a routine for that family (e.g., resource fit, balancing conflicts, meaning with respect to goals, and stability and predictability). Traditional measures of family adjustment are useful, but the valence of the items in such measures is predetermined and not considered in relationship to the whole family system. What might be a "good" score for one family on an item may not be relevant or in fact may be negative for another. For example, eating meals together might work for an easier child and be a sign of coparticipation for one family, but for a behaviorally difficult and unpredictable child, the practice may be impossible to sustain and be more disruptive; it may be better for sustainability to have the child eat first. Hence, a higher score on an item such as "family eats dinner together often" might or might not indicate less chaos and greater family sociality and well-being. More participation in support programs or parent training might fit with the time available, resources, and goals for one family, and perhaps benefit the child, but for another family, it might be too difficult to sustain, with too little impact on the child. Therefore, a higher score on an item such as "higher participation in support programs" might or might not be indicative of family well-being (Skinner & Weisner, 2007). For some children with disabilities, having more typically developing peers as friends might be important and assist them in fuller inclusion. For other children, however, having friends who also have disabilities and whom they meet more regularly at school or other programs could be more sustaining and more likely to be related to greater overall satisfaction (Matheson, Olsen, & Weisner, 2007)

Sustainability varied across the families in the disability study, including a group who were "multiply troubled." Many of these families would fit a chaotic environment description. For example, they reported and we observed the sense of being out of control, lack of structure or lack of family rituals and routines, and inability to engage with services. But others with similar difficulties and low SES were "improving/resilient" with effort, grit, and strong goals to persevere and improve their child's developmental course. Their problems in balancing life and conflicts were sometimes high, but their life satisfaction was also fairly high, and chaos was lower or absent.

Family Supports, Intervention Research, and Ecocultural Evidence

Understanding the cultural learning environment also matters for experimental research with applied and policy implications. For example, we developed a version of the EFI for working poor families and children in the United States as part of a prospective, longitudinal experimental intervention—the New Hope program (Bernheimer, Weisner, & Lowe, 2003; Duncan, Huston, & Weisner, 2007; Weisner, Gibson, Lowe, & Romich, 2002; Yoshikawa, Weisner, & Lowe, 2006). The New Hope program included wage supplements, child-care vouchers, health care subsidies, and a community service job if needed. To receive New Hope supports, adults had to be working full-time, that is, 30 hours a week or more. Over 1,300 adults in two zip codes in Milwaukee, Wisconsin, were recruited for a study of the impact of work and family supports for working poor families. Half were randomly selected as eligible for New Hope program supports; the control families, as well as New Hope families, were eligible for other available public and private programs. A survey, child assessments, teacher reports, and administrative records were also used in the mixed method design (Bos et al., 1999; Huston et al., 2001, 2003, 2005).

We randomly sampled a total of 44 families from the treatment and control groups and conducted a longitudinal ethnographic study, including the EFI. The EFI focused on how families organized their daily routine in the face of low incomes, often poor and unsafe neighborhood circumstances, limited community services, jobs that were often episodic and poorly paid with few or no benefits, and other difficult work and family circumstances in their lives. How parents learned about and used the New Hope program depended in part on their family sustainability and their prior beliefs and expectations regarding support programs (Gibson & Weisner, 2002). How parents used the child-care vouchers also depended on their beliefs and values about appropriate care, and how such formal care fit into their daily routines (Lowe & Weisner, 2004).

About 15% of families had multiple (more than two) barriers to employment, such as physical or mental health concerns, not having a high school degree, prior incarceration, several young children, drug or alcohol problems, little prior work experience, or very low and sporadic income. Although selected for the New Hope program, they were seldom able to engage with program services. Others led quite difficult, often chaotic lives, and so struggled to retain long enough work hours to sustain use of New Hope supports. Those with fewer barriers to work (one or perhaps two) were the most likely to use New Hope services, work longer periods, and show gains in income. Repetti and Wang (see chap. 12) reviewed the many ways that work issues can influence family well-being, stress, and chaos (parent–child conflict, unpredictable schedules, disrupted family routines, marital tension). Chaotic environments kept some of these economically poor families from benefiting from New Hope supports by preventing regular work. New Hope supports kept other families out of chaotic situations.

Interventions, no matter how well designed and implemented, will not have an impact unless a place can be found for them in the daily routines of the individuals, families, social settings, or organizations they are intended to change. In chaotic family circumstances with few meaningful routines, even the best-designed interventions are less likely to work; it is necessary to first reduce family chaos. The mixed method integration of the EFI into developmental, economic, work, and intervention outcome evidence led to important findings about how the interventions fit into the daily routines and goals of families. Contextually informed interventions will not only benefit efforts to improve chaotic circumstances for families and children but will do so in a way that fully recognizes the plurality of developmental beliefs, goals, and practices in the United States and around the world. Some of the families eligible for New Hope benefits indeed reduced chaotic circumstances and improved family and child well-being by using the New Hope supports. If researchers, practitioners, and parents themselves intend to use strong research evidence to change chaotic circumstances, or simply to support families and children and thereby increase their well-being, they all will benefit from evidence on the cultural learning environments and daily routines of those families and children.

Summary

Living in chaotic families and neighborhoods can thwart even the best efforts of parents and children to achieve some level of well-being in life. I have emphasized the importance of well-being for children, parents, families, and communities because it is a contextual, holistic outcome that I think can be widely applied across cultures. In addition to the specific indicators of chaos, and structural conditions associated with it, it is also useful to think about the life projects and goals—pluralistic and varied as they can be around the world—that matter deeply, and that are difficult or impossible to sustain in chaotic circumstances. The contrast to chaos is not the absence of negative conditions but rather something very positive: the well-being that comes from an active, meaningful, sustainable routine of everyday life for ourselves and others we care about.

References

Baumrind, D. (1989). Rearing competent children. In W. Damon (Ed.), *Child development today and tomorrow* (pp. 349–378). San Francisco: Jossey-Bass.

Ben-Arieh, A., & Goerge, R. M. (Eds.). (2005). *Indicators of children's well being: Understanding their role, usage and policy influence*. Dordrecht, Netherlands: Springer Publishing Company.

Bernheimer, L. P., & Weisner, T. S. (2007). "Let me just tell you what I do all day": The family story at the center of research and practice. *Infants & Young Children, 20*, 192–201.

Bernheimer, L. P., Weisner, T. S., & Lowe, E. (2003). Impacts of children with troubles on working poor families: Mixed-methods and experimental evidence. *Mental Retardation, 41*, 403–419.

Bornstein, M. H., Davidson, L., Keyes, C. L. M., & Moore, K. A. (Eds.). (2003). *Well-being: Positive development across the life course*. Mahwah, NJ: Erlbaum.

Bos, H., Huston, A., Granger, R., Duncan, G., Brock, T., McLoyd, V., et al. (1999). *New Hope for people with low incomes: Two-year results of a program to reduce poverty and welfare reform*. New York: Manpower Demonstration Research Corporation Press.

Bronfenbrenner, U. (1979). *The ecology of human development: Experiments by nature and design*. Cambridge, MA: Harvard University Press.

Bronfenbrenner, U. (Ed.). (2005). *Making human beings human. Bioecological perspectives on human development*. Thousand Oaks, CA: Sage.

Brown, B. V. (1997). Indicators of children's well-being: A review of current indicators based on data from the federal statistical system. In R. M. Hauser, B. Brown, & W. R. Prosser (Eds.), *Indicators of children's well-being* (pp. 3–35). New York: Russell Sage Foundation.

Chao, R. K. (1994). Beyond parental control and authoritarian parenting style: Understanding Chinese parenting through the cultural notion of training. *Child Development, 65*, 1111–1119.

Cole, M. (1996). *Cultural psychology: A once and future discipline*. Cambridge, MA: Harvard University Press.

Darling, N., & Steinberg, L. (1993). Parenting style as context: An integrative model. *Psychological Bulletin, 113*, 487–496.

DeLoache, J., & Gottlieb, A. (2000). *A world of babies: Imagined childcare guides for seven societies*. Cambridge, England: Cambridge University Press.

Duncan, G., Huston, A., & Weisner, T. S. (2007). *Higher ground: New Hope for working families and their children*. New York: Russell Sage Foundation.

Earls, F., & Carlson, M. (1995). Towards sustainable development for American families. *Daedalus: Journal of the American Academy of Arts and Sciences, 122*, 93–121.

Evans, G. W. (2006). Child development and the physical environment. *Annual Review of Psychology, 57*, 423–451.

Evans, G. W., Gonnella, C., Marcynyszyn, L., Gentile, L., & Selpekar, N. (2005). The role of chaos in poverty and children's socioemotional adjustment. *Psychological Science, 167*, 560–565.

Gallimore, R., Bernheimer, L. P., & Weisner, T. S. (1999). Family life is more than managing crisis: Broadening the agenda of research on families adapting to childhood disability. In R. Gallimore, L. P. Bernheimer, D. L. MacMillan, D. L. Speece, & S. Vaughn (Eds.), *Developmental perspectives on high incidence handicapping conditions* (pp. 55–80). Mahwah, NJ: Erlbaum.

Gallimore, R., Coots, J. J., Weisner, T. S., Garnier, H. E., & Guthrie, D. (1996). Family responses to children with early developmental delays II: Accommodation intensity and activity in early and middle childhood. *American Journal on Mental Retardation, 101*, 215–232.

Gallimore, R., Goldenberg, C., & Weisner, T. (1993). The social construction and subjective reality of activity settings: Implications for community psychology. *American Journal of Community Psychology, 21*, 537–559.

Gibson, C., & Weisner, T. S. (2002). "Rational" and ecocultural circumstances of program take-up among low-income working parents. *Human Organization, 61*, 154–166.

Goldstein, D. M. (2003). *Laughter out of place: Race, class, violence, and sexuality in a Rio shantytown*. Berkeley: University of California Press.

Gottlieb, A. (2004). *The afterlife is where we come from: The culture of infancy in West Africa*. Chicago: University of Chicago Press.

Harkness, S., & Super, C. M. (Eds.). (1996). *Parents' cultural belief systems*. New York: Guilford Press.

Harwood, R., Miller, J., & Irizarry, N. L. (1995). *Culture and attachment. Perceptions of the child in context*. New York: Guilford Press.

Huston, A., Duncan, G. J., Granger, R., Bos, J., McLoyd, V., Mistry, R., et al. (2001). Work-based antipoverty programs for parents can enhance the school performance and social behavior of children. *Child Development, 72*, 318–336.

Huston, A. C., Duncan, G. J., McLoyd, V. C., Crosby, D. A., Ripke, M. N., Weisner, T. S., & Eldred, C. A. (2005). Impacts on children of a policy to promote employment and reduce poverty: New Hope after five years. *Developmental Psychology, 41*, 902–918.

Huston, A., Miller, C., Richburg-Hayes, L., Duncan, G. J., Eldred, C. A., Weisner, T. S., et al. (2003). *New Hope for families and children: Five-year results of a program to reduce poverty and reform welfare*. New York: MDRC.

Isiugo-Abanihe, U. C. (1985). Child fosterage in West Africa. *Population and Development Review, 11*(1), 53–73.

Kagitçibasi, C. (1996). *Family and human development across cultures. A view from the other side*. Mahwah, NJ: Erlbaum.

Keller, H. (2007). *Cultures of infancy*. Mahwah, NJ: LEA Press.

Lareau, A. (2003). *Unequal childhoods: Class, race, and family life*. Berkeley: University of California Press.

LeVine, R. (2007). Ethnographic studies of childhood: A historical overview. *American Anthropologist, 109*, 247–260.

LeVine, R. A., Miller, P. M., & West, M. M. (Eds.). (1988). *Parental behaviors in diverse societies*. San Francisco: Jossey-Bass.

LeVine, R. A., & New, R. (Eds.). (2008). *Anthropology and child development. A cross-cultural reader*. Malden, MA: Wiley-Blackwell.

LeVine, R. A., & Norman, K. (2001). The infant's acquisition of culture: Early attachment reexamined in anthropological perspective. In C. C. Moore & H. F. Mathews (Eds.), *The psychology of cultural experience* (pp. 83–104). New York: Cambridge University Press.

Lowe, E., & Weisner, T. S. (2004). "You have to push it—Who's gonna raise your kids?": Situating child care in the daily routines of low-income families. *Children and Youth Services Review, 26*, 143–171.

Matheson, C., Olson, R., & Weisner, T. S. (2007). A good friend is hard to find: Friendship among adolescents with disabilities. *American Journal of Mental Retardation, 112*, 319–329.

McKenna, J. J., & McDade, T. (2005). Why babies should never sleep alone: A review of the co-sleeping controversy in relation to SIDS, bedsharing and breast feeding. *Pediatric Respiratory Reviews, 6*, 134–152.

Nihira, K., Weisner, T. S., & Bernheimer, L. P. (1994). Ecocultural assessment in families of children with developmental delays: Construct and concurrent validities. *American Journal of Mental Retardation, 98,* 551–566.

Okami, P., Weisner, T. S., & Olmstead, R. (2002). Outcome correlates of parent-child bedsharing: An 18-year longitudinal study. *Journal of Developmental and Behavioral Pediatrics, 23,* 244–253.

Repetti, R., Taylor, S. E., & Seeman, T. E. (2002). Risky families: Family, social environments and the mental and physical health of offspring. *Psychological Bulletin, 128,* 330–366.

Rogoff, B. (2003). *The cultural nature of human development.* Oxford, England: Oxford University Press.

Rothbaum, F., Pott, M., Azuma, H., Miyake, K., & Weisz, J. (2000). The development of close relationships in Japan and the United States: Paths of symbiotic harmony and generative tension. *Child Development, 71,* 1121–1142.

Serpell, R. (1993). *The significance of schooling. Life-journeys in an African society.* New York: Cambridge University Press.

Seymour, S. C. (1999). *Women, family, and child care in India: A world in transition.* Cambridge, England: Cambridge University Press.

Shweder, R. A., Balle-Jensen, L., & Goldstein, W. (1995). Who sleeps by whom revisited: A method for extracting the moral goods implicit in practice. In J. J. Goodnow, P. J. Miller, & F. Kessel (Eds.), *Cultural practices as contexts for development* (pp. 21–39). San Francisco: Jossey-Bass.

Shweder, R., Minow, M., & Markus, H. R. (Eds.). (2002). *Engaging cultural differences.* New York: Russell Sage Foundation.

Skinner, D., & Weisner, T. S. (2007). Sociocultural studies of families of children with intellectual disabilities. *MRDD Research Reviews, 13,* 302–312.

Super, C., & Harkness, S. (1986). The developmental niche: A conceptualization at the interface of child and culture. *International Journal of Behavior Development, 9,* 1–25.

Taylor, S. E., Repetti, R. L., & Seeman, T. (1997). Health psychology: What is an unhealthy environment and how does it get under the skin? *Annual Review of Psychology, 48,* 411–447.

Verhoef, H. (2005). 'A child has many mothers': Views of child fostering in northwestern Cameroon. *Childhood, 12,* 369–390.

Vleminckx, K., & Smeeding, T. M. (Eds.). (2001). *Child well-being, child poverty and child policy in modern nations.* Bristol, England: Policy Press.

Wachs, T. D., & Corapci, F. (2003). Environmental chaos, development and parenting across cultures. In C. Raeff & J. Benson (Eds.), *Social and cognitive development in the context of individual, social, and cultural processes* (pp. 54–83). London: Routledge.

Weisner, T. S. (1996). The 5-7 transition as an ecocultural project. In A. J. Sameroff & M. M. Haith (Eds.), *The five to seven year shift: The age of reason and responsibility* (pp. 295–326). Chicago: University of Chicago Press.

Weisner, T. S. (2001). Anthropological aspects of childhood. In *The international encyclopedia of the social sciences* (Vol. 3, pp. 1697–1701). Oxford, England: Elsevier.

Weisner, T. S. (2002). Ecocultural understanding of children's developmental pathways. *Human Development, 45,* 275–281.

Weisner, T. S. (2005). Attachment as a cultural and ecological problem with pluralistic solutions. *Human Development, 48,* 89–94.

Weisner, T. S. (2008). The Urie Bronfenbrenner Top 19: Looking back at his bioecological perspective. Review of *Making human beings human. Bioecological perspectives on human development,* by Urie Bronfenbrenner (Ed.). *Mind, Culture, & Activity, 15,* 258–262.

Weisner, T. S. (2009). Well-being and sustainability of the daily routine of life. In G. Mathews & C. Izquerdo (Eds.), *The good life: Well-being in anthropological perspective* (pp. 228–247). New York: Berghahn Press.

Weisner, T. S., Gibson, C., Lowe, E. D., & Romich, J. (2002). Understanding working poor families in the New Hope Program. *Poverty Research Newsletter, 6*(4), 3–5.

Weisner, T. S., Matheson, C., Coots, J., & Bernheimer, L. P. (2005). Sustainability of daily routines as a family outcome. In A. Maynard & M. Martini (Eds.), *Learning in cultural context: Family, peers and school* (pp. 47–74). New York: Kluwer Academic/Plenum Publishers.

Whiting, B. (1980). Culture and social behavior: A model for the development of social behavior. *Ethos, 8*, 95–116.

Whiting, B., & Edwards, C. P. (1988). *Children of different worlds: The formation of social behavior.* Cambridge, MA: Harvard University Press.

Yoshikawa, H., Weisner, T. S., & Lowe, E. (Eds.). (2006). *Making it work: Low-wage employment, family life and child development.* New York: Russell Sage Foundation.

14

Chaos and the Macrosetting: The Role of Poverty and Socioeconomic Status

Gary W. Evans, John Eckenrode, and Lyscha A. Marcynyszyn

> You can never plan a week or a day ahead . . . one minute you're fine and the next you just can't deal.
> —Low-income mother describing daily life
> (Roy, Tubbs, & Burton, 2004, p. 174)

There are many reasons why the lives of children from low-income households are more chaotic than are those from middle- and high-income households. Low-income parents suffer from a plethora of physical and social stressors. Poor parents juggle overlapping time obligations and have fewer resources than do wealthier parents to deal with the multitude of demands and obligations they face. Low-income parents are less likely to have a dependable car; they cannot afford reliable flexible child care or after-school care; their children are enrolled less often in structured child or youth programs; and low-income parents are less likely to have a partner, who could share household management and parenting responsibilities. Residential and school relocations, which erode social networks, are more common, and family disruptions and turmoil are more frequent among low-income families (Evans, 2004). In this chapter we document linkages between poverty and low socioeconomic status (SES) and chaos, review studies examining chaos as a mediator of poverty's impacts on children, and offer preliminary thoughts about why chaos is problematic for children's development.

Chaos, Poverty, and Socioeconomic Status

Components of chaos associated with poverty and SES include household crowding, noise levels, household routines and rituals, residential and school relocation, and parental partner instability. For each of these components of chaos, we have constructed a summary table. These tables are not included because of length restrictions but can be found at http://www.macses.ucsf.edu/Research/Social%20Environment/notebook/chaos.html. Chaos is operationalized in a few studies as a composite variable, with these results summarized in Table 7. Herein we characterize the overall direction and strength of relations among various indices of chaos and poverty and SES. Before we describe the evidence linking chaos to income or SES, however, it is important to emphasize that chaos is not equivalent to income or SES. In almost every study of chaos and child development, as indicated in chapters 3, 4, 6, 9, 10, and 11 of this volume as well as by the data described in this chapter, statistically significant associations between chaos and child outcomes remain with statistical controls for income or SES. In other words, the impacts of chaos on child development cannot be explained by income or class. Furthermore, as we show in the Chaos as a Mechanism for Social Class Effects on Human Development section later in this chapter, evidence for a pathway of SES → chaos → child development includes data showing that the chaos → child development link is significant when SES is in the model.

Because most research on chaos has focused on its impact on child development, investigators have used various means to minimize the potential contaminating role of income or SES in these relationships. In many cases this has likely led to downwardly biased estimates of the role of income or SES in chaos. The most typical analytic approach has been to incorporate SES statistical controls in analyses. Many of these same investigators have also tried to minimize variability in SES or income in their samples to better isolate the effects of chaos. Moreover, many chaos and child development studies have truncated variance in income or class because of sampling restrictions. Only about 20% of American families with children are poor, plus they are more difficult to recruit and retain in empirical studies. Thus, low-income families are underrepresented in social science research, including work on chaos. Finally, nearly all of the data come from the United States or Western Europe, where the potential range of chaos is dramatically less than that found in economically underdeveloped societies. Each of these limitations downwardly biases estimates of chaos sequelae of SES. On the other hand, self-selection and other omitted variables could produce spuriously high chaos–SES associations. Because most of the evidence on chaos and child development emanates from cross-sectional studies, this concern is applicable. One or more predisposing conditions (e.g., maternal depression) could produce inflated chaos–SES relations. It is noteworthy that most psychological research on chaos and child development has focused on selection bias and other factors that could inflate the association (see chap. 10) even though Type II errors are equally if not more salient. The following subsections summarize the evidence for ecological covariation among each of the constituent parts of chaos.

Crowding

Crowding contributes to chaotic living and school settings. Crowded environments are overstimulating, confusing, unpredictable, and uncontrollable. The salient index of crowding is people per room (Evans, 2006). High interior density is problematic because it interferes with the regulation of social interaction. Most child crowding studies show associations between crowding and outcomes with statistical controls for SES (Evans, 2001, 2006). There is a consistent modest correlation between crowding and poverty status. In studies using either income or poverty status the correlations range from a low of .32 (Light, 1973) to a high of .53 (Evans, Lepore, Shejwal, & Palsane, 1998), with a mean correlation of .44 (see Table 1). The Evans study examined lower and working-class families in India, whereas the Light study examined a large sample of U.S. households. When categorical outcomes (e.g., less than or greater than one person per room) are used, the percentage difference ranges from 5% (Federman et al., 1996) to 30% (Davie, Butler, & Goldstein, 1972), with a mean differential of 16.3% between poor and not-poor households (see Table 1). None of the studies found contradictory data. Low-income and low-SES families live under more crowded conditions.

Noise

Noise is measured by decibels, a logarithmic scale of sound intensity. A 10-decibel increase doubles perceived loudness. Common noise sources include transportation sources, especially airplanes and vehicular traffic, activities of other people, music, and various appliances. Noise elevates physiological arousal as well as interferes with relaxation and sleep (Cohen, Evans, Stokols, & Krantz, 1986). It also disrupts concentration and causes greater effort to maintain attention. The most robust adverse developmental outcome associated with chronic noise is reading deficits (Evans & Hygge, 2007). Noise can also produce fatigue and negative affect, including irritability and hostility (Evans, 2001, 2006). Nearly all of the noise studies incorporate statistical controls for SES, and a few longitudinal studies show effects of similar magnitude.

Studies of noise and SES relative to crowding are fewer, but they converge on a link between SES and noise exposure (see Table 2). Among a restricted range of lower-middle- and middle-income families, noise levels correlated –.14 with income (Heft, 1979), whereas a larger, more representative study of urban households found a correlation of –.61 between income and community noise levels (Environmental Protection Agency, 1977). In studies comparing noise exposures between various groups by income, poor children are exposed to between 5 and 10 more decibels on average (Evans & English, 2002; Haines, Stansfeld, Head, & Job, 2002).

Routines and Rituals

One of the key elements of stability in children's lives is the degree of structure and predictability in daily routines. Routines convey what needs to be done by whom and occur repeatedly over time (Fiese, 2006; see also chap. 4, this volume). Rituals happen less often and convey through practice and symbol a sense of community and belonging. Households with more structured and regular daily activities engage in more family rituals.

Routines tend to be positively related to SES, with correlations ranging from –.15 (Crouter, Head, McHale, & Tucker, 2004) to .47 (Boyce et al., 1977) for income, with an average correlation between income and routines of $r = .12$

(see Table 3). One reason for the high variability in SES correlates of routines and rituals is sampling restrictions. Samples with greater heterogeneity of income uncover higher correlations between SES and routines and rituals. For example, in a nationally representative sample of nearly 30,000 households, Bradley, Corwyn, McAdoo, and Garcia-Coll (2001) found that families below the poverty line were about 15% less likely than other families to have a regular mealtime routine in their home. In another large, heterogeneous sample, 44% of families headed by a high school dropout had regular meal, nap, and bedtime routines for their infants and toddlers, whereas 58% of parents who were high school graduates maintained such routines (Britto, Fuligni, & Brooks-Gunn, 2002). Very few studies have examined rituals and SES, so we cannot speak with confidence about the role of SES and child exposure to rituals.

Some parents undoubtedly do their best to maintain routines and rituals in the face of poverty. Although they benefit children, such efforts may exact costs for caregivers. For example, adolescent adherence to medical regimens for asthma does not vary with social class, but lower SES parents experience greater burden to maintain the level of desired asthmatic care for their teenage children compared with more advantaged parents (Fiese, Wamboldt, & Anbar, 2005). Flexibility and support may be in relatively short supply for parents with fewer economic or SES resources to draw on.

Changes in parental work hours, particularly if frequent or unpredictable, can make it difficult to maintain structure and routines in daily life. Opportunities to schedule other activities for children, particularly when they are young, that depend on parental involvement for transportation are also more challenging if parental work schedules are unstable (see chaps. 2 and 12). Two of the hallmarks of lower wage jobs, at least in the United States, are less regular schedules and change. In a nationally representative sample of 50,000 American households, about 10% of the total workforce have work schedules that vary weekly. Such jobs are 27% more likely to occur among high school dropouts (Goldin, 2001). Among all American workers, 26% of high school dropouts worked nonday shifts or variable schedules compared with 10% of college graduates (Presser, 2003). Forty percent of female heads of households with a child under 14 worked nonstandard hours or on weekends, or both, compared with 31% of their counterparts with a partner in the house (Presser, 2003). Data are not presented by income, but American laborers are 47% more likely to work variable hours on a weekly basis compared with those with other types of occupation. Professionals, by comparison, are 35% less likely to work variable hours (Goldin, 2001). In a study of 900 women who had worked from the time their child was born until their child was 3 years old, 69% of low-income mothers had worked at some time on a nontraditional schedule compared with 55% of mothers above the poverty line (Han, 2005). Lower wage American workers are also much more likely to work during evening or night times compared with typical daytime work hours (Hamermesh, 1999).

Residential Relocation

A major contributor to chaos in children's lives is changes in home or school location. If relocations are frequent, children may become reluctant to establish new friendships or to rely on adults for guidance and support, knowing that they may soon have to break those ties and start over again (Adam, 2004). When school changes occur, children may also be confronted with unfamiliar scholastic demands in addition to changes in peers and teachers.

Low-income children are more apt to change residences (see Table 4). In a 1988 national survey of more than 10,000 American schoolchildren between 1st and 12th grade, 39% of children had moved three or more times. Of these children, 48% were from families below the poverty line, whereas 37% were from families above the poverty line (Simpson & Fowler, 1994). Examining children ages 3 to 17 from that same data set, Long (1992) noted that children living below the poverty line were 2.4 times more likely than their nonpoor counterparts to move more than the average number of times. Census data on residential locations reveal that 24% of households below the poverty line moved between 2002 and 2003, whereas only 13% of nonpoor families relocated during that same year (U.S. Census, 2004). In a Canadian national sample, the difference in family income between 5- to 11-year-olds who had never relocated and children in the same age range who had located three or more times was greater than $10,000 (Kohen, Hertzman, & Wiens, 1998). A U.S. national study of high-frequency moves during childhood (six or more relocations) found that nearly twice as many low-income children as nonpoor children had moved frequently (Wood, Halfon, Scarlata, Newacheck, & Nessim, 1993). One aspect of residential relocation that has not been well documented is the extent of forced relocation or eviction. Two studies, both with modest-sized samples, revealed that families below the poverty line are 3 to 4 times more likely to move involuntarily than are those above the poverty line (Federman et al., 1996; Mayer & Jencks, 1989). See chapters 8 and 15 of this volume for more discussion on involuntary relocation and children's welfare.

School Relocations

Data on school relocations have to be considered carefully because some transitions are normative (e.g., elementary to middle school), whereas changes within one type of school are not. See Table 5 for a summary of SES and school relocation studies. The largest study of school relocations, conducted by the U.S. General Accounting Office (1994), looked at school changes between first and third grade. Three times as many low-income children (30%) had moved three or more times during this period compared with 10% of those who were not poor. Canadian national data on 5- to 11-year-olds found that children who had been to three or more schools were from households with nearly $20,000 less annual income compared with children who never changed schools (Kohen et al., 1998). Teachers, like pupils, also relocate and are more likely to do so if they work in a low-income school. Rutter et al. (1974), for example, documented 43% versus 26% teacher turnover rates in lower social class schools compared with working-class schools in London. In the United States the turnover rate of teachers in public schools is about 50% higher in low-income relative to middle-income schools (15.2% vs. 10% turnover rates; Ingersoll, 2001).

Maternal Partner Change

Family turmoil associated with poverty often leads to dissolution of romantic partnerships. Changes in household composition, particularly among adults with whom young children develop attachments, are clearly highly disruptive for children. The levels of divorce and changes in parental partners are strongly linked to SES (see Table 6). For example, in the United States, the divorce rate is almost 5 times higher in the lowest income quintile (25.4%) than it is among the upper income quintile for households with children (5.7%; Evans, 2004). Similar findings have been uncovered in U.K. national surveys as well (Kiernan & Mueller, 1999; Reid, 1989). Between 1992 and 1995, in a national sample of British adults the odds of first marriage dissolution were doubled if before the marriage either partner had received welfare and tripled if either partner had been unemployed prior to the marriage (Kiernan & Mueller, 1999). In a study of over 2,000 American children, the percentage of 3-year-olds whose mother had experienced three or more partner changes during the child's lifetime was more than double for those living below versus above the poverty line (Osborne & McLanahan, 2007).

Composite Indices

Some studies have used composite indices of chaos that do not disaggregate specific components of the measures. The most common composite index is CHAOS (Confusion, Hubbub, and Order Scale; Matheny, Wachs, Ludwig, & Phillips, 1995), which contains parental ratings of noise, confusion, crowding, and hectic pace in the household. A range of relations are stronger for SES heterogeneous samples (see Table 7). The average correlation between income and composite indices of chaos is .26, ranging from .09 (Dumas et al., 2005) to .40 (Marcynyszyn, Evans, & Eckenrode, 2008), in families with some variance in income. Households with lower levels of SES rather than income are also more chaotic, with an average correlation of .20 and a range of .02 (Marcynyszyn et al., 2008) to .28 (Dumas et al., 2005; Pike, Iervolino, Elev, Price, & Plomin, 2006). In addition, children below the poverty line are more than 4 times as likely (13%) to experience turbulence in their families than are those 3 or more times above the poverty line (3%; Moore, Vandivere, & Ehrle, 2000). Turbulence is a composite index indicative of residential, school, or parental job changes plus serious illness in the immediate family.

Chaos as a Mechanism for Social Class Effects on Human Development

We have shown that poverty and low SES are associated with higher levels of chaos. Moreover, many of the chapters in this volume document adverse effects of chaotic living conditions on children's cognitive, socioemotional, and physical well-being. Most of these studies also show that the effects of chaos on child outcomes are independent of SES. Recall that most studies of chaos statistically control for SES or use homogeneous samples of all low-income or all middle-income families. This pattern of interrelationships suggests the potential for chaotic environments to function as underlying mechanisms, conveying some of the covariation between income or SES and children's development.

We are aware of only two research programs that have directly examined the mediational pathway income (SES) → chaos → child development. Brody and Flor (1997) found that perceived financial hardship was associated with greater externalization and internalization symptoms as well as lower cognitive development among 6- to 9-year-old, low-income African American children. Family routines as assessed by the Family Routine Index (Jensen, James, Boyce, & Hartnett, 1983) conveyed some of these effects. Both the direct and indirect effects of perceived financial strain on children's cognitive development and well-being were significant. Children in households with greater financial burden experienced more chaotic household routines, which in turn partially mediated some of the ill effects of financial strain on their socioemotional and cognitive development.

Evans and English (2002) tested the mediating role of multiple stressor exposure in the link between poverty and developmental outcomes. Multiple stressor exposure was operationalized by coding six continuous indicators of risk into categorical codes of "risk" and "not at risk." The continuous measures of risk included residential noise and crowding, housing problems, family turmoil, child separation from parents, and exposure to violence. For each stressor, children exposed to greater than 1 standard deviation above the mean for the particular stressor were defined as being at risk. Approximately half of the sample of 8- to 10-year-old children lived in households at or below the federal poverty line, and half were from middle-income families. Multiple stressor exposure consisted of the sum (0–6) of the six dichotomous exposure metrics. Relations between poverty and elevated resting blood pressure and overnight stress hormones (cortisol and epinephrine) of low- relative to middle-income children were mediated by multiple stressor exposure. For psychological distress, partial mediation was uncovered, indicating both direct and indirect effects of poverty on this outcome. For self-regulatory behavior (Mischel's delay of gratification paradigm; Mischel, Shoda, & Rodriguez, 1989), the pattern of mediation more closely matched physiological stress outcomes indicative of full mediation. In a more recent study, Evans and Kim (2007) found prospective, longitudinal evidence that multiple stressor exposure mediated linkages between poverty and physiological stress in this same sample, 4 years later during early adolescence. Evans Gonnella, Marcynyszyn, Gentile, and Salpekar (2005) investigated the role of chaos in their longitudinal sample as well. They modified the original CHAOS (Matheny et al., 1995) slightly by adding measures of structure and routines. This modified CHAOS fully mediated the links between income and psychological distress, self-regulatory behavior, and learned helplessness during early adolescence, residualizing earlier measures of the same set of outcomes. The latter measure consisted of persistence on a challenging geometric puzzle.

Both Brody's and Evans's respective research programs provide preliminary evidence that some of the covariation between low-income and adverse socioemotional outcomes, including psychological distress, learned helplessness, and self-regulatory behavior, is mediated by higher levels of chaos in low-income households compared with middle-income families. Moreover, these effects help explain some of the well-documented negative outcomes of exposure to poverty or low SES. A strength of this pattern of results is multimethodological indices of socioemotional development in conjunction with both cross-sectional and longitudinal evidence. Moreover, the findings from these two research programs are supplemented by Evans's work showing parallel mediational results for physiological markers of stress. Finally, there are several replications of the poverty → chaos → socioemotional outcomes pathway across the two different research programs. An important limitation of both programs of research is the samples. Evans's sample is modest in size, nearly all Caucasian, with children living in rural areas of upstate New York. Brody's sample consists of African American, predominantly low-income families headed by a single parent, living in rural Georgia.

In sum, chaos is not simply a marker or surrogate for poverty or low SES. If it were, analyses of developmental sequelae of chaos would not be significant when income or SES are incorporated as statistical controls. Yet many studies, as reviewed in chapters 3, 4, 6, 9, 10, and 11, plus data from the studies described in this chapter reveal that statistically significant associations between chaos and child outcomes maintain with statistical controls for income or SES. Moreover, the mediational results from the research programs of Brody and of Evans also show that chaos predicts developmental outcomes, statistically controlling for income or SES. That is, chaos mediates the impacts of income (SES) on child outcomes rather than the other way around (MacKinnon, 2008). The covariation between chaos and child development cannot be explained by income or SES, but some of the covariation between income or SES and child development is mediated by chaos.

Chaos and Human Development

In this section we offer preliminary ideas about why chaos is harmful to children's development. Bronfenbrenner's bioecological model of human development provides a valuable theoretical framework to address this question. One of the key elements of the bioecological model is proximal process. Proximal process involves a transfer of energy between the developing human being and the persons, objects, and symbols in the immediate environment (Bronfenbrenner & Evans, 2000; Bronfenbrenner & Morris, 1998). For proximal processes to be effective they must take place regularly, over an extended period, and involve progressively more complex, reciprocal interactions between the developing human and his or her immediate surroundings (Bronfenbrenner & Evans, 2000; Bronfenbrenner & Morris, 1998). A fundamental reason why chaos is harmful to children is because it interferes with effective proximal processes. The predictability and sustained nature of increasingly complex interactions become more difficult to maintain in chaotic households. High levels of background stimulation (noise, crowding) and irregular and unstructured activities and schedules in conjunction with changes in familiar social and physical environments (adult household composition, residential or school changes) do not comport with proximal processes. Chaos shortens the duration of exchanges of energy between the developing child and his or her immediate surroundings. The child is less able to count on or predict that certain types of interactions will happen. At a most basic level, individuals cannot develop socially cohesive, meaningful relationships with people unless they see them regularly and can count on them being around. Regular interaction with family and friends is critical for the development and maintenance of healthy social networks.

Not only are proximal processes less likely to occur in chaotic settings, but children and their caregivers living under chaotic conditions may respond to their surroundings in ways that exacerbate aversive consequences of chaos. Warm and responsive familial interactions suffer under chaotic living conditions, which in turn could undermine the development and maintenance of secure attachment. Crowding and noise, for example, each interfere with the development and maintenance of warm, supportive parent–child interactions (Evans, 2001, 2006; Wachs & Corapci, 2003). Parents in both higher density and noisier homes are more socially withdrawn from their children, less patient with them, and less responsive. They also are less likely to speak with their children. In a research program on crowding, Evans and Lepore have shown in both the laboratory and the field that high-density living conditions interfere with the development of socially supportive relationships among adults. Evidence includes cross-sectional and prospective, longitudinal data in both the field and the laboratory (Evans, Palsane, Lepore, & Martin, 1989; Evans & Lepore, 1993; Evans, Rhee, Forbes, Allen, & Lepore, 2000; Lepore, Evans, & Schneider, 1991). Less social cohesion and more social conflict have been shown in higher density homes of elementary-school-age children as well (Evans et al., 1998). All of these crowding studies incorporated statistical controls for SES. Several studies in both the field and the laboratory have shown that noise leads to less altruism among adults (Cohen & Spacapan, 1984; Evans & Cohen, 2004). Composite indices of chaos such as Matheny's CHAOS have also been linked to less responsive parenting (Corapci & Wachs, 2002; Matheny et al., 1995) as well as less parental warmth and harsher, more hostile parent–child interactions (Coldwell, Pike, & Dunn, 2006). Composite indices of instability (e.g., residential change, maternal partner change) are also associated with greater family conflict (Ackerman, Kogos, Youngstrom, Schoff, & Izard, 1999) as well as parenting difficulties (e.g., rejection, overcontrolling behavior; Forman & Davies, 2003). Similar findings have been found for maternal romantic partner changes and parenting (Martinez & Forgatch, 2002). However, not only parents are affected by chaos. Residential instability can also compromise sibling relationships. Head Start children who moved more often had more conflicted relationships with their older siblings (Stoneman, Brody, Churchill, & Winn, 1999). Changes in day-care facilities are positively correlated with more restrictive parenting coupled with less nurturing family characteristics (Howes & Stewart, 1987). When parents are overstimulated and stressed by chaotic surroundings, it may be adaptive, initially, for them to tune out or withdraw from the negative conditions. However, an unintended side effect of such withdrawal may be less engagement in proximal processes (e.g., parental responsiveness, warmth).

Lack of structure coupled with unpredictability may also have consequences for feelings of mastery and self-efficacy. It is difficult to learn that one's actions can influence one's surroundings if many person–environment exchanges are unpredictable or uncontrollable (Evans & Stecker, 2004; White, 1959). Routines and rituals are a source of feedback from the environment about the consequences of the individual's and family's plans and expectations. Children need to know what is going to happen, when and where, and what preparations precede outcomes such as meals, religious services, and celebration of important milestones and holidays. Routines provide scaffolding to learn how to translate goals into actions (Fiese, 2006; see also chap. 4, this volume). At a more basic level, lack of structure and routine robs the developing child of fundamental building blocks of comprehension of temporal sequences and, ultimately, cause and effect (Heft, 1985). Corapci and Wachs (2002) noted a negative association between chaotic living conditions and mothers' sense of parental efficacy. Evans et al. (2005) showed in a prospective, longitudinal analysis that more chaotic living conditions predicted greater susceptibility to learned helplessness among young adolescents.

Lack of routines and structure may also undermine self-regulatory ability. It is hard enough for children to learn to manage their own behaviors and emotions; to do so against a backdrop of unpredictable and shifting physical and social circumstances asks a lot. Expectations for appropriate behaviors are reinforced by family routines and rituals (Fiese, 2006). In a sample of 6- to 9-year-olds, Brody and Flor (1997) showed that less organized and less structured daily life was negatively related to self-regulatory behavior. Both parental and teacher ratings of self-regulatory behaviors were negatively correlated with family routines. Thirteen-year-olds living in more chaotic households were rated lower in self-regulatory behavior by one of their schoolteachers (Evans et al., 2005). Chaos was evaluated by the adolescent's mother. Part of the development of self-regulatory skills presumably involves opportunities to exercise person-initiated interactions with the surrounding social and physical environment. Chaotic households may provide fewer opportunities for this to occur, with many person–environment transactions dominated by environmental demands rather than initiated by the individual (Metcalfe & Mischel, 1999; Mischel & Ayduk, 2005).

Chaotic living environments can interfere with children's physical health. For example, families that maintain more structured, regular medical regimens with asthmatic children achieve better asthma control (Fiese et al., 2005). As Fiese suggested, it is possible to extend this idea beyond medication routines. Healthy eating is fostered by the planning and preparation of meals. The maintenance of exercise programs is fostered by having a regular place and time to exercise. All people, but especially children, sleep better if they have a regular bedtime routine and place to sleep. Young children need guidance and ground rules to help them learn to do homework, to read on a regular basis, and to master new skills, whether in music, art, or athletics.

In chapter 7 of this volume, Wachs described several person characteristics that appear to interact with chaos to affect development. Here we briefly note social and physical environmental conditions that could alter the developmental sequelae of chaos during early childhood. The negative relationship between harsh, unsupportive parenting and behavior problems in 6- to 8-year-old British children was exacerbated when it occurred against a backdrop of more chaotic living conditions. Households with less structure, more noise and crowding, and more frenetic levels of activity appeared to accentuate the link between poor parenting practices and behavioral problems during middle childhood (Coldwell, Pike, & Dunn, 2006). Eight- to 12-year-old inner-city African American children who lived in households with greater routines and structure were partially protected from the harmful impacts of daily stressors and hassles on their psychological well-being (Kleiwer & Kung, 1998). Among first through fifth graders whose parents were divorced, behavioral adjustment and academic performance were enhanced by more regular bedtimes (Guidubaldi, Cleminshaw, Perry, Nastasi, & Lightel, 1986).

Summary and Conclusions

Most developmental research focuses on the intensity or level of individual and environmental variables and processes as they impinge on children. The study of chaos offers a reminder that temporal qualities of organism–environment transactions warrant more consideration (Bronfenbrenner & Evans, 2000; Lepore, 1995; see also chap. 8, this volume). Most research on temporal dimensions of environmental exposure in the realm of development has focused on developmental timing, largely driven by biological considerations of plasticity. The duration, frequency, regularity, and contingency of interactions between children and their surroundings may be just as important as the level of environmental exposures. We have argued that chaotic living conditions can interfere with proximal processes that are integral to the development of healthy, well-adjusted children. Regular, sustained, increasingly complex person–environment interactions are more difficult to maintain against a backdrop of chaotic living contexts. As the chapters in this volume show, chaos can be harmful to children's development.

Abundant evidence shows that poverty is bad for children's development. One of the key and perhaps unique features of childhood poverty is the confluence of multiple physical and social risk factors that converge among low-income families. Not only are poorer children more likely to experience adverse living conditions, but such conditions often cohere in a manner that undermines predictability and contingency, interferes with structure and routines, and overstimulates the organism. Poverty breeds chaos, which in turn harms children and their families. To be clear, we are not arguing that chaos is equated with poverty. As indicated earlier, most empirical studies on chaos and developmental outcomes statistically control for income or SES. Nor are we arguing that all of the adverse impacts of poverty on children are due to chaotic living conditions. Instead, we believe chaos is one of several plausible pathways through which poverty harms children's development.

References

Ackerman, B. P., Kogos, J., Youngstrom, E., Schoff, K., & Izard, C. (1999). Family instability and the problem behaviors of children from economically disadvantaged families. *Developmental Psychology, 35*, 258–268.

Adam, E. K. (2004). Beyond quality: Parental and residential stability and children's adjustment. *Current Directions in Psychological Science, 13,* 210–213.

Boyce, W. T., Jensen, E. W., Cassel, J. D., Collier, A. M., Smith, A. H., & Ramey, C. T. (1977). Influence of life events and family routines on childhood respiratory tract illnesses. *Pediatrics, 60,* 609–615.

Bradley, R. H., Corwyn, R. F., McAdoo, H. P., & Garcia-Coll, C. (2001). The home environments of children in the United States Part I: Variations by age, ethnicity, and poverty status. *Child Development, 72,* 1844–1867.

Britto, P. R., Fuligni, A. S., & Brooks-Gunn, J. (2002). Reading, rhymes, and routines: American parents and their young children. In N. Halfon (Ed.), *Childrearing in America* (pp. 117–145). New York: Cambridge University Press.

Brody, G. H., & Flor, D. L. (1997). Maternal psychological functioning, family processes, and child adjustment in rural, single-parent, African American families. *Developmental Psychology, 33,* 1000–1011.

Bronfenbrenner, U., & Evans, G. W. (2000). Developmental science in the 21st century: Emerging theoretical models, research designs, and empirical findings. *Social Development, 9,* 115–125.

Bronfenbrenner, U., & Morris, P. (1998). The ecology of developmental processes. In W. Damon & R. Lerner (Eds.), *Handbook of child psychology: Vol. 1. Theoretical models of human development* (5th ed., pp. 992–1028). New York: Wiley.

Cohen, S., Evans, G. W., Stokols, D., & Krantz, D. S. (1986). *Behavior, health, and environmental stress.* New York: Plenum Press.

Cohen, S., & Spacapan, S. (1984). The social psychology of noise. In D. M. Jones & A. J. Chapman (Eds.), *Noise and society* (pp. 221–245). New York: Wiley.

Coldwell, J., Pike, A., & Dunn, J. (2006). Household chaos—links with parenting and child behaviour. *Journal of Child Psychology and Psychiatry, 47,* 1116–1122.

Corapci, F., & Wachs, T. D. (2002). Does parental mood or efficacy mediate the influence of environmental chaos upon parenting behavior? *Merrill Palmer Quarterly, 48,* 182–201.

Crouter, A. C., Head, M. R., McHale, S. M., & Tucker, C. J. (2004). Family time and the psychosocial adjustment of adolescent siblings and their parents. *Journal of Marriage and Family, 66,* 147–162.

Davie, R., Butler, N., & Goldstein, H. (1972). *From birth to seven: The second report of the national child development study.* London: The National Children's Bureau.

Dumas, J. E., Nissley, J., Nordstrom, A., Smith, E. P., Prinz, R. J., & Levine, D. W. (2005). Home chaos: Sociodemographic, parenting, interactional, and child correlates. *Journal of Clinical Child and Adolescent Psychology, 34,* 93–104.

Environmental Protection Agency. (1977). *The urban noise survey* (EPA 550/9-77-100). Washington, DC: Author.

Evans, G. W. (2001). Environmental stress and health. In A. Baum, T. Revenson, & J. E. Singer (Eds.), *Handbook of health psychology* (pp. 365–385). Hillsdale, NJ: Erlbaum.

Evans, G. W. (2004). The environment of childhood poverty. *American Psychologist, 59,* 77–92.

Evans, G. W. (2006). Child development and the physical environment. *Annual Review of Psychology, 57,* 423–451.

Evans, G. W., & Cohen, S. (2004). The adaptive costs of coping with suboptimal environmental conditions. In C. Speilberger (Ed.), *Encyclopedia of applied psychology* (pp. 815–824). Los Angeles: Sage.

Evans, G. W., & English, K. (2002). The environment of poverty: Multiple stressor exposure, psychophysiological stress, and socioemotional adjustment. *Child Development, 73,* 1238–1248.

Evans, G. W., Gonnella, C., Marcynyszyn, L. A., Gentile, L., & Salpekar, N. (2005). The role of chaos in poverty and children's socioemotional adjustment. *Psychological Science, 16,* 560–565.

Evans, G. W., & Hygge, S. (2007). Noise and performance in children and adults. In L. Luxon & D. Prasher (Eds.), *Noise and its effects* (pp. 549–566). London: Wiley.

Evans, G. W., & Kim, P. (2007). Childhood poverty and health: Cumulative risk exposure and stress dysregulation. *Psychological Science, 18,* 953–957.

Evans, G. W., & Lepore, S. J. (1993). Household crowding and social support: A quasi-experimental analysis. *Journal of Personality and Social Psychology, 65,* 308–316.

Evans, G. W., Lepore, S. J., Shejwal, B. R., & Palsane, M. N. (1998). Chronic residential crowding and children's well being: An ecological perspective. *Child Development, 69*, 1514–1523.

Evans, G. W., Palsane, M. N., Lepore, S. J., & Martin, J. (1989). Residential density and psychological health: The mediating effects of social support. *Journal of Personality and Social Psychology, 57*, 994–999.

Evans, G. W., Rhee, E., Forbes, C., Allen, K. M., & Lepore, S. J. (2000). The meaning and efficacy of social withdrawal as a strategy for coping with chronic crowding. *Journal of Environmental Psychology, 20*, 204–210.

Evans, G. W., & Stecker, R. (2004). The motivational consequences of environmental stress. *Journal of Environmental Psychology, 24*, 143–165.

Federman, M., Garner, T. I., Short, K., Cutter, W. B. I., Kiely, J., Levine, D., et al. (1996). What does it mean to be poor in America? *Monthly Labor Review, 119*(5), 3–17.

Fiese, B. H. (2006). *Family routines and rituals*. New Haven, CT: Yale University Press.

Fiese, B. H., Wamboldt, F. S., & Anbar, R. D. (2005). Family asthma management routines: Connections to medical adherence and quality of life. *Journal of Pediatrics, 146*, 171–176.

Forman, E. M., & Davies, P. T. (2003). Family instability and young adolescent maladjustment: The mediating effects of parenting quality and adolescent appraisals of family security. *Journal of Clinical Child and Adolescent Psychology, 32*, 94–105.

Goldin, L. (2001). Flexible work schedules: What are we trading off to get them? *Monthly Labor Review, 124*(3), 50–67.

Guidubaldi, J., Cleminshaw, H. K., Perry, J. D., Nastasi, B. K., & Lightel, J. (1986). The role of selected family environment factors in children's post-divorce adjustment. *Family Relations, 35*, 141–151.

Haines, M. M., Stansfeld, S. A., Head, J., & Job, R. F. S. (2002). Multi-level modeling of aircraft noise on performance tests in schools around Heathrow London airport. *International Journal of Epidemiology and Community Health, 56*, 139–144.

Hamermesh, D. S. (1999). Changing inequality in work injuries and work timing. *Monthly Labor Review, 122*(10), 22–30.

Han, W. J. (2005). Maternal nonstandard work schedules and child cognitive outcomes. *Child Development, 76*, 137–154.

Heft, H. (1979). Background and focal environmental conditions of the home and attention in young children. *Journal of Applied Social Psychology, 9*, 47–69.

Heft, H. (1985). High residential density and perceptual-cognitive development: An examination of the effects of crowding and noise in the home. In J. F. Wohlwill & W. van Vliet (Eds.), *Habitats for children* (pp. 36–76). Hillsdale, NY: Erlbaum.

Howes, C., & Stewart, P. (1987). Child's play with adults, toys, and peers: An examination of family and child-care influences. *Developmental Psychology, 23*, 423–430.

Ingersoll, R. M. (2001). Teacher turnover and teacher shortages: An organizational analysis. *American Educational Research Journal, 38*, 499–534.

Jensen, E. W., James, S. A., Boyce, W. T., & Hartnett, S. A. (1983). The family routines inventory: Development and validation. *Social Science and Medicine, 17*, 201–211.

Kiernan, K., & Mueller, G. (1999). Who divorces? In S. McRae (Ed.), *Changing Britain* (pp. 377–403). Oxford, England: Oxford University Press.

Kleiwer, W., & Kung, E. (1998). Family moderators of the relation between hassles and behavior problems in inner city youth. *Journal of Clinical Child Psychology, 27*, 278–292.

Kohen, D. E., Hertzman, C., & Wiens, M. (1998). *Environmental changes and children's competencies* (W-98-25-E). Hull, Quebec, Canada: Applied Research Branch, Human Resources Development Canada..

Lepore, S. J. (1995). Measurement of chronic stressors. In S. Cohen, R. C. Kessler, & L. U. Gordon (Eds.), *Measuring stress* (pp. 102–120). New York: Oxford University Press.

Lepore, S., Evans, G. W., & Schneider, M. (1991). The dynamic role of social support in the link between chronic stress and psychological distress. *Journal of Personality and Social Psychology, 61*, 899–909.

Light, R. (1973). Abused and neglected children in America: A study of alternative policies. *Harvard Educational Review, 43*, 556–598.

Long, L. (1992). International perspectives on the residential mobility of America's children. *Journal of Marriage and Family, 54*, 861–869.

MacKinnon, D. (2008). *Introduction to statistical mediation analysis*. London: Psychology Press.

Marcynyszyn, L. A., Evans, G. W., & Eckenrode, J. J. (2008). Family instability during early and middle adolescence. *Journal of Applied Developmental Psychology, 29*, 380–392.

Martinez, C. R., & Forgatch, M. S. (2002). Adjusting to change: Linking family structure transitions with parenting and boys' adjustment. *Journal of Family Psychology, 16*, 107–117.

Matheny, A., Wachs, T. D., Ludwig, J., & Phillips, K. (1995). Bringing order out of chaos: Psychometric characteristics of the confusion, hubbub, and order scale. *Journal of Applied Developmental Psychology, 16*, 429–444.

Mayer, S. E., & Jencks, C. (1989). Poverty and the distribution of material hardship. *Journal of Human Resources, 24*, 88–114.

Metcalfe, J., & Mischel, W. (1999). A hot/cool system analysis of delay of gratification: Dynamics of willpower. *Psychological Review, 106*, 3–19.

Mischel, W., & Ayduk, O. (2005). Willpower in a cognitive-affective processing system. In R. F. Baumeister & K. D. Vohs (Eds.), *Handbook of self-regulation* (pp. 99–129). New York: Guilford Press.

Mischel, W., Shoda, Y., & Rodriguez, M. (1989, May 26). Delay of gratification in children. *Science, 244*, 933–938.

Moore, K. A., Vandivere, S., & Ehrle, J. (2000). *Turbulence and child well-being* (No. B-16). Washington, DC: Urban Institute Press.

Osborne, C., & McLanahan, S. (2007). Partnership instability and child well being. *Journal of Marriage and Family, 69*, 1065–1083.

Pike, A., Iervolino, A. C., Eley, T. C., Price, T. S., & Plomin, R. (2006). Environmental risk and young children's cognitive and behavioral development. *International Journal of Behavioral Development, 30*, 55–66.

Presser, H. B. (2003). *Working in a 24/7 economy*. New York: Russell Sage Foundation.

Reid, I. (1989). *Social class differences in Britain*. Glasgow, Scotland: Fontana.

Roy, K. M., Tubbs, C. Y., & Burton, L. M. (2004). Don't have no time: Daily rhythms and the organization of time for low-income families. *Family Relations, 53*, 168–178.

Rutter, M., Yule, B., Quinton, D., Rowlands, O., Yule, W., & Berger, M. (1974). Attainment and adjustment in two geographical areas: III—some factors accounting for area differences. *British Journal of Psychiatrics, 125*, 520–533.

Simpson, G. A., & Fowler, G. (1994). Geographic mobility and children's emotional/behavioral adjustment. *Pediatrics, 93*, 303–309.

Stoneman, Z., Brody, G. H., Churchill, S. L., & Winn, L. L. (1999). Effects of residential instability on Head Start children and their relationships with older siblings: Influences of child emotionality and conflict between family caregivers. *Child Development, 70*, 1246–1262.

U.S. Census. (2004, March). *Geographical mobility: 2002 to 2003. A Current Population Report* (No. P-20-549). Washington, DC: U.S. Census Bureau.

U. S. General Accounting Office. (1994). *Elementary school children: Many change schools frequently, harming their education* (GAO/HEHS-94-45). Washington, DC: Author.

Wachs, T. D., & Corapci, F. (2003). Environmental chaos, development and parenting across cultures. In C. Raeff & J. Benson (Eds.), *Social and cognitive development in the context of individual, social, and cultural processes* (pp. 54–83). New York: Routledge.

White, R. (1959). Motivation reconsidered: The concept of competence. *Psychological Review, 66*, 297–333.

Wood, D., Halfon, N., Scarlata, D., Newacheck, P., & Nessim, S. (1993). Impact of family relocation on children's growth, development, school function, and behavior. *Journal of the American Medical Association, 270*, 1334–1338.

15

An Ecological Framework for the Refugee Experience: What Is the Impact on Child Development?

Stuart L. Lustig

According to the United Nations Children's Fund (UNICEF), wars and conflicts in the past decade have killed an estimated 2 million children, disabled 6 million, separated 1 million from their parents, and left another 20 million homeless (UNICEF, 2007). The staggering number of young people whose lives have been irrevocably altered by war and its aftermath raises important questions about the developmental trajectories of these children and the environments in which they attempt to live, function, and grow up. The impact on normal developmental processes is significant, as summarized by several scholarly reviews on the mental well-being of child refugees and war-affected youth (Barenbaum, Ruchkin & Schwab-Stone, 2004; Jensen & Shaw, 1993; Lustig, Kea-Keating et al., 2004; Stichick, 2001). *Chaos* is a term Urie Bronfenbrenner has aptly applied to the experience of American youth, noting that

> America has yet to confront the reality that the growing chaos in the lives of our children, youth, and families pervades too many of the principal settings in which we live our daily lives: our homes, health care systems, child care arrangements, peer groups, schools, neighborhoods, workplaces and means of transportation and communication among all of them. (Bronfenbrenner, 2001b)

The lives of child refugees, whose entire social ecology has been decimated, are that much more engulfed by the chaos that is the unifying theme of this chapter. In this chapter, I (a) briefly review Bronfenbrenner's model of nested ecological structures in the context of the child refugee experience, with specific reference to the microsystem and the macrosystem (for more details on the overall model, see chap. 1); (b) define the refugee experience, outlining the chaotic components from different levels of a child's ecology that may impact development; (c) identify aspects of development among child refugees in particular

---

An earlier version of this chapter was presented on October 26, 2007, at the First Bronfenbrenner Conference on the Ecology of Human Development sponsored by the Bronfenbrenner Life Course Center and the College of Human Ecology at Cornell University. I, the chapter author, acknowledge the very helpful comments of Tom Weisner.

that are adversely affected by the chaos of war, within the context of Bronfenbrenner's model; and (d) identify current and future areas of study within the ecology of refugees' experiences.

Bioecology of Human Development

Thirty years ago Bronfenbrenner acknowledged his "intellectual debt of considerable size" to Kurt Lewin in his acceptance speech upon receiving the Kurt Lewin Memorial Award from the Society for the Psychological Studies of Social Issues (Bronfenbrenner, 1978, p. 199). In that speech, as elsewhere, he outlined a theoretical framework of the ecological environment as a nested and interconnected structure (Bronfenbrenner, 2001a, 2005).

Bronfenbrenner has described the ecological environment in which all children develop as a "set of nested structures, each inside the next like a set of Russian dolls" (Bronfenbrenner, 1979, p. 3). Within this hierarchically organized system, activities or circumstances at one level synergistically affect those of another level. The innermost level is the *microsystem*, which refers to environmental contexts in which the child is in direct contact with persons, activities, objects, and symbols that can act to influence the child's development. The psychological development of child refugees, perhaps more than most children, is a product of this chaotic system of nested ecological structures, and the upheaval and uncertainty of the refugee experience fundamentally threaten the microsystem. When cities, towns, and villages are burned to the ground, and unfortunate family members who were unable to successfully fight back or escape have died, the immediate cast of characters in a child's life is altered. Disappearances of family members, due to the actions of tyrannical governments, also destroy the bonds between close friends and family. It is at dangerous times such as these that stability would be most auspicious for a child's development. At the microsystem level, the term *proximal processes* refers to the reciprocal transactions between the child and persons, objects, and symbols in the immediate microsystem that directly influence children's development (Bronfenbrenner, 2001a). During the child's early years (i.e., ages 0–5), parenting is the primary proximal process. For refugee children, whose parents may be compromised in their caretaking abilities, proximal processes may be diminished in number and effectiveness.

Immediately beyond the microsystem is the *mesosystem*, which comprises relations among two or more settings (essentially different microsystems) in which the child becomes an active participant. Certainly for refugee children, who may be living in chaotic environments that are often in flux, mesosystems can be severely fragmented.

If one visualizes concentric circles, with the innermost being the microsystem, which is contained within the mesosystem, the next circle is the *exosystem*, defined as a setting that does not itself contain a developing person but in which events occur that affect the setting containing the person. Exosystems impact the well-being of those who come into contact with the developing child, and they may also exert policies or decisions that impact the child or child's family. For refugee families, changes in the parents' work environment, or the availability of a certain social service in the town where the family lives, can have a profound impact on the child's development. For example, Elbedour, Bensel, and Bastien (1993) suggested that strong social structures and resources for parents may help to mitigate the effects of war on children. More recent studies of this phenomenon are reviewed in the section on research.

Finally, Bronfenbrenner has described the outermost ring of the ecological environment as the *macrosystem*, which is defined as an overarching pattern of ideology and organization of the social institutions common to a particular culture or subculture. This societal blueprint influences the appearance of different lower order systems throughout the ecosystem, such that schoolrooms or slums have similar characteristics, wherever they may be. Child refugees, who move from country to country, however, may not derive any benefit afforded by belonging to the hegemonic macrosystem that may otherwise confer stability on those consistently living within it. Consider the destabilizing impact of one's family suddenly uprooting, often violently, from everything familiar (friends, neighborhoods, cherished places, favorite activities) and transplanting themselves, sometimes as a unit but potentially in fragments, in an entirely unknown world where everything is foreign: the food, the language, the geography, and the systems that regulate social interaction. Aspects of the macrosystem that may affect chaos at lower levels of the environment, especially for already uprooted refugee children, include economic disruption, cultural characteristics, and sectarian or political upheaval, all of which are common to the refugee experience as described later.

In chapter 2 of this volume, Lichter and Wethington asserted that "chaos at the macrosystem level or structural level has been increasingly replaced over the past century by chaos at the microsystem level (i.e., in children's family environments)." Though this assertion is perhaps true for most children, the refugee experience is by definition one of many tectonic shifts in typical macrosystemic structures of stability—community, culture, political, and educational systems—in light of the migration that refugees make around the world. In fact, the refugee experience has been conceptualized chronologically, with stages enumerated as premigration, migratory period (including time in a refugee camp), and settlement in the new country (Lustig, Kea-Keating, et al., 2004). However, even when resettled, migrants still remain sensitive to disruptions. For example, Sudanese youth resettled to the Boston area at the beginning of this decade expressed significant apprehension about their safety and how they would be perceived by others following the terrorist attacks on September 11, 2001. However, macrosystems can also reduce the impact of chaotic environments. Elbedour et al. (1993) referenced the macrosystemic, protective aspects of ideological commitment among those living in war zones (e.g., Punamaki, 1988) and of religious beliefs, as exemplified by the hope that a Jewish Auschwitz death camp inmate derived from the fact that the numbers branded on his forearm summed to 18 (representing the Hebrew word for "life"; Spilka, 1989). Thus, the political, cultural, or religious belief systems to which people adhere can be disruptive or comforting.

The Refugee Experience: A Product of Chaos

Throughout this volume, the authors have struggled with the definition of chaos and even, to a certain extent, whether chaos can be a good thing that fosters growth and autonomy (see chaps. 2 and 13). Here, however, chaos is conceptualized as what I call the *refugee experience*, which applies to children as well as to adult refugees who are rendered childlike in the process. The refugee experiences a series of interrelated events, interactions, and challenges that Bronfenbrenner's ecological model of development helps clarify. Aspects of the refugee experience may vary widely in the details from place to place and from individual to individual but are characterized in all cases by certain chaos-generating physical and emotional universals: deprivation, upheaval, fear, uncertainty, and loss.

As opposed to economic migrants who cross borders to seek better opportunities abroad, refugees are fleeing their countries and are unable or unwilling to return to their home country because of persecution or a well-founded fear of persecution on account of race, religion, nationality, membership in a particular social group, or political opinion (Anker, 1999). As stated by Miller et al. (2002), *valence of movement* and *urgency of departure* distinguish immigrants from refugees, both of whom may come from settings characterized by the structural violence of chronic poverty. Immigrants move toward the dream of a better life, generally after a well-considered decision-making process. Refugees move away from situations of persecution for the purpose of immediate safety, and the decision is usually made quite hastily. As such, deprivation often occurs early in the experience of a refugee. Those who are running from targeted persecution by their own governments flee, often shrouded in secrecy to escape undetected in the face of threats to personal safety, and immediately become destitute. Those fleeing war zones may travel in groups, also trying to move below the radar screens of invading armies (Eggers, 2006). Refugees usually travel with few personal belongings and many times do not even have key documents with them. They form droves of self-made paupers who must throw themselves at the mercy of others; often a refugee camp is their best and only means of survival. Refugee camps often have sparse resources for densely packed populations (the impact of crowding on child development was also addressed in chaps. 3 and 14, this volume).

Upheaval is unavoidable. The world as refugees used to know it, however perilous, was at least predictable in its peril. The new road, leading away, allows nothing to be taken. Everything has changed in an instant. Their world, previously with some order, is now chaos. Uncertainty, about both those left behind and what the future may bring, is an ever-present experience. Visibility on this new road is extremely limited. What lies beyond the next curve is completely unknown. The length of the road, or whatever awaits at the destination, if there is a defined one, is also unknown. Fear of this unknown, even fear about the availability of basic human survival needs (food, shelter, clothing), is inevitable. In addition, whether one travels through a war zone or crosses a border in the trunk of a smuggler's car, the journey itself may be one's last. Fear of bandits at a refugee camp, fear of officials at a foreign border, and fear of deportation should one arrive within the relative safety of another country are all experienced daily among refugees. Finally, despite the will to live that propels many from the familiar environment of home, this transplantation to a new environment is accompanied by the loss of all that is familiar: one's family; one's favorite places and pastimes; and familiar cuisine, language, and cultural values. This loss, termed *cultural bereavement*, is difficult to mitigate, although some evidence suggests that linkages with one's cultural group can help young refugees survive their new environments (Eisenbruch, 1991a).

Impact of the Refugee Experience on Normal Child Development

Disturbances at all levels of the social ecology can impede the developmental processes of children, which normally proceed simultaneously along several lines and encompass gains in physical growth, cognition, social skills, sexual growth, and moral thought. These trajectories have been described by many theorists of child development, and no single correct theory explains all of child development. In general, however, children benefit from stability as they develop a sense of themselves as competent and of the world as safe and nurturing (Bowlby, 1988).

As a result of an inherently unstable upbringing that fractures normal developmental processes, many refugee children are forced to become adults too soon. Key aspects of development, such as education, work, or interactions with family, are sacrificed to stay alive. Some refugee adolescents who arrive in the United States have cared for infants and young children at a young age themselves or have acted as heads of household. Those who have served as child soldiers have been involved in the kinds of destructive activities unknown to most young people and have a distorted sense of authority, likely believing that it comes from wielding a weapon. Child refugees may arrive with parents who were unable to protect or support them at critical developmental junctures. Meanwhile, these parents may lack language or vocational skills useful in the new country, which again places refugee youth in a caretaking role. Parents may feel they lack the respect necessary to regain their role as patriarch or matriarch of the family, and refugee children may not learn to trust in others, instead harboring a cynical view of the abilities of adults around them who are supposed to model competency. Adolescent refugees, in particular, may sense this self-doubt and guilt, which leads to blame or pity of the parents. Compounding the complexity of the adolescent refugees' experience in this country is the nonuniversal nature of adolescence; in many cultures, in which young and old work side by side (high-continuity cultures), young people revere elders whose knowledge they do not regard as obsolete.

Theoretical Perspectives on the Refugee Experience

The remainder of this section touches on how chaos can impact various aspects of development among refugees, including sexual development (Freud), ego psychology (Erikson), object relations (Mahler), attachment, and cognitive and moral development.

FREUDIAN PERSPECTIVES. From a Freudian perspective, within the child's unconscious, the same-sex parent is initially an adversary during toddlerhood with whom the child competes for exclusive attention from the parent of the opposite sex. In most cases, the child represses this libidinal desire by around age 5 or 6, but this oedipal conflict reawakens during puberty, when it is reworked. The same-sex parent is an object not only of competition but also of identification, and the child finds nonfamilial love objects with whom to develop a sexually gratifying relationship. However, for refugee children or resettled refugee parents, the oedipal adversary, namely, the parent, may appear ineffectual, embittered, and defeated. Such parents, instead of serving as useful objects of competition and identification, can become objects of disappointment and scorn. They are perceived as unworthy role models. Their children, feeling their own helplessness and that of the parents as a result of challenges in the social ecology at various levels, may unconsciously stunt their own social and sexual development so as not to exceed their parents' abilities, or, on the other hand, they may completely disregard any parental attempts to monitor or regulate their behavior.

ERIKSONIAN PERSPECTIVES. Another useful psychological perspective through which to view chaos in the lives of refugees is the developmental theory of Erik Erikson. As an offshoot of ego psychology, Erikson's perspective on identify development (Erikson, 1980) would suggest that the trauma experienced by many refugees may detract from key developmental phases (Garbarino & Kostelny, 1996). For example, in the initial Eriksonian stage (0–18 months) of trust versus mistrust, when children raised in the consistent presence of caretakers normally develop a sense of safety, refugee children may be separated from nurturing adults. Though very young children are able to trust a number of attachment figures simultaneously, this trust depends on an ongoing relationship with the key figures, typically mother and father. From the outset, then, children at this developmental stage are exposed to an unstable environment, in which parents are suddenly not present for a period of time; the children may thus develop a maladaptively dour view of the world and have great difficulty subsequently attaching to caretakers (Karen, 1998). In the second stage (18 months–3 years), autonomy versus shame and guilt, refugee children may exhibit war-induced regressions in previously acquired milestones in gaining autonomy, such as toilet training or language development. This regression is typical of all children whose circumstances do not adequately foster a sense of confidence and support, and can be particularly acute when the stressor is severe, such as war-related chaos and separations from family.

The third stage (3–5 years) of initiative versus guilt normally involves limit testing of individual freedom and group responsibility. This stage and also stage four (5–13 years) of industry versus inferiority can be severely disrupted by absence of schooling and vocational development for many refugee children, environments in which these social skills develop. Stage five (13–21 years) of identity versus role confusion is particularly problematic. Young people yearn to belong to a group, but they are disconnected from history and culture, adrift in space as well as time. They belong in some ways to two cultures, in other ways to neither, perhaps belonging most to the refugee culture of fleeing and struggling to survive in the wilderness or in refugee or detention camps. Conflict between parents and refugee children about which cultural identity is most desirable can cause serious schisms in families.

OBJECT RELATIONS AND ATTACHMENT THEORY. An additional developmental field of study concerns object relations, or attachment theory. Margaret Mahler conceptualized separation–individuation phases of attachment (following symbiosis from 0 to 6 months) as differentiation (6–10 months), practicing (10–16 months), rapprochement (16–24 months), and consolidation and object constancy (Mahler, 1975). Her observations of infants passing through these phases chronicle healthy infants' abilities to gradually develop physical autonomy from the parent (which coincide with the abilities to scoot, crawl, and walk), albeit with significant ongoing interaction with the parent in the process. The ability to tolerate these separations from caregivers has been the subject of attachment theory. Attachment theorists, most notably Bowlby (1988), have emphasized the relational context of child development; for example, infants' responses to separations from and reunions with caregivers depend on the quality of the preexisting bond between them. The underlying neuroendocrinology of attachment styles in day-care settings (where separations and reunions are common) has been discussed by Repetti, Taylor, and Saxbe (2007). If refugee children had disordered attachment styles early in life, one would expect to find neurobiological correlates to the challenges of the refugee experience in this population as well (for additional discussion of the neurobiological consequences of chaos, see chap. 8, this volume). One such challenge is described by Lee (1988), who explores the concept of adolescence as a second separation or individuation, another time when youngsters gain independence from parents. However, for child refugees, the previous separations endured by many young refugees make these adolescent tasks more difficult and can reawaken the pain of earlier separations, especially in the context of trauma. Furthermore, Lee argued that in Southeast Asian culture it is not uncommon for extended families to live under one roof or in very close proximity. American mores about independence may cause doubt and guilt among Southeast Asian parents on this continent and may cause conflict between parents and children.

THEORIES OF COGNITIVE AND MORAL DEVELOPMENT. From the standpoint of cognitive development, young refugees may egocentrically attribute the atrocities of war to their own destructive machinations. Thus, children should be protected from war, argued Freud and Burlingham (1943), not only because of the trauma or horror but also because the circumstance of war parallels their internal destructive impulses, thereby interfering with normal developmental tasks of repression and sublimation. Older children, however, may correctly interpret war-related events as external. These children may react with anxiety or depression. It is not uncommon for those who survive to choose to enter the helping professions in adulthood, as a way of either consciously giving back to those in need or unconsciously reworking their own traumatic experiences.

Finally, the refugee experience may also affect moral development. In addition to tangible losses such as home, possessions, friends, and sometimes family members, refugee children may lose trust in authority figures who are unable to provide for their basic needs or who themselves are engaged in perpetrating atrocities. Macksoud, Dyregrov, and Raundalen (1993) postulated that these youngsters lose their moral perspective because they come to believe that looting is not the same as stealing, or that killing for political reasons may, in fact, be justified. Nevertheless, among these same children, delinquency in the refugee camp was reportedly rare.

Studying the Ecology of the Refugee Experience

Before I review a sample of research on refugee children within their ecological environment, it is worth noting, first of all, that research in war zones is fraught with problems, including the possibility of immediate physical danger, unstable study populations, and lack of consistent resources. Thus, although refugees come from developing countries, as described by Barenbaum et al. (2004), most research efforts that study the impact of development within an ecological framework occur in the safer, economically privileged developed countries where the hegemonic cultures are often very different from those of refugees' homelands. This geographical relocation can potentially attenuate symptoms made worse by prior proximity to trauma. However, relocation could be associated with difficulties in either grieving the loss of the native culture (i.e., cultural bereavement) or adjusting to the new culture. This forced abandonment of everything one holds dear can augment psychiatric symptoms. From a methodological perspective, the immediate impact of war on children's developmental trajectories would perhaps be best studied in situ to avoid the confounds of these additional issues over time. Nevertheless, such an ideal study design is often not possible because of the obvious safety and logistical difficulties of doing research in a war zone. In addition, many local governments and agencies are suspicious of the motives, ethics, and long-term commitments of visiting researchers, who may be perceived as merely gathering data for their own interests back home, not to mention their projects' sustainability. Thus, investigations of migrating child refugees often involve studying the impact of migration itself on other outcome variables, even if this is done unintentionally.

The inherent difficulties of research among refugees notwithstanding, environmental chaos among all children, both refugees and nonrefugees alike, is perhaps most easily studied at the microsystem level (e.g., home and school). Elbedour et al.'s (1993) eloquent review examined the impact of war in children through the lens of an ecological model. The authors identified the following as important microsystem influences on children's well-being: the extent to which parents successfully cope with the anxiety and challenges of war, the extent to which they model and transmit a sense of calm to their children, and the ability of the family to remain intact.

In recent years, studies of child refugees have broadened beyond a narrow focus on psychiatric symptomatology to consider the bidirectional impact of various ecological levels and functioning. For example, one study examined not only psychiatric symptoms but also functioning within the microsystem level of the school system as a product of a school-based intervention. In this trauma-focused intervention for immigrant schoolchildren in Los Angeles, sixth-grade students exposed to violence were randomly assigned to a 10-session standardized cognitive–behavioral therapy early intervention group ($n = 61$) or to a waitlist delayed-intervention comparison group ($n = 65$). The intervention was associated with significantly lower scores on measures of posttraumatic stress disorder, depression, and psychosocial dysfunction, although improvements in classroom behavior were not noted (Stein et al., 2003). Despite basing the study within the children's microsystem, the authors commented on the challenges of social isolation, economic deprivation, and cultural bereavement when attempting to extend clinic-based interventions beyond the confines of the clinical setting.

Instead of assessing children's functioning within the microsystem, another study actually targeted for intervention the mothers of internally displaced, traumatized children who were involved in proximal processes within the children's microsystem (Dybdahl, 2001). The intervention was a 5-month, manualized, psychoeducational program for mothers of displaced children in Bosnia and Herzegovina, and included therapeutic discussions, psychoeducation about trauma, and guidance on facilitating parent–child interactions and communication. Outcome variables included children's psychosocial functioning and maternal health. Compared with mothers in the control group who only received medical care and participated in scheduled evaluations (which were also received by mothers comprising the intervention group), mothers in the intervention group had reduced symptoms as measured by the Impact of Events Scale and also reported being happier. Though children of mothers in the control group changed little, mothers and psychologists documented multiple changes among children in the intervention group: fewer problems; improved cognitive abilities; and greater gain in height, weight, and hemoglobin counts. This study evokes the earlier, well-known work of Beardslee et al. (1997) on a psychoeducational preventive intervention of treating maternal depression to prevent symptom development among (nonrefugee) children.

As stated earlier, chaos in the lives of refugees occurs at many levels. Another component of the microsystem in which refugee children function, namely, their school settings, has been identified as not only a location for interventions, as with the Los Angeles study of immigrant children (Stein et al., 2003), but also a target of interventions. O'Shea, Hodes, Down, and Bramley (2000) examined the impact of a range of interventions for refugee pupils and their families (who often had significant macrosystemic challenges such as poverty and unofficial legal status) attending a London junior high school. Students, their relatives, and their teachers were all potential recipients of services. A pre-post design with no control group showed a trend toward symptom reduction using the Strengths and Difficulties Questionnaire completed by teachers. Although identification of the school setting was a promising step toward integrating various ecological levels that impact refugee children, changes among parents, teachers, or the school setting were not evaluated.

Among child refugees, access to services, or interest in them, is problematic (National Child Traumatic Stress Network, 2005). Legal, financial, residential, or occupational difficulties may contribute to difficulties in accessing health services. Meanwhile, cultural beliefs that prevail at the macrosystem level may render certain interventions, such as psychiatric or psychological services, irrelevant at best, and highly stigmatized at worst. Psychiatric disorders may not be widely recognized or acknowledged beyond American or European nosologies. Other cultures' idioms of distress and healing may instead frame these problems as medical, spiritual, or moral, and therefore delegate them to spiritual or faith healers, synthetic churches, or medical healers of various types. In an attempt to move away from a narrow focus on psychopathology, an intervention to increase social support and access to services targeted the exosystem (e.g., the support network of parents caring for children) of Kosovar families living in Chicago. Weine et al. (2003) facilitated various multifamily groups held weekly at a community agency that were led by bilingual or bicultural workers following a manualized curriculum. The intervention was effective at engaging families, increasing access to mental health services, and reducing symptoms of depression and improving family communication, even though the focus was not specifically on symptom reduction. Although this intervention focused on adults, the previous studies suggest that an exosystemic focus could significantly benefit the children of refugees. Connection with one's cultural roots, and the increased social support and access to services that this connection can provide, ultimately help structure the otherwise chaotic environments of refugees (see chap. 4, this volume, which discusses, as an example of structure, the importance of family rituals).

The ability to beneficially alter the macrosystem was the focus of one study that evaluated the presence of cultural bereavement, a concept mentioned earlier. Unaccompanied adolescent Cambodian refugees fostered in Cambodian group care in Australia were compared with a similar group placed with non-Cambodian foster families in the United States. Results showed that cultural bereavement was greater among the latter group (Eisenbruch, 1991b). Contributing factors may have included greater isolation from other Cambodians in the U.S. group and more tolerance and encouragement of indigenous practices in Australia and access to familiar religious figures and traditional healers. Similarly, unaccompanied Indochinese refugee adolescents resettled in the United States with ethnic foster families were significantly less depressed and had higher grade point averages than did those in foster care with Caucasian families or in group homes (Porte & Toney-Purta, 1987). These two studies demonstrate the salubrious effects of maximizing access to one's original cultural roots within the macrosystem.

An adjustment to macrosystemic challenges of acculturative stress, isolation, alienation, and a general sense of being different is also represented for young refugees by the concept of belonging, which was evaluated in a recent study by Kia-Keating and Ellis (2007). In this study of Somali refugees resettled in the northeast United States, school belonging, as measured by the Psychological Sense of School Membership questionnaire, was associated with lower depression and higher self-efficacy, explaining over 25% of the variance of the latter, regardless of the level of past exposure to adversities.

Attention to the macrosystem, specifically the geopolitical problems that govern so many aspects of refugees' lives, is the focus of testimonial psychotherapy, which is telling one's story of survival with an emphasis on creating a written account that could be used for purposes of education and advocacy. In fact, though testimonial psychotherapy may have therapeutic benefit, participants need not identify themselves as mentally ill or in need of mental health services. Among Sudanese refugee adolescents resettled in Boston, testimonials were therefore acceptable and feasible (Lustig, Weine, Saxe, & Beardslee, 2004). As recently summarized, acceptability and feasibility were common attributes of testimonial and narrative therapies among child refugees (Lustig & Tennakoon, 2008). In Weine's study of testimonial psychotherapy among Bosnian refugees (Weine, Dzubur, Kelanvic, Pavkovic, & Gibbons, 1998), the success of testimony was attributed in part to the collective meaning-making in a group that deems social cohesion (e.g., absence of chaos) to be of utmost importance.

Concluding Thoughts

Refugee children, perhaps more than any other vulnerable group, are susceptible to actions at all levels of their social ecology. Because significant chaos for these children can occur at any level, often unpredictably, theirs is a world of ongoing turbulence during their journey on the path to potential stability. The studies of refugees' well-being reviewed here attempt to measure the influence of change within various ecological spheres, and some studies even focus on attempts to salubriously target the spheres themselves. However, the interactions among ecological levels and refugee populations, and the mutually determined impact on these levels, or on refugee youth themselves, have yet to be systematically studied. The task is made more complex by the challenges of research with this population. However, one can hope that gradually increasing awareness of the impact of trauma and structural violence on young people will, in the coming years, foster ongoing dedication to studying and promoting the well-being of these disenfranchised children of war within their complex ecological framework. They deserve nothing less.

References

Anker, D. E. (1999). *Law of asylum in the United States* (3rd ed.). Boston: Refugee Law Center.

Barenbaum, J., Ruchkin, V., & Schwab-Stone, M. S. (2004). The psychosocial aspects of children exposed to war: Practice and policy initiatives. *Journal of Child Psychology and Psychiatry, 45*, 4–62.

Beardslee, W. R., Versage, E. M., Wright, E. J., Salt, P., Rothberg, P., Drezner, K., & Gladstone, T. (1997). Examination of preventive interventions for families with depression: Evidence of change. *Development and Psychopathology, 9*, 109–130.

Bowlby, J. (1988). *A secure base: Parent-child attachment and healthy human development*. New York: Basic Books.

Bronfenbrenner, U. (1978). Lewinian space and ecological substance. *Journal of Social Issues, 33*, 199–212.

Bronfenbrenner, U. (1979). *The ecology of human development: Experiments by nature and design*. Cambridge, MA: Harvard University Press.

Bronfenbrenner, U. (2001a). The bioecological theory of human development. In N. J. Smelser & P. B. Baltes (Eds.), *International encyclopedia of the social and behavioral sciences* (pp. 6963–6970). New York: Elsevier.

Bronfenbrenner, U. (2001b). Growing chaos in the lives of children, youth, and families: How can we turn it around? In J. C. Westman (Ed.), *Parenthood in America* (pp. 197–210). Madison: University of Wisconsin Press.

Bronfenbrenner, U. (2005). *Making human beings human: Biocecological perspectives on human development*. Thousand Oaks, CA: Sage.

Dybdahl, R. (2001). Children and mothers in war: An outcome study of a psychosocial intervention program. *Child Development, 72*, 1214–1230.

Eggers, D. (2006). *What is the what.* San Francisco: McSweeney's.

Eisenbruch, M. (1991a). Cultural bereavement and homesickness. In S. Fischer & C. L. Cooper (Eds.), *On the move: The psychology of change and transition* (pp. 191–205). Chichester, England: Wiley.

Eisenbruch, M. (1991b). From post-traumatic stress disorder to cultural bereavement: Diagnosis of Southeast Asian refugees. *Social Science and Medicine, 33*, 673–680.

Elbedour, S., Bensel, R. T., & Bastien, D. T. (1993). Ecological integrated model of children of war: individual and social psychology. *Child Abuse and Neglect, 17*, 805–819.

Erikson, E. H. (1980). *Identity and the life cycle.* New York: Norton.

Freud, A., & Burlingham, D. (1943). *War and children.* New York: Medical War Books.

Garbarino, J., & Kostelny, K. (1996). What do we need to know to understand children in war and community violence? In R. J. Apfel & B. Simon (Eds.), *Minefields in their hearts: The mental health of children in war and communal violence.* New Haven, CT: Yale University Press.

Jensen, P., & Shaw, J. (1993). Children as victims of war: Current knowledge and future research needs. *Journal of the American Academy of Child and Adolescent Psychiatry, 32*, 697–708.

Karen, R. (1998). *Becoming attached: First relationships and how they shape our capacity to love.* New York: Oxford University Press.

Kia-Keating, M., & Ellis, B. H. (2007). Belonging and connection to school in resettlement: Young refugees, school belonging, and psychosocial adjustment. *Clinical Child Psychology and Psychiatry, 12*, 29–43.

Lee, E. (1988). Cultural factors in working with Southeast Asian refugee adolescents. *Journal of Adolescence, 11*, 167–179.

Lustig, S. L., Kea-Keating, M., Grant-Knight, W., Geltman, P., Ellis, H., Kinzie, J. D., et al. (2004). Review of child and adolescent refugee mental health. *Journal of the American Academy of Child and Adolescent Psychiatry, 43*, 24–36.

Lustig, S. L., & Tennakoon, L. (2008). Testimonials, narratives and drawings: Child refugees as witnesses. *Child and Adolescent Psychiatric Clinics of North America, 17*, 569–584.

Lustig, S. L., Weine, S. M., Saxe, G. N., & Beardslee, W. R. (2004). Testimonial psychotherapy for adolescent refugees: a case series. *Transcultural Psychiatry, 41*, 31–45.

Macksoud, M., Dyregrov, A., & Raundalen, M. (1993). Traumatic war experiences and their effects on children. In J. P. Wilson & B. Raphael (Eds.), *International handbook of traumatic stress syndromes* (pp. 625–634). New York: Plenum Press.

Mahler, M. (1975). *The psychological birth of the human infant: Symbiosis and individuation.* New York: Basic Books.

Miller, K., Worthington, G., Muzurovic, J., Fasmina, J., Tipping, S., & Goldman, A. (2002). Bosnian refugees and the stressors of exile: A narrative study. *American Journal of Orthopsychiatry, 72*, 341–354.

National Child Traumatic Stress Network, Refugee Trauma Task Force. (2005). *Mental health interventions for refugee children in resettlement: White paper II.* Retrieved May 11, 2009, from http://www.nctsnet.org/nctsn_assets/pdfs/promising_practices/MH_Interventions_for_Refugee_Children.pdf

O'Shea, B., Hodes, M., Down, G., & Bramley, J. (2000). A school-based mental health service for refugee children. *Clinical Child Psychology and Psychiatry, 5*, 189–201.

Porte, Z., & Toney-Purta, J. (1987). Depression and academic achievement among Indochinese refugee unaccompanied minors in ethnic and non-ethnic placements. *American Journal of Orthopsychiatry, 57*, 536–547.

Punamaki, R. L. (1988). Historical, political and individualistic determinants of coping modes and fears among Palestinian children. *International Journal of Psychology, 23*, 721–739.

Repetti, R., Taylor, S. E., & Saxbe, D. (2007). The influence of early socialization experiences on the development of biological systems. In: J. Grusec & P. Hastings (Eds.), *Handbook of specialization theory and research* (pp. 124–152). New York: Guilford Press.

Spilka, B. (1989). Functional and dysfunctional roles of religions: An attribution approach. *Journal of Psychology and Christianity, 8*, 5–15.

Stein, B. D., Jaycox, L. H., Kataoka, S. H., Wong, M., Tu, W., Elliott, M. N., & Fink, A. (2003). A mental health intervention for schoolchildren exposed to violence: A randomized controlled trial. *JAMA, 290,* 603–611.

Stichick, T. (2001). The psychosocial impact of armed conflict on children: Rethinking traditional paradigms in research and intervention. *Child and Adolescent Psychiatric Clinics of North America, 10,* 797–814.

UNICEF. (2007). *Child protection from violence, exploitation and abuse: Children in conflict and emergencies.* Retrieved May 11, 2009, from http://www.unicef.org/protection/index_armedconflict.html

Weine, S. M., Dzubur Kelanvic, A., Pavkovic, I., & Gibbons, R. (1998). Testimony psychotherapy with Bosnian refugees: A pilot study. *American Journal of Psychiatry, 155,* 1720–1726.

Weine, S., Raina, D., Zhubi, M., Delesi, M., Huseni, D., & Feetham, S. (2003). The TAFES multi-family group intervention for Kosovar refugees: A feasibility study. *Journal of Nervous and Mental Disease, 191,* 100–107.

Part VI

Conclusions

16

Dynamic Developmental Systems: Chaos and Order

Arnold Sameroff

Chaos is an excellent theme for academics in modern times, when multitasking at work, at home, and at play frequently elicits the feeling encapsulated in the title of an old Broadway musical: "Stop the world, I want to get off!" These negative feelings arise from a complex balancing act people perform among roles in organized institutional structures at home, at work, and at play. These settings may seem to have a life of their own but are highly dependent on the regulated functioning of their participants. When someone's home life becomes disrupted, it may intrude into his or her work life. For example, in many contemporary families child care is divided between two parents and several third parties but becomes disrupted if one partner becomes ill and the other has to fill in. This solution to the caregiving situation may reverberate to the workplace, where the absence of one or the other may mean that a function is not being performed, requiring someone else to fill in, affecting the regular role that person played. To the extent that the parents or their substitutes are in supervisory roles, their charges lose a source of regulation in their own occupations, which may affect the larger community when a product is not produced or a service not delivered (see chap. 12).

These examples reflect the ecological model that is the basis of much that is reported here. Bronfenbrenner's (1986) micro-, meso-, exo-, and macrosystems provide an important reorientation to the psychological study of human development. At a time when the child research enterprise was focused either on the unfolding of innate characteristics of the child or on the environment's reinforcement contingencies, Bronfenbrenner brought a sociological and anthropological perspective that added multiple levels of analysis to the requirements for predicting and influencing child development.

Modern times are based on highly regulated, ordered systems that connect multiple individuals to multiple roles that offer many opportunities for order but also for disorder. This disorder can have both contemporary and developmental effects and can be produced by either the individual or the context. The study of chaos must be integrated into a broader understanding of change over time in both individuals and settings. Time is part of the Bronfenbrenner model, but it is discussed primarily with regard to historical time and in terms of the accumulation of effects from either the absence or unpredictability of environmental supports, with a longer duration of chaos having more deleterious effects than a shorter duration. However, the notion of timing must be added to the study of time. Depending on what is developing at a specific point in time, chaos may have major, minor, or no effects. For example, Elder's (1995) studies of military experience document how the same historical events can facilitate the development of one cohort (teenagers) and interfere with the development of another (those in their 20s; for a detailed discussion of cumulative influences and timing issues, see chap. 8).

Models for Understanding Successful Development

Contemporary developmental science requires at least four models for understanding human growth: a personal change one, a contextual one, a regulation one, and a representational one (Sameroff, 2009a).

The Personal Change Model

The personal change model is necessary for understanding the progression of competencies from infancy on. It requires unpacking the changing complexity of the individual as he or she moves from the sensorimotor functioning of infancy into increasingly complicated levels of cognition; from early attachments with a few caregivers to relationships with peers, teachers, and individuals in the world beyond home and school; and from the early differentiation of self and other to the multifaceted personal and cultural identities of adolescence and adulthood.

The Contextual Model

The contextual model overlaps substantially with Bronfenbrenner's formulations. Before Bronfenbrenner, in the United States development was believed to be a linear progression, with each new step in behavior predictable from the preceding one, and directionality determined by the history of contingent reinforcements in a behaviorist model that could be replicated in the laboratory. But this approach was defeated as one exogenous variable after another, such as the ethnicity and gender of the subject or experimenter, was found to affect even laboratory behavior. The child psychologist now had to attend to context, and it was Bronfenbrenner who provided the theoretical basis that made attending to settings scientific. Many predecessors felt that families, schools, neighborhoods, and culture had influences on development, but Bronfenbrenner turned these ideas into a comprehensive framework with predictions of how these settings affect the child and also how they affect each other. Although his terminology of microsystems, mesosystems, macrosystems, exosystems, and chronosystems may not be universally accepted, his principles that the family, school, work, and community are all intertwined in explaining any particular child's development were part of a revolution in psychological science that is now universally acknowledged. After Bronfenbrenner, behavior in general, and development in particular, could not be separated from the social context. An individual's behavior could not be predicted independent of situational demands and constraints.

The growing child is increasingly involved with multiple social settings and institutions. The move was traditionally from participation wholly in the family microsystem into contact with the peer group and school system. Now, however, many infants are placed in out-of-home group child care in the first months of life. Similarly, where the effects of neighborhoods used to be primarily mediated by the family, street violence now directly impacts people of every age, including the very young. Each of these settings has its own system properties. Concerns about the development of the child are only one of many institutional functions. For example, the administration of a school setting needs to attend to financing, hiring, training of staff, and building maintenance before it can perform its putative function of caring for or educating children. In chapter 6, Maxwell's description of the school setting is an excellent detailing of the many functional dimensions involved.

Attention to the effects of changing settings on children over time must be augmented by attention to changing characteristics of individuals within a setting. Contemporary social models also take a life course perspective that includes the interlinked life trajectories of not only the child but also other family members (Elder, Johnson, & Crosnoe, 2003). Experiences for the child may be quite different if the mother is in her teens with limited education, or in her 30s after completing professional training and entry into the job force. Similarly, Lustig's (see chap. 15, this volume) discussion of the refugee experience highlights the disruptions in developmental timing where young children are filling adult roles as caregivers, breadwinners, or soldiers. The life course perspective incorporates the historical changes over the past century, as described by Lichter and Wethington (see chap. 2, this volume), that produced better health and less poverty while also maintaining or even expanding major social inequalities; in other words, life for many did not fully benefit from overall societal progress. Put in terms of concepts that are the focus of this volume, chaos may be reduced for some but increased for others.

The Regulation Model

The third model of regulation reflects the systems orientation of modern science (Sameroff, 1983). Most of the rhetoric in developmental research is about self-regulation, giving the illusion that regulation is a property of the individual, but self-regulation occurs only if there is a social surround that is engaged in other-regulation (Sameroff & Fiese, 2000). This regulation by others provides the increasingly complex social, emotional, and cognitive experiences to which the child must self-regulate and the safety net when self-regulation fails. Even early functional physiological self-regulation of sleep, crying, and attention are augmented by caregiving that provides children with regulatory experiences to help them quiet down on the one hand and become more attentive on the other. Vygotsky's (1978) zone of proximal development is analogous to other-regulation in cognitive development. Successful socialization and education are based on fitting experience to the developmental status of the child. As children create their understanding of the world, the world is made more complex through steps in a curriculum to move them along toward some societal goal of mature thought. Arithmetic is an excellent example: As soon as children learn to add they are required to learn to subtract, following which they are taught to multiply and divide. Each step is a regulation of the environment by the teacher to keep one step ahead of the child's mathematical regulation. Similarly, in the social realm increases in social responsibility are paced to the success of the child's adjustment to previous levels of responsibility (Rogoff, 2003).

More important, in the systems view no individuals are separate from relationships (Sameroff, 2009b). Each child is functioning in relation to characteristics of the physical and social world. For the most part other people are mediating this experience, from the first deitic episodes of the parent capturing the child's attention by pointing to objects in the surround to the university lecturer who uses multimedia presentations to engage a range of cognitive functions. One can consider many of these regulations in terms of affordances that capitalize on the perceptual abilities of the child, but many are developmentally paced to increase the complexity of experience for the child.

The Representational Model

In the last model, the representational one, an individual's here-and-now experiences in the world are given a timeless existence in thought. These representations are the cognitive structures where experience is encoded at abstracted levels that provide an interpretive structure for new experiences, as well as a sense of self and other. Representations are obviously not the same as what they represent. They have an adaptive function of bringing order to a variable world, producing a set of expectations of how things should fit together. In the social realm these representations include such things as an infant's working model of relationships (see chap. 4, on representations of family routines) or cultural practices (see chap. 13).

These four models for understanding development provide a framework for understanding the deleterious effects of chaos. The chapters in this volume tap into one or another model and document how the instability or incomprehensibility of experience interferes with or prevents adaptive current or later functioning. The contemporary effect of chaos is that functions are not performed, which leaves needs unmet. The developmental effect is that children growing in such circumstances may have limited or no models of effective regulation both for themselves and for their relationships with others.

Chaos

The vision of chaos discussed in the introductory chapter and detailed in chapter 14 by Evans et al. was somewhat simplified into categories of disorder or turbulence in chapter 10 by Brooks-Gunn et al.. Disorder arises from high levels of noise, excessive crowding, clutter, and lack of structure. Turbulence is related to instability of settings, instability of relationships, and unpredictability of routines. The negative effects of such environmental chaos are evidenced in poorer physical and mental health, reduced emotional maturity and social competence, and reduced cognitive competence and communication skills. These factors clearly lead to bad outcomes, but is it equally clear that their opposites would lead to good outcomes?

The other end of the spectrum for these variables includes silence for noisiness, being alone for crowdedness, rigid systems for lack of structure, unchanging settings for instability of settings, enmeshed relationships for unpredictable ones, and boredom for unpredictability. It would seem that extremes of either chaos or order may not foster effective or healthy outcomes. What we know about successful systems is that they are adaptive, capable of responding to a range of conditions, and still maintain competent functioning; in other words, they are well-regulated. Successful development requires regularities of experience for the formation of adapted systems but also requires new stimulation that may initially be experienced as noise, if growth is to occur.

An examination of the outcomes affected by chaos shows they are actually regulatory processes related to both internal and external experience. Physical health involves the regulation of a full range of biological functions that are a necessary platform for psychological processes. Psychological processes are systems for emotional and cognitive regulation served by social and communicative interactions. If one were to ask about the negative consequences of chaos, the answer would lie in how it affects these regulatory systems.

Interpreting Chaos Within Developmental Models

Although chaos sounds terrible and the chapters in this volume have listed a variety of negative outcomes associated with chaos, most of the consequences have not been through linear causality. The majority of the effects are mediated, such as when chaos affects parenting that in turn affects child outcomes; moderated, such as when children with certain biological or personality characteristics are more affected than are children without them; or accumulated, such as when chaos has an effect only when it occurs simultaneously in many settings of the social ecology. The conclusion to be drawn from all this research is that understanding the effects of chaos is as complex as understanding development itself.

Understanding what ingredients are necessary for developmental progress will throw light on how these ingredients are affected by chaos. Will chaos have its effect by removing these ingredients such that development does not take place? Will chaos have its effect by preventing the child from experiencing the ingredients that are available? Will chaos prevent the developmental process, or will it distort it? Moreover, as Lichter and Wethington suggested in chapter 2, dealing with chaos early in development could act as a vaccination such that the child will not be as affected by chaos later in development. The answers to all these questions require an articulated idea of what development is about and they definitely will vary depending on the particular developmental outcome and the particular developmental period. The four models outlined here provide a framework for understanding the effects of chaotic conditions on developmental success.

Personal Change Model

The time perspective of the personal change model is useful for examining the effects of chaos at specific ages as well as cascade effects, when chaotic effects during one period lead to chaotic effects during another period. The effects of chaotic conditions on regulatory systems vary in accord with their timing. Some experiences are important to the formation of regulatory systems, and others are related to their adaptive functioning once formed. The formation of regulatory systems requires consistencies of experience in repeated and then recognizable patterns. Communicative competence requires patterned experience to shape first the sound patterns for a specific language and then the words within a semantic framework. Social competence requires patterned social interactions initially in the service of emotional regulation—the soothing behavior of parents—and then in the service of using means to obtain ends, such as using parents to access desired outcomes.

To the extent that there are few regularities in experience (the unpredictability dimension of chaos), cognitive and social development is slowed or distorted. Early sensorimotor cognition is hard to derail because it is based on regularities in physical experiences such as touch, sound, or sight largely independent of the social surround, but even such basic perceptions can be distorted in children with mental challenges or children reared in stimulus-deprived institutions or in the extremely overwhelming, chaotic conditions of war and physical calamities (for a discussion on the impact of these types of conditions, see chap. 15). However, as the child reaches toddlerhood, and later in school where understanding the world requires explanations, the full range of chaotic disorder and turbulence comes into play, disrupting learning experiences as detailed in Maxwell's analysis of school effects (see chap. 6).

Early social development is typically associated with the establishment of secure attachments to caregiving figures that will be the basis for later social relationships. The study of attachment is an example of how difficult it is to pin down the effects of chaos. Almost all children form working models of relationships, so chaotic conditions do not prevent development except at the extremes. However, the instability of early experiences will affect the quality of these relationships and how they are represented. Anxious attachments are formed when experience with caregivers is unpredictable or unresponsive. Unresponsiveness is a reduction in regularity of parenting behavior, whereas unpredictability could be interpreted as an instance of chaos. But the real effect of chaos seems to be at the extreme where parental psychopathology comes into play, producing disorganized attachments through a variety of forms of emotional and physical abuse. As more longitudinal studies are completed early, insecurity with caregivers is being connected to later attachment anxiety and avoidance in peer and romantic relationships, but these are not necessarily pathology, only different forms of adaptation (Sroufe, Egeland, Carlson, & Collins, 2005). The chaos in later relationships is more clearly tied to the borderline personality disorders that mediate the link between earlier abuse and later social relationships.

Contextual Model

Through the ecological lens, the chapters in this volume have identified the way chaos affects many aspects of families, schools, cultures, and economies. Moreover, the discussion extended to how chaos in one setting contributes to chaos in another, as exosystems, mesosystems, or macrosystems. For example, Repetti and Wang's (see chap. 12) analysis of the workplace as an exosystem portrays how instability of work hours and autonomy in jobs affects parent emotional stability, which, in turn, could negatively affect child rearing. In addition, a parent's stress at work could affect the marital relationship, producing another pathway toward worse child rearing. However, the workplace could act to reduce stress and the associated chaos by providing maternal leaves and child care, possibly with resources and policies provided by governmental macrosystems.

Whether a setting increases or decreases chaos is important not only for research but also if one is seeking targets for intervention strategies to reduce the effects of chaos. Stability in primary caregivers is an important ingredient in successful socioemotional development. The historically increasing numbers of children being reared in single-parent households documented by Lichter and Wethington in chapter 2 suggest that this stability is being undermined over time. Good child care can be both a microsystem in providing additional stability to the parent's caregiving and a mesosystem by reducing parent stress, acting to offset chaotic family conditions. However, as Corapci and Bradley described in chapters 5 and 9, respectively, many poor-quality child-care settings undermine child development through low staff ratios, turnover of child-care workers, and poor facilities. These effects are further conditioned by macrosystem economic conditions that were described by Evans et al. in chapter 14. Good child care is expensive, and to the extent that economic resources are limited, as in lower socioeconomic status families, the available child care will be of a much lower quality and more chaotic than it will be where families are more affluent.

Regulation Model

Effective regulatory systems are guided by information as to what needs regulating. Negative feedback systems typically reduce discrepancies from a set point by either an increase or a decrease in activation. For effective classroom learning, the teacher must first wake up some students and quiet down other students to get their attention. Mesosystem effects can be seen when increasing student attention is made more difficult when they are tired from after-school jobs, irregular bedtime routines, or carousing late at night thanks to a lack of parental supervision. In a similar vein, getting the attention of overly aroused students is difficult when they have not developed good self-regulation skills because of stress at home or not having had experience at self-regulation earlier in development. Only afterward can instruction begin to stimulate cognitive development via other-regulation by teachers, through the use of curricula that move the child from one step to the next.

Chaos in and between settings interferes with the flow of information necessary to both self- and other-regulation. At one extreme is the destruction of regulatory systems by extreme conditions of war or disaster in which families, schools, neighborhoods, and other social settings are shattered. Less extreme chaotic conditions of high levels of noise, crowding, or instability of participants in the system prevent communication. This degraded communication may not stop development but should act to slow or distort growth processes.

Representational Model

Representations are encodings of experience. They are a more or less elaborated internal summary of the external world. In such a summation certain aspects are selected and others ignored. In the representation of a square, for example, the size, color, and texture of the square object are ignored. In an analogous way, when representations are made of a social object such as a parent, certain features are included in the representation and others are ignored. Research using the adult attachment interview (Main & Goldwyn, 1984) has found that such representations of parents are often idolized, with only positive aspects being included in the mental model.

Chaotic experience can be seen as both a hindrance and distorter in the formation of representations during development. In Weisner's summation of culture and chaos (see chap. 13, this volume), he argued that well-being is a result of meaningful cultural engagement with desirable everyday routines that have a script, goals, and values. Meaningfulness, a key component of his analysis, is found only in coherent representations. The development of this engagement is negatively affected by chaotic conditions. Once formed, these representations frequently provide a script for social engagement that may act as a buffer against later chaos. In dynamic family systems when one member of a family becomes disruptive, such as through alcoholism, activities of all other members are affected. An excellent example is Fiese and Winter's (see chap. 4) descriptions of how family routines provide a narrative representation for the rest of the family members that allows the whole to continue adaptive functioning despite the disruption of one of its parts. The order or disorder in a society's representation of itself certainly affects the adaptive functioning of its members. For example, native youth show much higher levels of suicide and other problem behavior when there are large inconsistencies in cultural continuity from one generation to another (Chandler, Lalonde, Sokol, & Hallett, 2003).

Person, Process, Context

An overarching problem for developmental research, in general, and the studies of chaos described in this volume, in particular, is that the effect sizes are usually small. Although statistical differences are found between individuals experiencing a variety of chaotic conditions on a variety of outcomes, there are few, if any, demonstrations of causal connections between disruptions in experience and behavior. One explanation of such weak effects is that there are so many influences on any developmental outcome that single variables such as noise or crowding cannot have a large effect. To demonstrate larger effects one must accumulate a number of chaotic variables into some form of aggregate chaos score. Such is the strategy suggested by Evans et al. (see chap. 14) and by Ackerman and Brown (see chap. 3). Their argument is that a single area of chaos will not have a major effect because it is only one of the many ingredients in each developmental achievement. Therefore, a cumulative aggregate composed of multiple chaotic elements would be a much better predictor of aberrant development.

The problem with cumulative scores is that they do not reveal the process by which a particular chaotic element might influence a particular developmental outcome, nor do they necessarily separate chaotic factors in microsystems from those in the meso- or macrosystems. An explanation of process would need to be framed by the developmental models outlined earlier. Those models have as a final common pathway Bronfenbrenner's (Bronfenbrenner & Crouter, 1983) person, process, context model, where outcomes may change when any of the three components change. The ultimate usefulness of this model could be achieved through a complete mapping of all the persons, processes, and contexts in which a function develops. Such a large-scale model was attempted by Wachs in chapter 7. He argues that the influence of chaos will depend, in part, on person characteristics such as temperament, age, gender, and prior history, as well as the larger macrosystem context within which the individual develops including measures such as war, poverty, and disaster. Dunn, Schaefer-McDaniel, and Ramsay (see chap. 11) contributed to this discussion by emphasizing the need to define specific aspects of development, specific age groups, and specific aspects of social settings to determine how chaos has its effect.

What is clear from such mappings of the literature is that direct effects are few and small because most indicators of chaos operate through mediating processes or are moderators of still other developmental processes. In light of the complexity of the constructs involved, one of the implications of the operation of mediating and moderating processes is that attempts to catalog the dimensions and influences of chaos are very likely to be oversimplified.

Reducing the Effects of Chaos

The goal of this volume is to offer an analysis of how chaos affects development, followed by strategies for reducing any negative effects. If one were to focus on the list of chaotic conditions presented in chapter 1—noise, excessive crowding, clutter and lack of structure, instability of settings, instability of relationships, and unpredictability of routines—then the most direct intervention strategies would be to reduce noise, crowding, clutter, instability, and unpredictability. Reduction rather than elimination should be the goal, because as Wachs and Evans pointed out in chapter 1, depending on individual and contextual characteristics, it is not clear what the demarcation is between amounts of chaos that produce a negative effect, a neutral effect, or even a potentially development-enhancing effect through stretching the adaptive regulations of the child into new realms.

Intervention itself, however, must be subject to an ecological analysis. If the effects of chaotic conditions are cumulative, then what are the costs and benefits of reducing one or another aspect? All social institutions have limited resources, so the decision of how to best invest in improving child development is generally difficult. Is it better to reduce noise or crowding? Is it better to foster cognitive growth or socioemotional adaptation? Such questions interact with the information provided in this volume. Suppose there were a higher social value on cognitive than social competence, but more was known about the effects of chaos on social development. How would one decide where to intervene? This volume has offered a major foundation for such discussions while at the same time pointing to the research necessary to further answer such questions.

References

Bronfenbrenner, U. (1986). Ecology of the family as a context for human development: Research perspectives. *Developmental Psychology, 22,* 723–742.

Bronfenbrenner, U., & Crouter, A. C. (1983). The evolution of environmental models in developmental research. In W. Kessen (Series Ed.) & P. H. Mussen (Vol. Ed.), *Handbook of child psychology: Vol. 1. History, theory, and methods* (4th ed., pp. 357–414). New York: Wiley.

Chandler, M. J., Lalonde, C. E., Sokol, B. W., & Hallett, D. (2003). Personal persistence, identity development, and suicide: A study of native and non-native North American adolescents. *Monographs of the Society for Research in Child Development, 68*(2, Serial No. 273).

Elder, G. H. (1995). The life course paradigm: Social change and individual development. In P. Moen, G. H. Elder, & K. Luscher (Eds.), *Examining lives in context: Perspectives on the ecology of human development* (pp. 101–139). Washington, DC: American Psychological Association.

Elder, G. H., Jr., Johnson, M. K., & Crosnoe, R. (2003). The emergence and development of life course theory. In J. T. Mortimer & M. J. Shanahan (Eds.), *Handbook of the life course* (pp. 3–19). New York: Kluwer Academic/Plenum Publishers.

Main, M., & Goldwyn, R. (1984). Predicting rejection of their infant from mother's representation of her own experience: Implications for the abused and abusing intergenerational cycle. Child Abuse and Neglect, 8, 203–217.

Rogoff, B. (2003). *The cultural nature of human development.* Oxford, England: Oxford University Press.

Sameroff, A. (1983). Developmental systems: Contexts and evolution. In P. H. Mussen (Series Ed.) & W. Kessen (Vol. Ed.), *Handbook of child psychology: Vol. 1. History, theories, and methods* (4th ed., pp. 238–294). New York: Wiley.

Sameroff, A. (2009a, April). *The development of developmental science.* Presidential address presented at the biennial meetings of the Society for Research on Child Development, Denver, CO.

Sameroff, A. (2009b). The transactional model. In A. Sameroff (Ed.), *The transactional model of development: How children and contexts shape each other* (pp. 3–22). Washington, DC: American Psychological Association.

Sameroff, A., & Fiese, B. H. (2000). Models of development and ecological risk. In C. H. Zeanah (Ed.), *Handbook of infant mental health* (2nd ed., pp. 3–19). New York: Guilford Press.

Sroufe, L. A., Egeland, B., Carlson, E. A., & Collins, W. A. (2005). *The development of the person: The Minnesota study of risk and adaptation from birth to adulthood.* New York: Guilford Press.

Vygotsky, L. S. (1978). *Mind in society: The development of higher psychological processes* (M. Cole, V. John-Steiner, S. Scribner, & E. Souberman, Eds.). Cambridge, MA: Harvard University Press.

Index

Academic achievement, 38, 40
   and classroom crowding, 86
   and class size, 87
   and noise, 86
   and transition into kindergarten, 121
Accidents, at-home, 200
Accommodation, to children with disabilities, 218–219
Ackerman, B. P., 53, 54, 57, 60–62
Adam, E. K., 54
Affordances, 182
African Americans, 53
Age, of child, 44, 98
Aggregate chaos, 37, 263
Airport noise, 56–57, 84
Air traffic controllers, 192
Alcoholics, 57–58
Alfred P. Sloan Foundation Center on Everyday Lives of Families, 204
Allostatic load, 106
Altruism, 233
Annoying noise, 84–85
Anxious attachment, 260
Army enlistment, 28
Asthma, 58, 228, 234
Attachment
   and child–adult ratios in child care, 71
   and child-care experience, 142
   cultural differences with, 213–214
   mutual emotional, 174
   neighborhood-level, 184
   to school, 116
   and social development, 260–261
Attachment theory, 245
Attention level, noise and, 75–76, 85
Auditory selective attention, 122
Authoritative parenting, 214
Autonomy
   and identity development, 244
   job, 193–194

Barenbaum, J., 246
Barnett, R. C., 196
Bastien, D. T., 241
BCLHDB (British Columbia Linked Health Database), 126
Beardslee, W. R., 247
Behavior problems
   and child care, 72, 142
   and classroom crowding, 86

   and family instability, 54, 61–62
   and residential moves, 53
Belonging, 248
Ben-Arieh, A., 214
Benefits, employer-provided, 197–200
Bensel, R. T., 241
Bianchi, S. M., 25, 51
Bioecological model. *See also* Proposition 2
   components of, 8–11
   nature and nurture in, 99
   and neighborhood chaos, 173–174
   propositions of, 173–174
   proximal processes of, 8, 9, 232–233
   and refugee experience, 240–241
Biological risk, 99
Biosocial risk, 99
Birth cohort studies, 125
Bonman, E., 84, 85, 87
Borderline personality disorders, 260–261
Bowlby, J., 245
Boyce, W. T., 43, 62
Bradley, R. H., 40, 53, 228
Bramley, J., 247
Brief Parenting Interventions, 139
British 1946 birth cohort, 117
British Columbia, 118, 120
British Columbia Linked Health Database (BCLHDB), 126
Brody, G. H., 61, 231, 232, 234
Broken windows, 157
Bronfenbrenner, Urie, xvii, 35, 50, 97, 173, 183, 191, 204, 217, 239–241, 255, 256, 263
Brooks-Gunn, J., 44, 176
Brown, E., 103
Brown, T., 55
Bryant, D. M., 140
Bumpas, M. F., 195
Burchinal, M. R., 72
Burlingame, D., 245
Burton, L. M., 52

Caldwell, B. M., 40
Cambodian refugees, 248
Canadian Medicare, 116, 157
Canadian National Longitudinal Survey of Children and Youth (NLSCY), 114, 157, 176
Caregivers
   communication between parents and, 138–139

multiple, 74–75, 213
  partnership of parents and, 139–140
Carlson, M., 214
Caspi, A., 43
Castro, D. C., 140
Cauce, A. M., 192
Cavanagh, S. E., 115
Center-based child care
  child–adult ratios in, 70–71
  closure of, 75
  defined, 68
Change, family, 244–245
CHAOS. *See* Confusion, Hubbub, and Order Scale
Chaos, 3–7. *See also specific* chaos; e.g., Family chaos
  and bioecological model, 9–11
  and child development, 232–235
  children's perception of, 92
  composite indices of, 230
  concept of, xvii
  defined, 5, 9
  dimensions of, 5–6, 9
  factors of, 258–259
  and family change, 22–25
  in historical context, 3–4
  in the home, 19–21
  and inequality, 26–28
  mechanisms of, 6–7, 10–11
  reducing effects of, 263–264
"Chaos amid stability," 29
Chaos research, 29–30
Chaos theory, 5n1
Chaotic environments
  and clutter/messiness, 157–158
  and crowding/density, 156–157
  cultural context of, 213–215
  deleterious effects of, 212–213
  and geographic mobility, 158
  and lack of predictability/routines, 158–159
  and noise/confusion, 157
  and supervision/monitoring, 159–160
  and well-being, 215
Chase-Landsdale, L., 176
Chase-Landsdale, P. L., 54
Cherlin, A., 114
Chicago, Illinois, 16, 177
Chicago School of Sociology, 16
Child–adult ratios
  in center-/home-based child care, 70–71
  and developmental outcomes, 71–72
  in foreign countries, 73
Child care. *See also* Caregivers; Family/child-care mesosystem
  access to quality, 67–68
  arranging for, 199–200
  continuity from home to, 141–143

  decision making about, 137
  goals of, 67
  group size and child–adult ratios for, 70–71
  information sources about, 68–69
  measurement of quality of, 68–70
  multiple arrangements for, 74–75, 213
  nonparental, 68
  outside the United States, 72–73
  parental involvement in, 140–141
  and parental stress, 233
  quality of, 67–69
  by siblings, 18, 200
  socially distributed, 213
  time spent in, 143–144
Child-care chaos, 67–79
  and crowding, 70–73
  future directions in research on, 76–78
  and instability/lack of regularity, 73–75
  and measurement of child-care quality, 68–70
  and noise, 75–76
Child development
  and crowding, 71–72
  measurement of, 262–263
  models of. *See* Developmental model(s)
  reducing effects of chaos on, 263–264
Childhood mental health, 118
Child labor, 16, 18
Child longitudinal studies, 125
Child maltreatment, 142, 164, 200
Child mortality, 18
Child poverty
  and cognitive development, 44
  in United States, 17
Children of the Great Depression (Glen Elder), 29
Children's Charter, 16
Children's lives, 15–25
  and child labor, 16
  and community, 16
  divergent destinies of, 25–28
  and geographic mobility, 16–18
  in historical context, 15–19
  and housing, 16
  and living arrangements, 22–23
  and maternal education, 25–26
  in recent historical context, 19–25
  temporal–spatial instability in, 114–118
Children with disabilities, 218–219
China, 72
Clarke-Stewart, K. A., 73
Classroom chatter, 84, 85
Class size, 87
Cleanliness, 161–162, 167
Clifford, R. M., 69
Clutter, 157, 161–162
Cognitive development, 35–45

and behavioral regulation, 41–42
child mediators of, 41–43
moderators of, 43–44
and motivational processes, 42–43
parenting mediators of, 38–41
and physical chaos, 39–40
and psychological chaos, 40–41
and socioeconomic status, 231
and stress physiology/reactivity, 42
Cognitive development theory, 245
Cohabitation, 23
Cohesion, social, 233
Coldwell, J., 19, 36, 60
Coll, C. G., 53
Collective efficacy, 160, 164, 165
Collective meaning-making, 249
College ambitions, 28
Communication
caregiver–parent, 138–139, 145
in classroom, 84, 85
at mealtimes, 52
Community(-ies)
gated, 180, 186
need for, 16
"purified," 179
temporal–spatial instability in, 118–120
unstable, 128–129
Community involvement, 115
Commuting, 195
Complexity, job, 193–194
Conflict
family, 53
marital, 58
social, 233
Confusion, 157. *See also* Noise
Confusion, Hubbub, and Order Scale (CHAOS), 19–20, 37, 69, 157, 160–161, 230, 231
Context dimension
of bioecological model, 8, 9
of microsystem chaos, 105–106
Contextual model of development, 256–257, 261
Continuity, from home to child care, 138, 141–143
Control
effortful, 38, 41, 62
over work schedule, 197
Coontz, Stephanie, 18
Corapci, F., 103, 214, 234
Coregulation, 180, 181
Cortisol levels, 72, 77
Corwyn, R. F., 53, 228
Cosleeping arrangements, 214
Costigan, C. L., 192
"Couple unemployment," 201
Cox, M. J., 192
Crime rates, 20
Crouter, A. C., 195

Crowding
and child-care chaos, 70–73, 144
and child development, 71–72
chronic, 57
and cognitive development, 39
and developmental outcomes, 71–72
and group size and child–adult ratios, 70–71
measurement of, 156–157
in non-U.S. child-care settings, 72–73
and school chaos, 86–87, 89
and socioeconomic status, 226–227, 233
and socioemotional development, 90–91
Cryer, D., 69, 78
Cultural attitudes
toward child care, 73
toward crowding, 106
Cultural beliefs, 247–248
Cultural bereavement, 243, 248
Cultural context
of environment, 213–215
well-being/chaos in, 215
Cultural identity, 244–245
Cumulative chaos, 106–107
Cumulative life course effects, 123
Cumulative risk indexes, 37–38
Cystic fibrosis, 54

Daily instability, in child care, 73–74
Daily routine(s), 211–212, 215–221
assessing sustainability of, 218–219
assessment of, 158–159
and child's well-being, 52–54
and cognitive development, 40
direct observation and meaning of, 54–55
as feedback, 234
integration of interventions into, 219–221
measurement of, 159
and nighttime waking, 58–59
and parental alcoholism, 57–58
processes of maintaining, 216–218
school attendance as, 88, 89
and socioeconomic status, 227–228
and stress, 235
universal problem of maintaining, 215–216
D'Anguilli, A., 122
Davies, P. T., 114
Density, 164. *See also* Crowding
De Schipper, J. C., 74
Developing countries, child care in, 72–73
Developmental biology, 124
Developmental model(s), 256–258
chaos within, 259–262
contextual model, 256–257, 261
Eriksonian model, 244–245
personal change model, 256, 260–261
regulation model, 257–258, 261–262
representational model, 258, 262

Disabilities, children with, 218–219
Disorder
  causes of, 258
  defined, 179
  family-level, 160–162, 167
  neighborhood-level, 163–164, 166, 178–181
  physical, 157
Divorce
  and maternal employment, 195n3
  preventing, 167
  rates of, 23
  and socioeconomic status, 230
  unemployment and risk of, 200
Doherty, G., 174
Down, G., 247
Dumas, J. E., 54, 61
Duncan, G. J., 44
Dunn, J., 19, 36, 60
Dunn, J. R., 185
Dynamic neighborhoods, 180–181
Dynamic systems theory, 5n1
Dyregrov, A., 246

Earls, F., 214
Early Child Care Research Network (ECCRN), 67
Early child development (ECD), 119
Early Development Instrument (EDI), 119
Early Head Start, 136
"Ear to the ground" hypothesis, 122
ECCRN (Early Child Care Research Network), 67
ECD (early child development), 119
ECERS–R (Early Childhood Environment Rating Scale—Revised), 69
Ecocultural Family Interview (EFI), 218–219
Economic conditions, 118
EDI (Early Development Instrument), 119
Education level
  maternal, 25–26
  and time at work, 195
Effortful control, 38, 41, 62
EFI (Ecocultural Family Interview), 218–219
Elbedour, S., 241, 246
Elder, G. H., 256
Elder, Glen, 29
Ellis, B. H., 248
El-Sheikh, M., 42
Emotion regulation, 74
Employer-provided benefits, 197–200
Employment. See Maternal employment; Parent employment
Employment programs, 167
Engagement, 216
English, K., 41, 231
Enmarker, I., 84, 85, 87
Enterprise Zones, 167

Environmental chaos. See also Chaotic environments
  defined, 4
  features of, 49, 68, 144
Environmental traffic, 69
Erikson, Erik, 244
Eriksonian developmental perspective, 244–245
ERPs (event-related potentials), 122
European child care, 73
Evans, G., 101, 102, 106
Evans, G. W., 19, 20, 35, 37, 39, 41–43, 50, 54, 56, 61, 62, 76, 78, 122, 144, 195, 227, 231–234
Evans, Gary, 180
Event-related potentials (ERPs), 122
Exclusion, social, 25
Executive functioning, 41
Exosystems, 8, 9, 240–241
Experience, 173, 178

Family change, 22–25
Family chaos, 49–63
  and changing family structure, 19–21
  child-care awareness of, 145
  cyclic variations in, 56–60
  and disorder, 160–162
  future directions in research on, 165–166
  multidimensional nature of, 50–56
  prevention of, 166–167
  process models of, 60–62
  and routines, 52–55
  and time spent with family, 50–52
  and turbulence, 162–163
Family child care
  child–adult ratios in, 71
  communications with, 145
  defined, 68
Family/child-care mesosystem, 135–148
  and chaos, 144–146
  and communication between parents and caregivers, 138–139
  and continuity from home to child care, 141–143
  in early 21st century, 135–137
  future directions in research on, 147–148
  and parental involvement in child care, 140–141
  and partnership of parents and caregivers, 139–140
  and pervasiveness of child care, 146–147
  and well-being of children, 143–144
Family conflict, 53
Family-friendly job benefits, 197–200
Family health, 197–200
Family instability, 114–115
Family mealtimes, 51–54

and children with disabilities, 219
and family chaos, 51–54
as measure of turbulence, 159
and nonstandard work hours, 197
Family Medical Leave Act (FMLA), 198, 199
Family size, 20, 147
Family time, 50–52, 102
Fathers
   and child care, 141
   and maternal employment, 195–196
   roles of, 25
Fatigue hypothesis, 103–105
Feedback, 234
Female heads of households, 228
Fiese, B. H., 234
Financial hardship, perceived, 231
Fivush, R., 55
Flor, D. L., 61, 231, 234
Fluidity. *See* Instability
FMLA. *See* Family Medical Leave Act
Fomby, P., 114
Forman, E. M., 114
Foster care, 89
Foundation Skills Assessment (FSA), 127
Fragile Families and Child Well-Being Study, 157
Freud, A., 245
Freud, Sigmund, 244
Freudian developmental perspective, 244
FSA (Foundation Skills Assessment), 127

Garcia-Coll, C., 228
Garron, D., 142
Gated communities, 180, 186
Gender differences, with microsystem chaos, 98
Genetics, and chaos, 99–100
Gentile, L., 231
Geographic mobility
   and children's lives, 16–18
   and cognitive development, 40–41
   and neighborhood chaos, 158
   of refugees, 246
   social problems caused by, 16–18
   and socioeconomic status, 228–229
Gibson, J. J., 182
Giddens, Anthony, 183, 184
Goals, 216
Goerge, R. M., 214
Gonnella, C., 231
Graham, K., 175
Great Depression, 15–16
Grych, J. H., 55

Haarlar, N., 101
Habitat for Humanity, 167
Habituation hypothesis, 103
Hambrick-Dixon, P. J., 76

Han, W., 137
Harkness, S., 214
Harms, T., 69
"Haves" and "have-nots," 26
Hazzard, A., 55
Head Start, 140
Health
   family, 197–200
   neighborhood effects on, 178
   and parent employment, 195
   physical, 234
Health behaviors, 59
Healthy eating, 53, 234
Heat Wave (E. Klinenberg), 181, 184
Heft, H., 7
HELP. *See* Human Early Learning Partnership
Helplessness, 90, 91
Helpless or hopeless approach (to academic challenge), 42
Heritability, 99–100
Hertzman, C., 176, 185
Hippocampus, 43
Hodes, M., 247
Hofferth, S. L., 51
HOME. *See* Home Observation for Measurement of the Environment Scale
Home-based child care
   child–adult ratios in, 71
   crowding in, 72
   defined, 68
Home Builders, 167
Home environment, 35–45
   and bioecological model, 35
   chaos in, 19–21
   and cognitive development, 38–45
   continuity from, to child care, 141–143
   directions of effects on chaos and, 38
   family-level variables of, 36
   and income poverty, 36–37
   and individual vs. aggregate representations, 37–38
   theoretical issues with, 36–38
Homelessness, 89
Home Observation for Measurement of the Environment (HOME) Scale, 39–40, 53, 157
Homeownership rates, 20
Hong Kong, 72
Household size, 20
Housing
   and children's lives, 16
   need for safe, 16
HPA (hypothalamic–pituitary–adrenal) axis, 115
Hsueh, J., 40
Human Early Learning Partnership (HELP), 118, 119

Huston, A., 115
Hypothalamic–pituitary–adrenal (HPA) axis, 115

Identity development, 244–245
IHDP (Infant Health and Development Program), 176
Immigrants
 poverty rates among, 27
 refugees vs., 242
Immune function, 144
Income gap, 25
Income level, 28, 41
Individualism, 19
Individual vulnerability, 98–99
Indonesia, 72
Inequality, 26–28
Infant Health and Development Program (IHDP), 176
Infant mortality, 18
Infectious disease, 18
Instability, 73–75
 and chaos, 5–6
 in child care, 73–75
 family, 114–115
 geographic, 158
 neighborhood, 158
 and problem behavior, 54, 61–62
 residential, 164
 and socioeconomic status, 233
 temporal–spatial. *See* Temporal–spatial instability
Interactions, 173, 180
Intermittent chaos, 57
Internal representations, 55
Interventions
 with child refugees, 246–249
 with family/neighborhood chaos, 166–167
Irritability, 91
Izard, C., 53

Janus, M., 174
Job loss, 200–201
Job overload, 192–193
Job selection, 193–194, 204
Johnston-Brooks, C. H., 144
Joshi, H., 176

Kagan, J., 174
Kewin, Kurt, 240
Kia-Keating, M., 248
Kim, P., 231
Kindergarten(s)
 Asian, 72
 transition into, 121
Klinenberg, E., 181, 184
Klockow, L. L., 55

Kogos, J., 53
Kohen, D., 176, 185
Kohn, M. L., 193
Kohn, Melvin, 193
Korea, 73

LaBarre, R., 142
Language skills, 75, 76, 116–117, 161
Lareau, Annette, 20
Latent life course effects, 123
Layoffs, 118
Learned helplessness, 61, 103–105, 234
Learning, 43
Learning environment, 39, 142
Leave
 maternity, 199n8
 paid and unpaid, 198–199
LECP (Life in Early Childhood Programs) Scale, 69
Lee, E., 245
Lee, V. E., 89
Lemery-Chalfant, K., 60
Lepore, S., 101, 102
Lepore, S. J., 233
Lewis, J., 144
Lichter, D. T., 26
Life course effects, 123
Life in Early Childhood Programs (LECP) Scale, 69
Light, R., 227
Lives, of children. *See* Children's lives
Living arrangements, of children, 22–23, 89
Long, L., 229
Longitudinal, person-specific data, 125–128
Low, C., 103
Ludwig, J., 144

Macksoud, M., 246
Macrosystems, 8, 9, 241
Maguire, M. C., 195
Mahler, Margaret, 245
Major depressive disorder, 23
Marbles in a bowl (experiment), 123
Marcynyszyn, L. A., 231
Marital conflict, 58
Marital relationship
 and long work hours, 195
 and nonstandard work hours, 197
 strengthening, 167
 and work stress, 193
Marriage, poverty and, 27–28
Mastery, sense of, 42, 233–234
Maternal education, 25–26
Maternal employment, 23–25, 51
 and child care, 137, 140
 child care and increasing, 67
 and paternal involvement, 195–196

Maternal mortality, 18
Maternal partner change, 230, 233
Maternity leave, 199n8
Matheny, A., 144, 145
Maxwell, L., 76, 78
Maxwell, L. E., 86, 142
Mayer, S., 21
McAdoo, H. P., 53, 228
McCall, R., 101
McClelland, M. M., 41
McCulloch, A., 176
McHale, S. M., 195
McLanahan, Sara, 25–26
Mealtimes. *See* Family mealtimes
Meaningfulness, 262
Meaning-making, 249
Medicaid, 21
Memory, 43, 85
Mental health, childhood, 118
Mesosystems, 8–10, 240
Messiness, 157
Microsystem chaos, 97–108
    context dimension of, 105–106
    defined, 97
    future directions in research on, 100–105
    and gender, 98
    and individual vulnerability, 98–99
    mediators of, 101–102
    moderation of genetic influences of, 99–100
    person dimension of, 97–101
    PPCT model of, 107–108
    process dimension of, 101–105
    time dimension of, 106–107
Microsystems, 8, 240, 246
Miller, K., 242
Milton, John, xvii
Mobility
    geographic. *See* Geographic mobility
    residential, 40–41
Monitoring, 159–160
Moral development, 245–246
"The more the merrier," 3
Mutual coregulation, 180, 181
Mutual emotional attachment, 174

Narratives, 54–56
National Association for the Education of Young Children, 71, 141
National Center for Health Statistics, 23
National Child Development Study (NCDS), 176
National Institute of Child Health and Human Development (NICHD), 67, 143
National Longitudinal Survey of Children and Youth (NLSCY), 176
National Longitudinal Survey of Youth (NLSY), 176

NCDS (National Child Development Study), 176
Negative emotionality hypothesis, 103–105
Negative emotion spillover, 192, 202–203
Neighborhood chaos, 173–186
    and disorder, 163–164, 178–181
    and ecology of child development, 173–174
    future directions in research on, 165–166, 184–186
    as lack of social capital, 181–182
    literature on, 182–184
    as poverty vs. disadvantage, 177–178
    prevention of, 166–167
    research on effects of, 175–177
    terminology, 174–175
    and turbulence, 164–165
Neighborhood disorder, 178
Neighborhoods
    and child development, 119–120
    defining, 174–175
    dynamic, 180–181
    off-diagonal, 128
New Hope program, 40, 117, 219–220
NICHD. *See* National Institute of Child Health and Human Development
Nighttime waking, 58–59
NLSCY. *See* Canadian National Longitudinal Survey of Children and Youth
NLSY (National Longitudinal Survey of Youth), 176
Noise
    annoying, 84–85
    and chaos, 5–7
    and child-care chaos, 75–76
    chronic exposure to, 56–57
    and cognitive development, 39
    and healthy eating, 53
    measurement of, 157
    and reading achievement, 161
    and school chaos, 83–86
    and socioeconomic status, 227, 233
    and socioemotional development, 91
Noncustodial fathers, 25
Nonmarital births, 26
Nonparental child care, 68

Object relations theory, 245
Occupational safety codes, 18
Off-diagonal neighborhoods, 128
Offord, D., 174
Oliver, L., 185
Ontario Child Health Study, 117
Ontological security, 183–184
Optimal stimulation hypothesis, 4
Order
    dis-. *See* Disorder
    measurement of. *See* Confusion, Hubbub, and Order Scale

school, 93
O'Shea, B., 247
Overload, work, 192
Overweight children, 54

Paid leave, 198–199
Palsane, M., 101
Panel Study of Income Dynamics, 161
Parental involvement
  in child care, 140–141
  and work hours, 195–196
Parent employment, 191–205. *See also* Maternal employment
  and autonomy/complexity, 193–194
  and child care, 199–200
  and family-friendly benefits, 197–200
  future directions in research on, 203–205
  integration of research findings on, 201–203
  and job loss/unemployment, 200–201
  model of effects of, on family, 202
  and overload/social stressors, 192–193
  and paid/unpaid leave, 198–199
  and socioeconomic status, 228
  and time at work, 194–197
  and workplace psychosocial characteristics, 191–194
Parenting behavior
  and child-care experience, 142
  cultural differences with, 214
  and environmental chaos, 101–102
  and socioeconomic status, 233
Parenting styles, 20–21
Parents
  alcoholism in, 57–58
  communication between caregivers and, 138–139
  efficacy of, 234
  partnership of caregivers and, 139–140
  psychopathology in, 260
  warmth of, 53
Partnership, of parents and caregivers, 139–140
Pathways life course effects, 123
Pathways model, 217
Peabody Picture Vocabulary Test—Revised (PPVT–R), 176
Peisner-Feinberg, E., 72
Peisner-Feinberg, E. S., 140
Perry-Jenkins, M., 197
Person, process, context, time (PPCT) model, 50, 107–108, 217, 263. *See also* Bioecological model
Personal change model of development, 256, 260–261
Person dimension
  of bioecological model, 8, 9, 97–98
  of microsystem chaos, 98–101

Petrill, S., 106
PHDCN (Project on Human Development in Chicago Neighborhoods), 157
Phillips, K., 144
Physical chaos, 39–40
Physical disorder, 157
Physical health, 234
Pike, A., 19, 36, 60
Plato, 3
Play, 142
Population-based data, 125
Poverty, neighborhood, 177–178
Poverty rates
  among immigrants, 27
  by household structure, 26, 27
  stability of U.S., 25
  in United States, 21
PPCT model. *See* Person, process, context, time model
PPVT–R (Peabody Picture Vocabulary Test—Revised), 176
Pratchett, Terry, xvii
Predictability. *See also* Daily routine(s)
  of caregiver, 260
  and mastery, 233–234
  neighborhood, 158–159
Prevention, of family/neighborhood chaos, 166–167
Process mechanisms, of microsystem chaos, 101–105
Process quality, child-care, 69, 71–73
Project on Global Working Families, 198
Project on Human Development in Chicago Neighborhoods (PHDCN), 157
Proposition 2 (of bioecological model), 97, 98, 100, 105–106, 173
Provost, M. A., 142
Proximal processes, 8, 9, 232–233, 240
Psychosocial chaos, 40–41
Psychosocial risk, 99
Public assistance, 21
Public health, 18, 167
"Purified community," myth of, 179

Qian, Z., 26
Quas, J. A., 43

Racial minorities
  and class size, 87
  family mealtimes among, 53
  inequality and chaos among, 26–28
Raley, S. B., 51
Raundalen, M., 246
Reading achievement, 161
Refugee camps, 242
Refugee experience, 239–249
  and biological model of development, 240–241

chaos with, 27
impact of, on child development, 243–249
as product of chaos, 242–243
stages of, 241
studying ecology of, 246–249
theoretical perspectives on, 243–246
Regulation. *See also* Self-regulation
behavioral, 41–42
emotion, 74
mutual co-, 180, 181
Regulation model of development, 257–258, 261–262
Reiser, M., 60
Remarriage, 23
Repetti, R., 245
Repetti, R. L., 192
Representational model of development, 258, 262
Residential instability, 164
Residential mobility, 40–41
Residential moves
and community involvement, 115
and neighborhood stability, 158
as opportunity vs. necessity, 114
and problem behavior, 53
rates of, 17
and school moves, 128
and socioeconomic status, 228–229
Responsiveness, of caregiver, 260
Rituals, 227–228
Ross, N., 175
Routines. *See* Daily routine(s)
Roy, K. M., 52
Ruhm, C. J., 198
Rutter, M., 229

Sales, J. M., 55
Salpekar, N., 231
Sandberg, J. F., 51
Sapolsky, R. M., 124
Sarfati, D., 55
Sawmill communities, in British Columbia, 115, 125–127
Saxbe, D., 245
Schoff, K., 53
School(s)
crowding/noise in, 166
transiency and attachment to, 116
School, physical condition of, 88–89
School attendance, 88, 89
School auditoriums, 91
School cafeterias, 91
School chaos, 83–94
and crowding, 86–87
individual differences with, 91–92
and noise, 83–86
reversing, 92–93

social factors of, 89–90
and socioemotional development, 90–91
and visual complexity, 87–89
School dropouts, 86, 228
Schooler, C., 193
School libraries, 91
School relocation, 229
School size, 88
Script, for normative conduct, 216
Seaton, E. K., 61
SECCYD (Study of Early Child Care and Youth Development), 68
Security
development of, 213–214
ontological, 183–184
Self-efficacy, 137, 233–234
Self-efficacy hypothesis, 103–105
Self-identity, 184
Self-regulation. *See also* Regulation model of development
and chaos, 7, 38
and cognitive development, 41
and mutual coregulation, 181
and routines/structure, 234
and school crowding, 91
and socioeconomic status, 231
Sennett, R., 179
Serotonin transporters, 43
SES. *See* Socioeconomic status
Sexual development, 244
Shejwal, B., 101
Shift work, 196–197, 199
Sibling relationships, 233
Siblings, as caregivers, 18, 200
Single mothers
and child care arrangements, 199
and nonstandard work hours, 196–197
and poverty, 26
and risk of job loss, 201
Skinner, M. L., 140
Sleep, family mealtime and, 53
Sleep hygiene, 59
Sleeping arrangements, 214
Sleep problems, 42, 58–59, 103
Smeeding, T. M., 214
Smith, J. B., 89
Smoking, initiation of, 200
Social affiliation, 122
Social capital, 181–182
Social cohesion, 233
Social conflict, 233
Social cues, response to, 54
Social development, 260–261
Social exclusion, 25
Socialization, of workers' values, 193
Socially distributed child care, 213
Social play, 142

Social Security, 21
Social space, 121
Social stressors, 192–193
Social withdrawal, 192
Social withdrawal hypothesis, 102–104
Socioeconomic status (SES), 225–232
  and auditory selective attention, 122
  and chaos, 5
  and child development, 230–232
  and composite indices of chaos, 230
  and crowding, 226–227
  and maternal partner change, 230
  and neighborhood, 177
  and noise, 227
  and residential relocation, 228–229
  and routines/rituals, 227–228
  and school relocation, 229
Socioemotional development, 40, 90–91
South Bronx (New York City), 177
Southeast Asian culture, 245
Space, social, 121
Speech perception, 84, 85
SSO (systematic social observations), 179
Stability, 261. *See also* Instability
Staff turnover, 75
State Children's Health Insurance Program, 21
Stepfamilies, 23
Stimulation, 3–4
"Stimulus shelters," 142
Stress
  and daily routines, 217–218, 235
  and socioeconomic status, 231
  at work, 192–193, 202–203
Stress reactivity, 43–44, 145
Structural quality, child-care, 69
Structure, lack of, 233–235
Study of Early Child Care, 143
Study of Early Child Care and Youth Development (SECCYD), 68
Substitute teachers, 90
Suburbia, movement toward, 20
Subway noise, 84
Sudanese refugees, 241, 249
Suicide, 126
Suomi, S. J., 124
Super, C., 214
Supervision, 159–160
Support, social, 101
Sure Start program, 117
Sustainability, of daily routines, 215–219
Systematic social observations (SSO), 179

"Tag team" work hours, 196
Taylor, R. D., 61
Taylor, S. E., 245
Teacher absenteeism, 90
Teacher education, 73, 76

Teacher turnover, 229
Teenage births, 26
Television, 53, 167
Temperament
  adaptable, 62
  and child-care adaptation, 74
  and environmental chaos, 99
  and self-efficacy hypothesis, 105
Temporal–spatial instability, 113–129
  in child's life, 114–118
  in community, 118–120
  community role in, 128–129
  data systems for studying, 124–128
  developmental biology of, 124
  future directions in research on, 124–129
  theoretical bases for importance of, 120–123
Testimonial psychotherapy, 248–249
Thailand, 72
Thomas, J. C., 114
Three-City Study, 52
Time
  in child care, 143–144
  with family, 50–52
  at work, 194–197, 228
Time allotment hypothesis, 102
Time dimension
  of bioecological model, 8, 255–256
  of microsystem chaos, 106–107
  of personal change model, 260
Traditional maturational perspective (of child development), 120
Tran, H., 75
Transactional perspective (of child development), 120
Tremblay, S., 175
The Truly Disadvantaged (W. J. Wilson), 159
Trust, 244
Tubbs, C. Y., 52
Tuning out hypothesis, 103
Turbulence
  causes of, 258–259
  family-level, 162–163, 167
  neighborhood-level, 164–166

Unemployment, 200–201
Unequal Childhoods (Annette Lareau), 20
United Nations Children's Fund (UNICEF), 239
United States
  child poverty in, 17
  1900 population of, 16
  poverty rates in, 21
University of Kansas Language Acquisition Project, 39
Unmarried mothers, 23
Unpaid leave, 197–199
Urbanization, 16
Urgency of departure, 242

U.S. Census Bureau, 16
U.S. General Accounting Office, 229
The Uses of Disorder (R. Sennett), 179

Valence of movement, 242
Valiente, C., 38, 60, 102, 105
Values
   and daily activities, 216
   socialization of workers', 193
Van IJzendoorn, M. H., 138
Verbal responsiveness, 39
Violence exposure, 55
Violent behavior, 114
Visual stimulation, in classroom, 87–89
Vleminckx, K., 214
Vulnerability, individual, 98–99
Vygotsky, L. S., 257

Wachs, T. D., 42, 69, 103, 144, 214, 234
Wachsmuth-Schlaefer, T., 55
War zones, research within, 246
Water supplies, on Indian reservations, 167
The Way We Never Were (Stephanie Coontz), 18
Weine, S., 248
Weine, S. M., 249
Weintraub, M., 75
Well-being of child
   and child care, 143–144
   in cultural context, 215
   and daily routines, 52–54, 216
   and family/child-care mesosystem, 143–144
   and family-level disorder/turbulence, 160–163
   and neighborhood-level disorder/turbulence, 163–165
Wener, R. E., 195
Whalen, C. K., 144
White House Conference on Child Health and Protection, 16
Wilson, W. J., 159, 167
Winnipeg, Manitoba, 116
Withdrawal, social, 102–104, 192
Wohlwill, J., 3, 7
Wohlwill, Joachim, xvii
Wolin, S. J., 57–58
Women, status of, 23
Work hours
   long, 194–196, 228
   nonstandard, 196–197, 228
Work overload, 192
Workplace, 191–200
   autonomy in, 193–194
   family-friendly benefits in, 197–200
   injuries and death in, 18
   and job complexity, 193–194
   psychosocial characteristics of, 191–194
   time spent in, 194–197

Yoshikawa, H., 40
Youngstrom, E., 53

About the Editors

Gary W. Evans, PhD, Elizabeth Lee Vincent Professor of Human Ecology in the Departments of Design and Environmental Analysis and of Human Development at Cornell University in Ithaca, New York, studies how the physical environment affects human health and well-being among children. His specific areas of expertise include childhood poverty, environmental stress, and children's environments. He holds a doctorate in environmental psychology, with postdoctoral training in psychoneuroendocrinology and human development.

Dr. Evans is the author of 5 books and over 300 scholarly articles and book chapters. Evans is a member of the John D. and Catherine T. MacArthur Foundation Network on Socioeconomic Status and Health; the Board on Children, Youth, and Families of the National Academy of Sciences; and the Board of Scientific Counselors, National Center for Environmental Health/Agency for Toxic Substance and Disease Registry, Centers for Disease Control and Prevention. He has been awarded two Fulbright Research Fellowships and is the recipient of a Senior National Research Service Award from the National Institutes of Health.

Theodore D. Wachs, PhD, is professor of psychological sciences at Purdue University in West Lafayette, Indiana. He received his doctorate in psychology from George Peabody College in 1968. In 1995–1996 he was a Golestan Fellow at the Netherlands Institute for Advanced Studies in the Humanities and Social Sciences, and in 2003 he was a Fulbright Distinguished Scholar at the Centre for International Child Health at the University of London, London, England. Currently, he is a member of the editorial boards of the *International Journal of Behavioral Development* and the *Journal of Applied Developmental Psychology*. His research focuses on the role of chaotic family environments on development, micronutrient deficiencies in infancy and cognitive and socioemotional development, and temperament in infancy and childhood. In addition to his research in the United States, he has also been involved in research projects in Egypt, Jamaica, and Peru. He has authored or coedited 8 books, numerous book chapters, and over 90 research and review articles in scientific and professional journals.